HOW WE BECAME POSTHUMAN

HOW WE BECAME POSTHUMAN

Virtual Bodies in Cybernetics, Literature, and Informatics

N. KATHERINE HAYLES

The University of Chicago Press
Chicago & London

N. KATHERINE HAYLES is professor of English at the University of California, Los Angeles. She holds degrees in both chemistry and English. She is the author of *The Cosmic Web: Scientific Field Models and Literary Strategies in the Twentieth Century* (1984) and *Chaos Bound: Orderly Disorder in Contemporary Literature and Science* (1990) and is the editor of *Chaos and Order: Complex Dynamics in Literature and Science* (1991), the last published by the University of Chicago Press.

The University of Chicago Press, Chicago 60637
The University of Chicago Press, Ltd., London
© 1999 by The University of Chicago
All rights reserved. Published 1999
Printed in the United States of America
08 07 06 05 04 03 02 01 00   3 4 5

ISBN (cloth): 0-226-32145-2
ISBN (paper): 0-226-32146-0

Library of Congress Cataloging-in-Publication Data

Hayles, N. Katherine.
    How we became posthuman : virtual bodies in cybernetics,
literature, and informatics / N. Katherine Hayles.
        p.   cm.
    Includes bibliographical references and index.
    ISBN: 0-226-32145-2 (cloth : alk. paper). — ISBN: 0-226-32146-0
(pbk. : alk. paper)
    1. Artificial intelligence.   2. Cybernetics.   3. Computer science.
4. Virtual reality.   5. Virtual reality in literature.   I. Title.
Q335.H394   1999
003'.5—dc21   98-36459
                                                          CIP

♾ The paper used in this publication meets the minimum requirements of the
American National Standard for the Information Sciences—Permanence of Paper for
Printed Library Materials, ANSI Z39.48-1992.

For Nicholas
one of the world's great technology archivists
and much more besides

Contents

A c k n o w l e d g m e n t s

The notion of distributed cognition, central to the posthuman as it is defined in this book, makes acknowledging intellectual and practical contributions to this project an inevitability as well as a pleasure. The arguments have benefited from conversations and correspondence with many friends and colleagues, among them Evelyn Fox Keller, Felicity Nussbaum, Rob Latham, Adalaide Morris, Brooks Landon, Peter Galison, Timothy Lenoir, Sandra Harding, Sharon Traweek, and Marjorie Luesebrink. Mark Poster and an anonymous reader for the University of Chicago Press gave valuable suggestions for revisions and rethinking parts of the argument. Tom Ray, Rodney Brooks, and Mark Tilden graciously spoke with me about their artificial life projects, and Stefan Helmreich shared with me an early version of his book on artificial life. Many of my students gave valuable feedback and criticism of early versions of my ideas, including Carol Wald, Jim Berkley, Kevin Fisher, Evan Nisonson, Mark Sander, Linda Whitford, and Jill Galvin.

I am also very grateful for the institutional support I have received, including a fellowship from the Guggenheim Foundation, a fellowship from the Stanford Humanities Center, a Presidential Research Fellowship from the University of California, support from the Council on Research at the University of California at Los Angeles, and a leave of absence and research support from the University of Iowa. I could not have completed this project without this generous support.

I owe a debt of gratitude as well to Routledge Press for allowing me to reprint "Narratives of Artificial Life," from *FutureNatural: Nature, Science, Culture,* edited by George Robertson, Melinda Mash, Lisa Tickner, John Bird, Barry Curtis, and Tim Putnam, pp. 145–46, © 1996 (appearing in revised form as chapter 9); and "Designs on the Body: Cybernetics, Norbert Wiener, and the Play of Metaphor," from *History of the Human Sciences* 3 (1990): 212–28 (appearing in revised from as a portion of chapter 4). Johns Hopkins University Press has graciously allowed me to reprint three articles appearing in *Configurations: A Journal of Literature, Science, and Technology*—"The Materiality of Informatics," *Configurations* 1 (1993): 147–70 (appearing in revised form as a portion of chapter 8); "Boundary Disputes: Homeostasis, Reflexivity, and the Foundations of Cybernetics," ibid. 3 (1994): 441–67 (appearing in revised form as part of chapter 3); and "The Posthuman Body: Inscription and Incorporation in *Galatea 2.2* and *Snow Crash*," ibid. 5 (1997): 241–66 (appearing as part of chapter 10). MIT Press has given permission to reprint "Virtual Bodies and Flickering Signifiers," from *October* 66 (Fall 1993): 69–91 (appearing in slightly revised form as chapter 2). The University of North Carolina Press has given permission to reprint a portion of "Voices Out of Bodies, Bodies Out of Voices," from *Sound States: Innovative Poetics and Acoustical Technologies*, edited by Adalaide Morris, pp. 74–78, 86–96, © 1997 by The University of North Carolina Press (appearing in revised form as a part of chapter 8). *The Journal of the Fantastic in the Arts* has given permission to reprint "Schizoid Android: Cybernetics and the Mid-60s Novels of Dick," *JFIA* 8 (1997): 419–42 (appearing in slightly revised form as chapter 6).

Finally, my greatest debt is to my family, who have listened patiently to my ideas over the years, and to my husband, Nick Gessler, from whom I have learned more than I can say.

You are alone in the room, except for two computer terminals flickering in the dim light. You use the terminals to communicate with two entities in another room, whom you cannot see. Relying solely on their responses to your questions, you must decide which is the man, which the woman. Or, in another version of the famous "imitation game" proposed by Alan Turing in his classic 1950 paper "Computer Machinery and Intelligence," you use the responses to decide which is the human, which the machine.[1] One of the entities wants to help you guess correctly. His/her/its best strategy, Turing suggested, may be to answer your questions truthfully. The other entity wants to mislead you. He/she/it will try to reproduce through the words that appear on your terminal the characteristics of the other entity. Your job is to pose questions that can distinguish verbal performance from embodied reality. If you cannot tell the intelligent machine from the intelligent human, your failure proves, Turing argued, that machines can think.

Here, at the inaugural moment of the computer age, the erasure of embodiment is performed so that "intelligence" becomes a property of the formal manipulation of symbols rather than enaction in the human lifeworld. The Turing test was to set the agenda for artificial intelligence for the next three decades. In the push to achieve machines that can think, researchers performed again and again the erasure of embodiment at the heart of the Turing test. All that mattered was the formal generation and manipulation of informational patterns. Aiding this process was a definition of information, formalized by Claude Shannon and Norbert Wiener, that conceptualized information as an entity distinct from the substrates carrying it. From this formulation, it was a small step to think of information as a kind of bodiless fluid that could flow between different substrates without loss of meaning or form. Writing nearly four decades after Turing, Hans

Moravec proposed that human identity is essentially an informational pattern rather than an embodied enaction. The proposition can be demonstrated, he suggested, by downloading human consciousness into a computer, and he imagined a scenario designed to show that this was in principle possible. The Moravec test, if I may call it that, is the logical successor to the Turing test. Whereas the Turing test was designed to show that machines can perform the thinking previously considered to be an exclusive capacity of the human mind, the Moravec test was designed to show that machines can become the repository of human consciousness—that machines can, for all practical purposes, become human beings. You are the cyborg, and the cyborg is you.

In the progression from Turing to Moravec, the part of the Turing test that historically has been foregrounded is the distinction between thinking human and thinking machine. Often forgotten is the first example Turing offered of distinguishing between a man and a woman. If your failure to distinguish correctly between human and machine proves that machines can think, what does it prove if you fail to distinguish woman from man? Why does gender appear in this primal scene of humans meeting their evolutionary successors, intelligent machines? What do gendered bodies have to do with the erasure of embodiment and the subsequent merging of machine and human intelligence in the figure of the cyborg?

In his thoughtful and perceptive intellectual biography of Turing, Andrew Hodges suggests that Turing's predilection was always to deal with the world as if it were a formal puzzle.[2] To a remarkable extent, Hodges says, Turing was blind to the distinction between saying and doing. Turing fundamentally did not understand that "questions involving sex, society, politics or secrets would demonstrate how what it was possible for people to *say* might be limited not by puzzle-solving intelligence but by the restrictions on what might be *done*" (pp. 423–24). In a fine insight, Hodges suggests that "the discrete state machine, communicating by teleprinter alone, was like an ideal for [Turing's] own life, in which he would be left alone in a room of his own, to deal with the outside world solely by rational argument. It was the embodiment of a perfect J. S. Mill liberal, concentrating upon the free will and free speech of the individual" (p. 425). Turing's later embroilment with the police and court system over the question of his homosexuality played out, in a different key, the assumptions embodied in the Turing test. His conviction and the court-ordered hormone treatments for his homosexuality tragically demonstrated the importance of *doing* over *saying* in the coercive order of a homophobic society with the power to enforce its will upon the bodies of its citizens.

The perceptiveness of Hodges's biography notwithstanding, he gives a strange interpretation of Turing's inclusion of gender in the imitation game. Gender, according to Hodges, "was in fact a red herring, and one of the few passages of the paper that was not expressed with perfect lucidity. The whole point of this game was that a successful imitation of a woman's responses by a man would *not* prove anything. Gender depended on facts which were *not* reducible to sequences of symbols" (p. 415). In the paper itself, however, nowhere does Turing suggest that gender is meant as a counterexample; instead, he makes the two cases rhetorically parallel, indicating through symmetry, if nothing else, that the gender and the human/machine examples are meant to prove the same thing. Is this simply bad writing, as Hodges argues, an inability to express an intended opposition between the construction of gender and the construction of thought? Or, on the contrary, does the writing express a parallelism too explosive and subversive for Hodges to acknowledge?

If so, now we have two mysteries instead of one. Why does Turing include gender, and why does Hodges want to read this inclusion as indicating that, so far as gender is concerned, verbal performance cannot be equated with embodied reality? One way to frame these mysteries is to see them as attempts to transgress and reinforce the boundaries of the subject, respectively. By including gender, Turing implied that renegotiating the boundary between human and machine would involve more than transforming the question of "who can think" into "what can think." It would also necessarily bring into question other characteristics of the liberal subject, for it made the crucial move of distinguishing between the enacted body, present in the flesh on one side of the computer screen, and the represented body, produced through the verbal and semiotic markers constituting it in an electronic environment. This construction necessarily makes the subject into a cyborg, for the enacted and represented bodies are brought into conjunction through the technology that connects them. If you distinguish correctly which is the man and which the woman, you in effect reunite the enacted and the represented bodies into a single gender identity. The very existence of the test, however, implies that you may also make the wrong choice. Thus the test functions to create the possibility of a disjunction between the enacted and the represented bodies, regardless which choice you make. What the Turing test "proves" is that the overlay between the enacted and the represented bodies is no longer a natural inevitability but a contingent production, mediated by a technology that has become so entwined with the production of identity that it can no longer meaningfully be separated from the human subject. To pose the question of "what can think" inevitably also changes, in a reverse feedback loop, the terms of "who can think."

On this view, Hodges's reading of the gender test as nonsignifying with respect to identity can be seen as an attempt to safeguard the boundaries of the subject from precisely this kind of transformation, to insist that the existence of thinking machines will not necessarily affect what being human means. That Hodges's reading is a misreading indicates he is willing to practice violence upon the text to wrench meaning away from the direction toward which the Turing test points, back to safer ground where embodiment secures the univocality of gender. I think he is wrong about embodiment's securing the univocality of gender and wrong about its securing human identity, but right about the importance of putting embodiment back into the picture. What embodiment secures is not the distinction between male and female or between humans who can think and machines which cannot. Rather, embodiment makes clear that thought is a much broader cognitive function depending for its specificities on the embodied form enacting it. This realization, with all its exfoliating implications, is so broad in its effects and so deep in its consequences that it is transforming the liberal subject, regarded as the model of the human since the Enlightenment, into the posthuman.

Think of the Turing test as a magic trick. Like all good magic tricks, the test relies on getting you to accept at an early stage assumptions that will determine how you interpret what you see later. The important intervention comes not when you try to determine which is the man, the woman, or the machine. Rather, the important intervention comes much earlier, when the test puts you into a cybernetic circuit that splices your will, desire, and perception into a distributed cognitive system in which represented bodies are joined with enacted bodies through mutating and flexible machine interfaces. As you gaze at the flickering signifiers scrolling down the computer screens, no matter what identifications you assign to the embodied entities that you cannot see, you have already become posthuman.

TOWARD EMBODIED VIRTUALITY

We need first to understand that the human form—including human desire and all its external representations—may be changing radically, and thus must be re-visioned. We need to understand that five hundred years of humanism may be coming to an end as humanism transforms itself into something that we must helplessly call post-humanism.

Ihab Hassan, "Prometheus as Performer: Towards a Posthumanist Culture?"

This book began with a roboticist's dream that struck me as a nightmare. I was reading Hans Moravec's *Mind Children: The Future of Robot and Human Intelligence,* enjoying the ingenious variety of his robots, when I happened upon the passage where he argues it will soon be possible to download human consciousness into a computer.[1] To illustrate, he invents a fantasy scenario in which a robot surgeon purees the human brain in a kind of cranial liposuction, reading the information in each molecular layer as it is stripped away and transferring the information into a computer. At the end of the operation, the cranial cavity is empty, and the patient, now inhabiting the metallic body of the computer, wakens to find his consciousness exactly the same as it was before.

How, I asked myself, was it possible for someone of Moravec's obvious intelligence to believe that mind could be separated from body? Even assuming such a separation was possible, how could anyone think that consciousness in an entirely different medium would remain unchanged, as if it had no connection with embodiment? Shocked into awareness, I began noticing he was far from alone. As early as the 1950s, Norbert Wiener proposed it was theoretically possible to telegraph a human being, a suggestion underlaid by the same assumptions informing Moravec's scenario.[2] The producers of *Star Trek* operate from similar premises when they imagine that the body can be dematerialized into an informational pattern and rematerialized, without change, at a remote location. Nor is the idea confined to what Beth Loffreda has called "pulp science."[3] Much of the discourse on molecular biology treats information as the essential code the body expresses, a practice that has certain affinities with Moravec's ideas.[4] In fact, a defining characteristic of the present cultural moment is the belief that information can circulate unchanged among different material substrates. It is not for nothing that "Beam me up, Scotty," has become a cultural icon for the global informational society.

Following this thread, I was led into a maze of developments that turned into a six-year odyssey of researching archives in the history of cybernetics, interviewing scientists in computational biology and artificial life, reading cultural and literary texts concerned with information technologies, visiting laboratories engaged in research on virtual reality, and grappling with technical articles in cybernetics, information theory, autopoiesis, computer simulation, and cognitive science. Slowly this unruly mass of material began taking shape as three interrelated stories. The first centers on how *information lost its body,* that is, how it came to be conceptualized as an entity separate from the material forms in which it is thought to be embedded. The second story concerns how *the cyborg was created as a technological artifact and cultural icon* in the years following World War II. The third, deeply implicated with the first two, is the unfolding story of how a historically specific construction called *the human is giving way to a different construction called the posthuman.*

Interrelations between the three stories are extensive. Central to the construction of the cyborg are informational pathways connecting the organic body to its prosthetic extensions. This presumes a conception of information as a (disembodied) entity that can flow between carbon-based organic components and silicon-based electronic components to make protein and silicon operate as a single system. When information loses its body, equating humans and computers is especially easy, for the materiality in which the thinking mind is instantiated appears incidental to its essential nature. Moreover, the idea of the feedback loop implies that the boundaries of the autonomous subject are up for grabs, since feedback loops can flow not only *within* the subject but also *between* the subject and the environment. From Norbert Wiener on, the flow of information through feedback loops has been associated with the deconstruction of the liberal humanist subject, the version of the "human" with which I will be concerned. Although the "posthuman" differs in its articulations, a common theme is the union of the human with the intelligent machine.

What is the posthuman? Think of it as a point of view characterized by the following assumptions. (I do not mean this list to be exclusive or definitive. Rather, it names elements found at a variety of sites. It is meant to be suggestive rather than prescriptive.)[5] First, the posthuman view privileges informational pattern over material instantiation, so that embodiment in a biological substrate is seen as an accident of history rather than an inevitability of life. Second, the posthuman view considers consciousness, regarded as the seat of human identity in the Western tradition long before Descartes thought he was a mind thinking, as an epiphenomenon, as an evolutionary upstart trying to claim that it is the whole show when in actuality it is only a minor sideshow. Third, the posthuman view thinks of the body as the original prosthesis we all learn to manipulate, so that extending or replacing the body with other prostheses becomes a continuation of a process that began before we were born. Fourth, and most important, by these and other means, the posthuman view configures human being so that it can be seamlessly articulated with intelligent machines. In the posthuman, there are no essential differences or absolute demarcations between bodily existence and computer simulation, cybernetic mechanism and biological organism, robot teleology and human goals.

To elucidate the significant shift in underlying assumptions about subjectivity signaled by the posthuman, we can recall one of the definitive texts characterizing the liberal humanist subject: C. B. Macpherson's analysis of possessive individualism. "Its possessive quality is found in its conception of the individual as essentially the proprietor of his own person or capacities, *owing nothing to society for them.* . . . The human essence *is freedom from the wills of others,* and freedom is a function of possession."[6] The italicized phrases mark convenient points of departure for measuring the distance between the human and the posthuman. "Owing nothing to society" comes from arguments Hobbes and Locke constructed about humans in a "state of nature" before market relations arose. Because ownership of oneself is thought to predate market relations and owe nothing to them, it forms a foundation upon which those relations can be built, as when one sells one's labor for wages. As Macpherson points out, however, this imagined "state of nature" is a retrospective creation of a market society. The liberal self is *produced* by market relations and does not in fact predate them. This paradox (as Macpherson calls it) is resolved in the posthuman by doing away with the "natural" self. The posthuman subject is an amalgam, a collection of heterogeneous components, a material-informational entity whose boundaries undergo continuous construction and reconstruction. Consider the six-million-dollar man, a paradigmatic citizen of the posthuman regime. As his name implies, the parts of the self are indeed owned, but they are owned precisely because they were purchased, not because ownership is a natural condition preexisting market relations. Similarly, the presumption that there is an agency, desire, or will belonging to the self and clearly distinguished from the "wills of others" is undercut in the posthuman, for the posthuman's collective heterogeneous quality implies a distributed cognition located in disparate parts that may be in only tenuous communication with one another. We have only to recall Robocop's memory flashes that interfere with his programmed directives to understand how the distributed cognition of the posthuman complicates individual agency. If "human essence is freedom from the wills of others," the posthuman is "post" not because it is necessarily unfree but because there is no a priori way to identify a self-will that can be clearly distinguished from an other-will. Although these examples foreground the cybernetic aspect of the posthuman, it is important to recognize that the construction of the posthuman does not require the subject to be a literal cyborg. Whether or not interventions have been made on the body, new models of subjectivity emerging from such fields as cognitive science and artificial life imply that even a biologically unaltered *Homo sapiens* counts as posthuman. The defining characteristics involve the construction of subjectivity, not the presence of nonbiological components.

What to make of this shift from the human to the posthuman, which both evokes terror and excites pleasure? The liberal humanist subject has, of course, been cogently criticized from a number of perspectives. Feminist theorists have pointed out that it has historically been constructed as a white European male, presuming a universality that has worked to suppress and disenfranchise women's voices; postcolonial theorists have taken issue not only with the universality of the (white male) liberal subject but also with the very idea of a unified, consistent identity, focusing instead on hybridity; and postmodern theorists such as Gilles Deleuze and Felix Guattari have linked it with capitalism, arguing for the liberatory potential of a dispersed subjectivity distributed among diverse desiring machines they call "body without organs."[7] Although the deconstruction of the liberal humanist subject in cybernetics has some affinities with these perspectives, it proceeded primarily along lines that sought to understand human being as a set of informational processes. Because information had lost its body, this construction implied that embodiment is not essential to human being. Embodiment has been systematically downplayed or erased in the cybernetic construction of the posthuman in ways that have not occurred in other critiques of the liberal humanist subject, especially in feminist and postcolonial theories.

Indeed, one could argue that the erasure of embodiment is a feature common to *both* the liberal humanist subject and the cybernetic posthuman. Identified with the rational mind, the liberal subject *possessed* a body but was not usually represented as *being* a body. Only because the body is not identified with the self is it possible to claim for the liberal subject its notorious universality, a claim that depends on erasing markers of bodily difference, including sex, race, and ethnicity.[8] Gillian Brown, in her influential study of the relation between humanism and anorexia, shows that the anoretic's struggle to "decrement" the body is possible precisely because the body is understood as an object for control and mastery rather than as an intrinsic part of the self. Quoting an anoretic's remark—"You make out of your body your very own kingdom where you are the tyrant, the absolute dictator"—Brown states, "Anorexia is thus a fight for self-control, a flight from the slavery food threatens; self-sustaining self-possession independent of bodily desires is the anoretic's crucial goal.[9] In taking the self-possession implied by liberal humanism to the extreme, the anoretic creates a physical image that, in its skeletal emaciation, serves as material testimony that the locus of the liberal humanist subject lies in the mind, not the body. Although in many ways the posthuman deconstructs the liberal humanist subject, it thus shares with its predecessor an emphasis on cognition rather than embodiment. William Gibson makes the point vividly in *Neuromancer* when the narrator characterizes the posthuman body as "data made flesh."[10] To the extent that the posthuman constructs embodiment as the instantiation of thought/information, it continues the liberal tradition rather than disrupts it.

In tracing these continuities and discontinuities between a "natural" self and a cybernetic posthuman, I am not trying to recuperate the liberal subject. Although I think that serious consideration needs to be given to how certain characteristics associated with the liberal subject, especially agency and choice, can be articulated within a posthuman context, I do not mourn the passing of a concept so deeply entwined with projects of domination and oppression. Rather, I view the present moment as a critical juncture when interventions might be made to keep disembodiment from being rewritten, once again, into prevailing concepts of subjectivity. I see the deconstruction of the liberal humanist subject as an opportunity to put back into the picture the flesh that continues to be erased in contemporary discussions about cybernetic subjects. Hence my focus on how information lost its body, for this story is central to creating what Arthur Kroker has called the "flesh-eating 90s."[11] If my nightmare is a culture inhabited by posthumans who regard their bodies as fashion accessories rather than the ground of being, my dream is a version of the posthuman that embraces the possibilities of information technologies without being seduced by fantasies of unlimited power and disembodied immortality, that recognizes and celebrates finitude as a condition of human being, and that understands human life is embedded in a material world of great complexity, one on which we depend for our continued survival.

Perhaps it will now be clear that I mean my title, *How We Became Posthuman,* to connote multiple ironies, which do not prevent it from also being taken seriously. Taken straight, this title points to models of subjectivity sufficiently different from the liberal subject that if one assigns the term "human" to this subject, it makes sense to call the successor "posthuman." Some of the historical processes leading to this transformation are documented here, and in this sense the book makes good on its title. Yet my argument will repeatedly demonstrate that these changes were never complete transformations or sharp breaks; without exception, they reinscribed traditional ideas and assumptions even as they articulated something new. The changes announced by the title thus mean something more complex than "That was then, this is now." Rather, "human" and "posthuman" coexist in shifting configurations that vary with historically specific contexts. Given these complexities, the past tense in the title—"became"—is intended both to offer the reader the pleasurable shock of a double take and to reference ironically apocalyptic visions such as Moravec's prediction of a "postbiological" future for the human race.

Amplifying the ambiguities of the past tense are the ambiguities of the plural. In one sense, "we" refers to the readers of this book—readers who, by becoming aware of these new models of subjectivity (if they are not already familiar with them), may begin thinking of their actions in ways that have more in common with the posthuman than the human. Speaking for myself, I now find myself saying things like, "Well, my sleep agent wants to rest, but my food agent says I should go to the store." Each person who thinks this way begins to envision herself or himself as a posthuman collectivity, an "I" transformed into the "we" of autonomous agents operating together to make a self. The infectious power of this way of thinking gives "we" a performative dimension. People become posthuman because they think they are posthuman. In another sense "we," like "became," is meant ironically, positioning itself in opposition to the techno-ecstasies found in various magazines, such as *Mondo 2000,* which customarily speak of the transformation into the posthuman as if it were a universal human condition when in fact it affects only a small fraction of the world's population— a point to which I will return.

The larger trajectory of my narrative arcs from the initial moments when cybernetics was formulated as a discipline, through a period of reformulation known as "second-order cybernetics," to contemporary debates swirling around an emerging discipline known as "artificial life." Although the progression is chronological, this book is not meant to be a history of cybernetics. Many figures not discussed here played important roles in that history, and I have not attempted to detail their contributions. Rather, my selection of theories and researchers has been dictated by a desire to show *the complex interplays between embodied forms of subjectivity and arguments for disembodiment throughout the cybernetic tradition.* In broad outline, these interplays occurred in three distinct waves of development. The first, from 1945 to 1960, took homeostasis as a central concept; the second, going roughly from 1960 to 1980, revolved around reflexivity; and the third, stretching from 1980 to the present, highlights virtuality. Let me turn now to a brief sketch of these three periods.

During the foundational era of cybernetics, Norbert Wiener, John von Neumann, Claude Shannon, Warren McCulloch, and dozens of other distinguished researchers met at annual conferences sponsored by the Josiah Macy Foundation to formulate the central concepts that, in their high expectations, would coalesce into a theory of communication and control applying equally to animals, humans, and machines. Retrospectively called the Macy Conferences on Cybernetics, these meetings, held from 1943 to 1954, were instrumental in forging a new paradigm.[12] To succeed, they needed a theory of information (Shannon's bailiwick), a model of neural functioning that showed how neurons worked as information-processing systems (McCulloch's lifework), computers that processed binary code and that could conceivably reproduce themselves, thus reinforcing the analogy with biological systems (von Neumann's specialty), and a visionary who could articulate the larger implications of the cybernetic paradigm and make clear its cosmic significance (Wiener's contribution). The result of this breathtaking enterprise was nothing less than a new way of looking at human beings. Henceforth, humans were to be seen primarily as information-processing entities who are *essentially* similar to intelligent machines.

The revolutionary implications of this paradigm notwithstanding, Wiener did not intend to dismantle the liberal humanist subject. He was less interested in seeing humans as machines than he was in fashioning human and machine alike in the image of an autonomous, self-directed individual. In aligning cybernetics with liberal humanism, he was following a strain of thought that, since the Enlightenment, had argued that human beings could be trusted with freedom because they and the social structures they devised operated as self-regulating mechanisms.[13] For Wiener, cybernetics was a means to extend liberal humanism, not subvert it. The point was less to show that man was a machine than to demonstrate that a machine could function like a man.

Yet the cybernetic perspective had a certain inexorable logic that, especially when fed by wartime hysteria, also worked to undermine the very liberal subjectivity that Wiener wanted to preserve. These tensions were kept under control during the Macy period partly through a strong emphasis on *homeostasis*.[14] Traditionally, homeostasis had been understood as the ability of living organisms to maintain steady states when they are buffeted by fickle environments. When the temperature soars, sweat pours out of the human body so that its internal temperature can remain relatively stable. During the Macy period, the idea of homeostasis was extended to machines. Like animals, machines can maintain homeostasis using feedback loops. Feedback loops had long been exploited to increase the stability of mechanical systems, reaching a high level of development during the mid-to-late nineteenth century with the growing sophistication of steam engines and their accompanying control devices, such as governors. It was not until the 1930s and 1940s, however, that the feedback loop was explicitly theorized as a flow of information. Cybernetics was born when nineteenth-century control theory joined with the nascent theory of information.[15] Coined from the Greek word for "steersman," cybernetics signaled that three powerful actors—information, control, and communication—were now operating jointly to bring about an unprecedented synthesis of the organic and the mechanical.

Although the informational feedback loop was initially linked with homeostasis, it quickly led to the more threatening and subversive idea of *reflexivity*. A few years ago I co-taught, with a philosopher and a physicist, a course on reflexivity. As we discussed reflexivity in the writings of Aristotle, Fichte, Kierkegaard, Gödel, Turing, Borges, and Calvino, aided by the insightful analyses of Roger Penrose and Douglas Hofstader, I was struck not only by the concept's extraordinarily rich history but also by its tendency to mutate, so that virtually any formulation is sure to leave out some relevant instances. Instructed by the experience, I offer the following tentative definition, which I hope will prove adequate for our purposes here. *Reflexivity is the movement whereby that which has been used to generate a system is made, through a changed perspective, to become part of the system it generates.* When Kurt Gödel invented a method of coding that allowed statements of number theory also to function as statements *about* number theory, he entangled that which generates the system with the system. When M. C. Escher drew two hands drawing each other, he took that which is presumed to generate the picture—the sketching hand—and made it part of the picture it draws. When Jorge Luis Borges in "The Circular Ruins" imagines a narrator who creates a student through his dreaming only to discover that he himself is being dreamed by another, the system generating a reality is shown to be part of the reality it makes. As these examples illustrate, reflexivity has subversive effects because it confuses and entangles the boundaries we impose on the world in order to make sense of that world. Reflexivity tends notoriously toward infinite regress. The dreamer creates the student, but the dreamer in turn is dreamed by another, who in his turn is dreamed by someone else, and so on to infinity.

This definition of reflexivity has much in common with some of the most influential and provocative recent work in critical theory, cultural studies, and the social studies of science. Typically, these works make the reflexive move of showing that an attribute previously considered to have emerged from a set of preexisting conditions is in fact used to generate the conditions. In Nancy Armstrong's *Desire and Domestic Fiction: A Political History of the Novel,* for example, bourgeois femininity is shown to be constructed through the domestic fictions that represent it as already in place.[16] In Michael Warner's *The Letters of the Republic: Publication and the Public Sphere in Eighteenth-Century America,* the founding document of the United States, the Constitution, is shown to produce the very people whose existence it presupposes.[17] In Bruno Latour's *Science in Action: How to Follow Scientists and Engineers through Society,* scientific experiments are shown to produce the nature whose existence they predicate as their condition of possibility.[18] It is only a slight exaggeration to say that contemporary critical theory is produced by the reflexivity that it also produces (an observation that is, of course, also reflexive).

Reflexivity entered cybernetics primarily through discussions about the observer. By and large, first-wave cybernetics followed traditional scientific protocols in considering observers to be outside the system they observe. Yet cybernetics also had implications that subverted this premise. The objectivist view sees information flowing from the system to the observers, but feedback can also loop *through* the observers, drawing them in to become part of the system being observed. Although participants remarked on this aspect of the cybernetic paradigm throughout the Macy transcripts, they lacked a single word to describe it. To my knowledge, the word "reflexivity" does not appear in the transcripts. This meant they had no handle with which to grasp this slippery concept, no signifier that would help to constitute as well as to describe the changed perspective that reflexivity entails. Discussions of the idea remained diffuse. Most participants did not go beyond remarking on the shifting boundaries between observer and system that cybernetics puts into play. With some exceptions, deeper formulations of the problem failed to coalesce during the Macy discussions.

The most notable exception turned out to hurt more than it helped. Lawrence Kubie, a hard-line Freudian psychoanalyst, introduced a reflexive perspective when he argued that every utterance is doubly encoded, acting both as a statement about the outside world and as a mirror reflecting the speaker's psyche. If reflexivity was already a subversive concept, this interpretation made it doubly so, for it threatened to dissolve the premise of scientific objectivity shared by the physical scientists in the Macy group. Their reactions to Kubie's presentations show them shying away from reflexivity, preferring to shift the conversation onto more comfortable ground. Nevertheless, the idea hung in the air, and a few key thinkers—especially Margaret Mead, Gregory Bateson, and Heinz von Foerster—resolved to pursue it after the Macy Conferences ran out of steam.

The second wave of cybernetics grew out of attempts to incorporate reflexivity into the cybernetic paradigm at a fundamental level. The key issue was how systems are constituted as such, and the key problem was how to redefine homeostatic systems so that the observer can be taken into account. The second wave was initiated by, among others, Heinz von Foerster, the Austrian émigré who became coeditor of the Macy transcripts. This phase can be dated from 1960, when von Foerster wrote the first of the essays that were later collected in his influential book *Observing Systems*.[19] As von Foerster's punning title recognizes, the observer of systems can himself be constituted as a system to be observed. Von Foerster called the models he presented in these essays "second-order cybernetics" because they extended cybernetic principles to the cyberneticians themselves. The second wave reached its mature phase with the publication of Humberto Maturana and Francisco Varela's *Autopoiesis and Cognition: The Realization of the Living*.[20] Building on Maturana's work on reflexivity in sensory processing and Varela's on the dynamics of autonomous biological systems, the two authors expanded the reflexive turn into a fully articulated epistemology that sees the world as a set of informationally closed systems. Organisms respond to their environment in ways determined by their internal self-organization. Their one and only goal is continually to produce and reproduce the organization that defines them as systems. Hence, they not only are self-organizing but also are autopoietic, or self-making. Through Maturana and Varela's work and that of other influential theorists such as German sociologist Niklas Luhmann,[21] cybernetics by 1980 had spun off from the idea of reflexive feedback loops a theory of autopoiesis with sweeping epistemological implications.

In a sense, autopoiesis turns the cybernetic paradigm inside out. Its central premise—that systems are informationally closed—radically alters the idea of the informational feedback loop, for the loop no longer functions to connect a system to its environment. In the autopoietic view, no information crosses the boundary separating the system from its environment. We do not see a world "out there" that exists apart from us. Rather, we see only what our systemic organization allows us to see. The environment merely *triggers* changes determined by the system's own structural properties. Thus the center of interest for autopoiesis shifts from the cybernetics of the observed system to the cybernetics of the observer. Autopoiesis also changes the explanation of what circulates through the system to make it work as a system. The emphasis now is on the mutually constitutive interactions between the components of a system rather than on message, signal, or information. Indeed, one could say either that information does not exist in this paradigm or that it has sunk so deeply into the system as to become indistinguishable from the organizational properties defining the system as such.

The third wave swelled into existence when self-organization began to be understood not merely as the (re)production of internal organization but as the springboard to emergence. In the rapidly emerging field of artificial life, computer programs are designed to allow "creatures" (that is, discrete packets of computer codes) to evolve spontaneously in directions the programmer may not have anticipated. The intent is to evolve the *capacity* to evolve. Some researchers have argued that such self-evolving programs are not merely models of life but are themselves alive. What assumptions make this claim plausible? If one sees the universe as composed essentially of information, it makes sense that these "creatures" are life *forms* because they have the form of life, that is, an informational code. As a result, the theoretical bases used to categorize all life undergo a significant shift. As we shall see in chapters 9 and 10, when these theories are applied to human beings, *Homo sapiens* are so transfigured in conception and purpose that they can appropriately be called posthuman.

The emergence of the posthuman as an informational-material entity is paralleled and reinforced by a corresponding reinterpretation of the deep structures of the physical world. Some theorists, notably Edward Fredkin and Stephen Wolfram, claim that reality is a program run on a cosmic computer.[22] In this view, a universal informational code underlies the structure of matter, energy, spacetime—indeed, of everything that exists. The code is instantiated in cellular automata, elementary units that can occupy two states: on or off. Although the jury is still out on the cellular automata model, it may indeed prove to be a robust way to understand reality. Even now, a research team headed by Fredkin is working on showing how quantum mechanics can be derived from an underlying cellular automata model.

What happens to the embodied lifeworld of humans in this paradigm? In itself, the cellular automata model is not necessarily incompatible with recognizing that humans are embodied beings, for embodiment can flow from cellular automata as easily as from atoms. No one suggests that because atoms are mostly empty space, we can shuck the electron shells and do away with occupying space altogether. Yet the cultural contexts and technological histories in which cellular automata theories are embedded encourage a comparable fantasy—that because we are essentially information, we can do away with the body. Central to this argument is a conceptualization that sees information and materiality as distinct entities. This separation allows the construction of a hierarchy in which information is given the dominant position and materiality runs a distant second. As though we had learned nothing from Derrida about supplementarity, embodiment continues to be discussed as if it were a supplement to be purged from the dominant term of information, an accident of evolution we are now in a position to correct.

It is this materiality/information separation that I want to contest—not the cellular automata model, information theory, or a host of related theories in themselves. My strategy is to complicate the leap from embodied reality to abstract information by pointing to moments when the assumptions involved in this move were contested by other researchers in the field and so became especially visible. The point of highlighting such moments is to make clear how much had to be erased to arrive at such abstractions as bodiless information. Abstraction is of course an essential component in all theorizing, for no theory can account for the infinite multiplicity of our interactions with the real. But when we make moves that erase the world's multiplicity, we risk losing sight of the variegated leaves, fractal branchings, and particular bark textures that make up the forest. In the pages that follow, I will identify two moves in particular that played important roles in constructing the information/materiality hierarchy. Irreverently, I think of them as the Platonic backhand and forehand.

The Platonic backhand works by inferring from the world's noisy multiplicity a simplified abstraction. So far so good: this is what theorizing should do. The problem comes when the move circles around to constitute the abstraction as the originary form from which the world's multiplicity derives. Then complexity appears as a "fuzzing up" of an essential reality rather than as a manifestation of the world's holistic nature. Whereas the Platonic backhand has a history dating back to the Greeks, the Platonic forehand is more recent. To reach fully developed form, it required the assistance of powerful computers. This move starts from simplified abstractions and, using simulation techniques such as genetic algorithms, *evolves* a multiplicity sufficiently complex that it can be seen as a world of its own. The two moves thus make their play in opposite directions. The backhand goes from noisy multiplicity to reductive simplicity, whereas the forehand swings from simplicity to mulilicity. They share a common ideology—privileging the abstract as the Real and downplaying the importance of material instantiation. When they work together, they lay the groundwork for a new variation on an ancient game, in which disembodied information becomes the ultimate Platonic Form. If we can capture the Form of ones and zeros in a nonbiological medium—say, on a computer disk—why do we need the body's superfluous flesh?

Whether the enabling assumptions for this conception of information occur in information theory, cybernetics, or popular science books such as *Mind Children,* their appeal is clear. Information viewed as pattern and not tied to a particular instantiation is information free to travel across time and space. Hackers are not the only ones who believe that information wants to be free. The great dream and promise of information is that it can be free from the material constraints that govern the mortal world. Marvin Minsky precisely expressed this dream when, in a recent lecture, he suggested it will soon be possible to extract human memories from the brain and import them, intact and unchanged, to computer disks.[23] The clear implication is that if we can become the information we have constructed, we can achieve effective immortality.

In the face of such a powerful dream, it can be a shock to remember that for information to exist, it must *always* be instantiated in a medium, whether that medium is the page from the *Bell Laboratories Journal* on which Shannon's equations are printed, the computer-generated topological maps used by the Human Genome Project, or the cathode ray tube on which virtual worlds are imaged. The point is not only that abstracting information from a material base is an imaginary act but also, and more fundamentally, that conceiving of information as a thing separate from the medium instantiating it is a prior imaginary act that constructs a holistic phenomenon as an information/matter duality.[24]

The chapters that follow will show what had to be elided, suppressed, and forgotten to make information lose its body. This book is a "rememory" in the sense of Toni Morrison's *Beloved*: putting back together parts that have lost touch with one another and reaching out toward a complexity too unruly to fit into disembodied ones and zeros.

Seriation, Skeuomorphs, and Conceptual Constellations

The foregoing leads to a strategic definition of "virtuality." *Virtuality is the cultural perception that material objects are interpenetrated by informa-*

tion patterns. The definition plays off the duality at the heart of the condition of virtuality—materiality on the one hand, information on the other. Normally virtuality is associated with computer simulations that put the body into a feedback loop with a computer-generated image. For example, in virtual Ping-Pong, one swings a paddle wired into a computer, which calculates from the paddle's momentum and position where the ball would go. Instead of hitting a real ball, the player makes the appropriate motions with the paddle and watches the image of the ball on a computer monitor. Thus the game takes place partly in real life (RL) and partly in virtual reality (VR). Virtual reality technologies are fascinating because they make visually immediate the perception that a world of information exists parallel to the "real" world, the former intersecting the latter at many points and in many ways. Hence the definition's strategic quality, strategic because it seeks to connect virtual technologies with the sense, pervasive in the late twentieth century, that all material objects are interpenetrated by flows of information, from DNA code to the global reach of the World Wide Web.

Seeing the world as an interplay between informational patterns and material objects is a historically specific construction that emerged in the wake of World War II.[25] By 1948, the distinction had coalesced sufficiently for Wiener to articulate it as a criterion that any adequate theory of materiality would be forced to meet. "Information is information, not matter or energy. No materialism which does not admit this can survive at the present day."[26] Wiener knew as well as anyone else that to succeed, this conception of information required artifacts that could embody it and make it real. When I say virtuality is a cultural perception, I do not mean that it is merely a psychological phenomenon. It is instantiated in an array of powerful technologies. The perception of virtuality facilitates the development of virtual technologies, and the technologies reinforce the perception.

The feedback loops that run between technologies and perceptions, artifacts and ideas, have important implications for how historical change occurs. The development of cybernetics followed neither a Kuhnian model of incommensurable paradigms nor a Foucauldian model of sharp epistemic breaks.[27] In the history of cybernetics, ideas were rarely made up out of whole cloth. Rather, they were fabricated in a pattern of overlapping replication and innovation, a pattern that I call "seriation" (a term appropriated from archaeological anthropology). A brief explanation may clarify this concept. Within archaeological anthropology, changes in artifacts are customarily mapped through seriation charts. One constructs a seriation chart by parsing an artifact as a set of attributes that change over time. Suppose a researcher wants to construct a seriation chart for lamps. A key attribute is the element that gives off light. The first lamps, dating from thousands of years ago, used wicks for this element. Later, with the discovery of electricity, wicks gave way to filaments. The figures that customarily emerge from this kind of analysis are shaped like a tiger's iris—narrow at the top when an attribute first begins to be introduced, with a bulge in the middle during the heyday of the attribute, and tapered off at the bottom as the shift to a new model is completed. On a seriation chart for lamps, a line drawn at 1890 would show the figure for wicks waxing large with the figure for filaments intersected at the narrow tip of the top end. Fifty years later, the wick figure would be tapering off, and the filament figure would be widening into its middle section. Considered as a set, the figures depicting changes in the attributes of an artifact reveal patterns of overlapping innovation and replication. Some attributes change from one model to the next, but others remain the same.

As figure 1 illustrates, the conceptual shifts that took place during the development of cybernetics display a seriated pattern reminiscent of material changes in artifacts. Conceptual fields evolve similarly to material culture, in part because concept and artifact engage each other in continuous feedback loops. An artifact materially expresses the concept it embodies, but the process of its construction is far from passive. A glitch has to be fixed, a material exhibits unexpected properties, an emergent behavior surfaces—any of these challenges can give rise to a new concept, which results in another generation of artifact, which leads to the development of still other concepts. The reasoning suggests that we should be able to trace the development of a conceptual field by using a seriation chart analogous to the seriation charts used for artifacts.

In the course of the Macy Conferences, certain ideas came to be associated with each other. Through a cumulative process that continued across several years of discussions, these ideas were seen as mutually entailing each other until, like love and marriage, they were viewed by the participants as naturally going together. Such a constellation is the conceptual entity corresponding to an artifact, possessing an internal coherence that defines it as an operational unit. Its formation marks the beginning of a period; its disassembly and reconstruction signal the transition to a different period. Indeed, periods are recognizable as such largely because constellations possess this coherence. Rarely is a constellation discarded wholesale. Rather, some of the ideas composing it are discarded, others are modified, and new ones are introduced. Like the attributes composing an artifact, the ideas in a constellation change in a patchwork pattern of old and new.

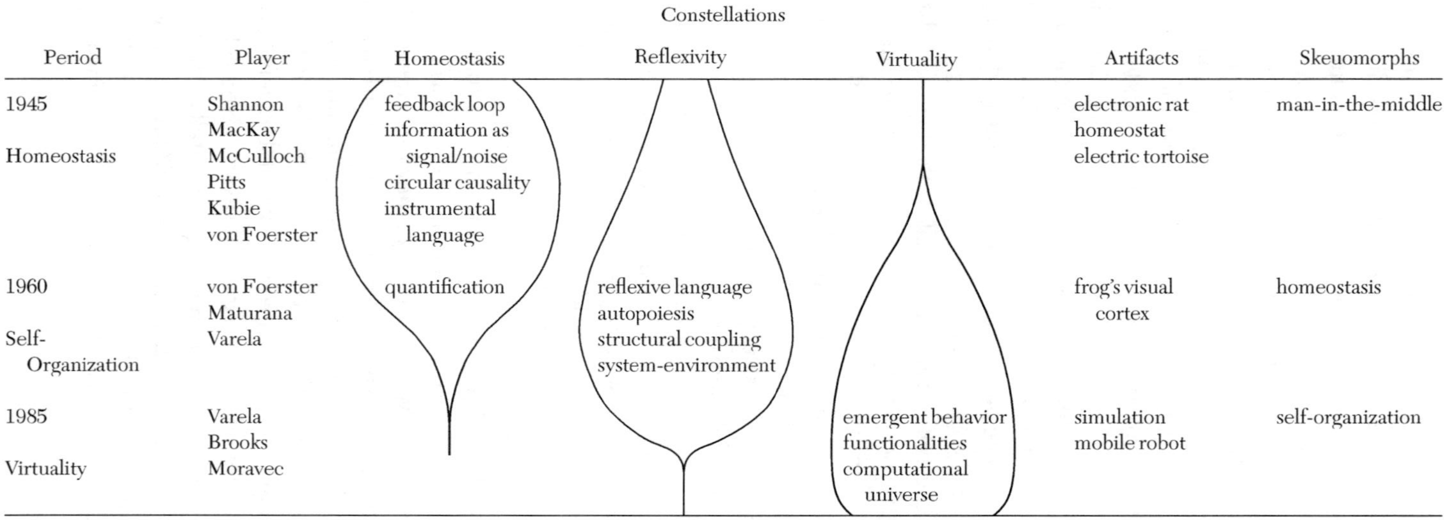

Constellations

Period	Player	Homeostasis	Reflexivity	Virtuality	Artifacts	Skeuomorphs
1945 Homeostasis	Shannon MacKay McCulloch Pitts Kubie von Foerster	feedback loop information as signal/noise circular causality instrumental language quantification			electronic rat homeostat electric tortoise	man-in-the-middle
1960 Self-Organization	von Foerster Maturana Varela		reflexive language autopoiesis structural coupling system-environment		frog's visual cortex	homeostasis
1985 Virtuality	Varela Brooks Moravec			emergent behavior functionalities computational universe	simulation mobile robot	self-organization

FIGURE 1   The three waves of cybernetics

Here I want to introduce another term from archaeological anthropology. A *skeuomorph* is a design feature that is no longer functional in itself but that refers back to a feature that was functional at an earlier time. The dashboard of my Toyota Camry, for example, is covered by vinyl molded to simulate stitching. The simulated stitching alludes back to a fabric that was in fact stitched, although the vinyl "stitching" is formed by an injection mold. Skeuomorphs visibly testify to the social or psychological necessity for innovation to be tempered by replication. Like anachronisms, their pejorative first cousins, skeuomorphs are not unusual. On the contrary, they are so deeply characteristic of the evolution of concepts and artifacts that it takes a great deal of conscious effort to avoid them. At SIGGRAPH, the annual computer trade show where dealers come to hawk their wares, hard and soft, there are almost as many skeuomorphs as morphs.

The complex psychological functions a skeuomorph performs can be illustrated by an installation exhibited at SIGGRAPH '93. Called the "Catholic Turing Test," the simulation invited the viewer to make a confession by choosing selections from the video screen; it even had a bench on which the viewer could kneel.[28] On one level, the installation alluded to the triumph of science over religion, for the role of divinely authorized interrogation and absolution had been taken over by a machine algorithm. On another level, the installation pointed to the intransigence of conditioned behavior, for the machine's form and function were determined by its religious predecessor. Like a Janus figure, the skeuomorph looks to past and future, simultaneously reinforcing and undermining both. It calls into a play a psychodynamic that finds the new more acceptable when it recalls the old that it is in the process of displacing and finds the traditional more comfortable when it is presented in a context that reminds us we can escape from it into the new.

In the history of cybernetics, skeuomorphs acted as threshold devices, smoothing the transition between one conceptual constellation and another. Homeostasis, a foundational concept during the first wave, functioned during the second wave as a skeuomorph. Although homeostasis remained an important concept in biology, by about 1960 it had ceased to be an initiating premise in cybernetics. Instead, it performed the work of a gesture or an allusion used to authenticate new elements in the emerging constellation of reflexivity. At the same time, it also exerted an inertial pull on the new elements, limiting how radically they could transform the constellation.

A similar phenomenon appears in the transition from the second to the third wave. Reflexivity, the key concept of the second wave, is displaced in the third wave by emergence. Like homeostasis, reflexivity does not altogether disappear but lingers on as an allusion that authenticates new elements. It performs a more complex role than mere nostalgia, however, for it also leaves its imprint on the new constellation of virtuality. The complex story formed by these seriated changes is told in chapters 3, 6, and 9, which discuss cybernetics, autopoiesis, and artificial life, respectively.

I have already suggested that living in a condition of virtuality implies we participate in the cultural perception that information and materiality are conceptually distinct and that information is in some sense more essential, more important, and more fundamental than materiality. The preamble to "A Magna Carta for the Knowledge Age," a document coauthored by Alvin Toffler at the behest of Newt Gingrich, concisely sums up the matter by proclaiming, "The central event of the 20th century is the overthrow of matter.(29) To see how this view began to acquire momentum, let us briefly flash back to 1948 when Claude Shannon, a brilliant theorist working at Bell Laboratories, defined a mathematical quantity he called information and proved several important theorems concerning it.[30]

Information Theory and Everyday Life

Shannon's theory defines information as a probability function with no dimensions, no materiality, and no necessary connection with meaning. It is a pattern, not a presence. (Chapter 3 talks about the development of information theory in more detail, and the relevant equations can be found there.) The theory makes a strong distinction between message and signal. Lacan to the contrary, a message does not always arrive at its destination. In information theoretic terms, no message is ever sent. What is sent is a signal. Only when the message is encoded in a signal for transmission through a medium—for example, when ink is printed on paper or when electrical pulses are sent racing along telegraph wires—does it assume material form. The very definition of "information," then, encodes the distinction between materiality and information that was also becoming important in molecular biology during this period.[31]

Why did Shannon define information as a pattern? The transcripts of the Macy Conferences indicate that the choice was driven by the twin engines of reliable quantification and theoretical generality. As we shall see in chapter 3, Shannon's formulation was not the only proposal on the table. Donald MacKay, a British researcher, argued for an alternative definition that linked information with change in a receiver's mindset and thus with meaning.[32] To be workable, MacKay's definition required that psychological states be quantifiable and measurable—an accomplishment that only now appears distantly possible with such imaging technologies as positron-emission tomography and that certainly was not in reach in the immediate post–World War II years. It is no mystery why Shannon's definition rather than MacKay's became the industry standard.

Shannon's approach had other advantages that turned out to incur large (and mounting) costs when his premise interacted with certain predispositions already at work within the culture. Abstracting information from a material base meant that information could become free-floating, unaffected by changes in context. The technical leverage this move gained was considerable, for by formalizing information into a mathematical function, Shannon was able to develop theorems, powerful in their generality, that hold true regardless of the medium in which the information is instantiated. Not everyone agreed this move was a good idea, however, despite its theoretical power. As Carolyn Marvin notes, a decontextualized construction of information has important ideological implications, including an Anglo-American ethnocentrism that regards digital information as more important than more context-bound analog information.[33] Even in Shannon's day, malcontents grumbled that divorcing information from context and thus from meaning had made the theory so narrowly formalized that it was not useful as a general theory of communication. Shannon himself frequently cautioned that the theory was meant to apply only to certain technical situations, not to communication in general.[34] In other circumstances, the theory might have become a dead end, a victim of its own excessive formalization and decontextualization. But not in the post–World War II era. The time was ripe for theories that reified information into a free-floating, decontextualized, quantifiable entity that could serve as the master key unlocking secrets of life and death.

Technical artifacts help to make an information theoretic view a part of everyday life. From ATMs to the Internet, from the morphing programs used in *Terminator II* to the sophisticated visualization programs used to guide microsurgeries, information is increasingly perceived as interpenetrating material forms. Especially for users who may not know the material processes involved, the impression is created that pattern is predominant over presence. From here it is a small step to perceiving information as more mobile, more important, more *essential* than material forms. When this impression becomes part of your cultural mindset, you have entered the condition of virtuality.

U.S. culture at present is in a highly heterogeneous state regarding the condition of virtuality. Some high-tech preserves (elite research centers such as Xerox Palo Alto Research Center and Bell Laboratories, most major research universities, and hundreds of corporations) have so thoroughly incorporated virtual technologies into their infrastructures that information is as much as part of the researchers' mindscapes as is electric lighting or synthetic plastics.[35] The thirty million Americans who are plugged into the Internet increasingly engage in virtual experiences enacting a division between the material body that exits on one side of the screen and the computer simulacra that seem to create a space inside the screen.[36] Yet for millions more, virtuality is not even a cloud on the horizon of their everyday worlds. Within a global context, the experience of virtuality becomes more exotic by several orders of magnitude. It is a useful corrective to remember that 70 percent of the world's population has never made a telephone call.

Nevertheless, I think it is a mistake to underestimate the importance of virtuality, for it wields an influence altogether disproportionate to the number of people immersed in it. It is no accident that the condition of virtuality is most pervasive and advanced where the centers of power are most concentrated. Theorists at the Pentagon, for example, see it as the theater in which future wars will be fought. They argue that coming conflicts will be decided not so much by overwhelming force as by "neocortical warfare," waged through the techno-sciences of information.[37] If we want to contest what these technologies signify, we need histories that show the erasures that went into creating the condition of virtuality, as well as visions arguing for the importance of embodiment. Once we understand the complex interplays that went into creating the condition of virtuality, we can demystify our progress toward virtuality and see it as the result of historically specific negotiations rather than of the irresistible force of technological determinism. At the same time, we can acquire resources with which to rethink the assumptions underlying virtuality, and we can recover a sense of the virtual that fully recognizes the importance of the embodied processes constituting the lifeworld of human beings.[38] In the phrase "virtual bodies," I intend to allude to the historical separation between information and materiality and also to recall the embodied processes that resist this division.

Virtuality and Contemporary Literature

I have already suggested that one way to think about the organization of this book is chronologically, since it follows the three waves of seriated changes in cybernetics. In this organization of the textual body, each of the three chronologically arranged divisions has an anchoring chapter discussing the scientific theories: on the Macy Conferences (chapter 3); on autopoiesis (chapter 6); and on artificial life (chapter 9), respectively. Each section also has a chapter showing specific applications of the theories: the work of Norbert Wiener (chapter 4); tape-recording technologies (chapter 8); and human-computer interactions (chapter 10). Also included in each of the three divisions are chapters on literary texts contemporaneous with the development of the scientific theories and cybernetic technologies (chapters 5, 7, and 10). I have selected literary texts that were clearly influenced by the development of cybernetics. Nevertheless, I want to resist the idea that influence flows from science into literature. The cross-currents are considerably more complex than a one-way model of influence would allow. In the *Neuromancer* trilogy, for example, William Gibson's vision of cyberspace had a considerable effect on the development of three-dimensional virtual reality imaging software.[39]

A second way to think about the organization of *How We Became Posthuman* is narratively. In this arrangement, the three divisions proceed not so much through chronological progression as through the narrative strands about the (lost) body of information, the cyborg body, and the posthuman body. Here the literary texts play a central role, for they display the passageways that enabled stories coming out of narrowly focused scientific theories to circulate more widely through the body politic. Many of the scientists understood very well that their negotiations involved premises broader than the formal scope of their theories strictly allowed. Because of the wedge that has been driven between science and values in U.S. culture, their statements on these wider implications necessarily occupied the position of ad hoc pronouncements rather than "scientific" arguments. Shaped by different conventions, the literary texts range across a spectrum of issues that the scientific texts only fitfully illuminate, including the ethical and cultural implications of cybernetic technologies.[40]

Literary texts are not, of course, merely passive conduits. They actively shape what the technologies mean and what the scientific theories signify in cultural contexts. They also embody assumptions similar to those that permeated the scientific theories at critical points. These assumptions included the idea that stability is a desirable social goal, that human beings and human social organizations are self-organizing structures, and that form is more essential than matter. The scientific theories used these assumptions as enabling presuppositions that helped to guide inquiry and shape research agendas. As the chapters on the scientific developments will show, culture circulates through science no less than science circulates through culture. The heart that keeps this circulatory system flowing is narrative—narratives about culture, narratives within culture, narratives about science, narratives within science. In my account of the scientific developments, I have sought to emphasize the role that narrative plays in articulating the posthuman as a technical-cultural concept. For example, chapter 4, on Wiener's scientific work, is interlaced with analyses of the narratives he tells to resolve conflicts between cybernetics and liberal humanism, and chapter 9, on artificial life, is organized by looking at this area of research as a narrative field.

What does this emphasis on narrative have to do with virtual bodies? Following Jean-François Lyotard, many theorists of postmodernity accept that the postmodern condition implies an incredulity toward metanarrative.[41] As we have seen, one way to construct virtuality is the way that Moravec and Minsky do—as a metanarrative about the transformation of the human into a disembodied posthuman. I think we should be skeptical about this metanarrative. To contest it, I want to use the resources of narrative itself, particularly its resistance to various forms of abstraction and disembodiment. With its chronological thrust, polymorphous digressions, located actions, and personified agents, narrative is a more *embodied* form of discourse than is analytically driven systems theory. By turning the technological determinism of bodiless information, the cyborg, and the posthuman into narratives about the negotiations that took place between particular people at particular times and places, I hope to replace a teleology of disembodiment with historically contingent stories about contests between competing factions, contests whose outcomes were far from obvious. Many factors affected the outcomes, from the needs of emerging technologies for reliable quantification to the personalities of the people involved. Though overdetermined, the disembodiment of information was not inevitable, any more than it is inevitable we continue to accept the idea that we are *essentially* informational patterns.

In this regard, the literary texts do more than explore the cultural implications of scientific theories and technological artifacts. Embedding ideas and artifacts in the situated specificities of narrative, the literary texts give these ideas and artifacts a local habitation and a name through discursive formulations whose effects are specific to that textual body. In exploring these effects, I want to demonstrate, on multiple levels and in many ways, that abstract pattern can never fully capture the embodied actuality, unless it is as prolix and noisy as the body itself. Shifting the emphasis from technological determinism to competing, contingent, embodied narratives about the scientific developments is one way to liberate the resources of narrative so that they work against the grain of abstraction running through the teleology of disembodiment. Another way is to read literary texts alongside scientific theories. In articulating the connections that run through these two discursive realms, I want to entangle abstract form and material particularity such that the reader will find it increasingly difficult to maintain the perception that they are separate and discrete entities. If, for cultural and historical reasons, I cannot start from a holistic perspective, I hope to mix things up enough so that the emphasis falls not on the separation of matter and information but on their inextricably complex compoundings and entwinings. For this project, the literary texts with their fashionings of embodied particularities are crucial.

The first literary text I discuss in detail is Bernard Wolfe's *Limbo*.[42] Written in the 1950s, *Limbo* has become something of an underground classic. It imagines a postwar society in which an ideology, Immob, has developed; the ideology equates aggression with the ability to move. "Pacifism equals passivity," Immob slogans declare. True believers volunteer to banish their mobility (and presumably their aggression) by having amputations, which have come to be regarded as signifiers of social power and influence. These amputees get bored with lying around, however, so a vigorous cybernetics industry has grown up to replace their missing limbs. As this brief summary suggests, *Limbo* is deeply influenced by cybernetics. But the technical achievements of cybernetics are not at the center of the text. Rather, they serve as a springboard to explore a variety of social, political, and psychological issues, ranging from the perceived threat that women's active sexuality poses for Immob men to global East-West tensions that explode into another world war at the end of the text. Although it is unusually didactic, *Limbo* does more than discuss cybernetics; it engages a full range of rhetorical and narrative devices that work both with and against its explicit pronouncements. The narrator seems only partially able to control his verbally extravagant narrative. There are, I will argue, deep connections between the narrator's struggle to maintain control of the narrative and the threat to "natural" body boundaries posed by the cybernetic paradigm. *Limbo* interrogates a dynamic that also appears in Norbert Wiener's work—the intense anxiety that erupts when the perceived boundaries of the body are breached. In addition, it illustrates how the body of the text gets implicated in the processes used to represent bodies within the text.

Several Philip K. Dick novels written from 1962 to 1966 (including *We Can Build You, Do Androids Dream of Electric Sheep?, Dr. Bloodmoney,* and *Ubik*) provide another set of texts through which the multiple implications of the posthuman can be explored.[43] Chronologically and thematically, Dick's novels of simulation cross the scientific theory of autopoiesis. Like Maturana, Varela, and other scientific researchers in the second wave of cybernetics, Dick is intensely concerned with epistemological questions and their relation to the cybernetic paradigm. The problem of where to locate the observer—in or out of the system being observed?—is conflated in his fiction with how to determine whether a creature is android or human. For Dick, the android is deeply bound up with the gender politics of his male protagonists' relations with female characters, who ambiguously figure either as sympathetic, life-giving "dark-haired girls" or emotionally cold, life-threatening schizoid women. Already fascinated with epistemological questions that reveal how shaky our constructions of reality can be, Dick is drawn to cybernetic themes because he understands that cybernetics radically destabilizes the ontological foundations of what counts as human. The gender politics he writes into his novels illustrate the potent connections between cybernetics and contemporary understandings of race, gender, and sexuality.

The chapter on contemporary speculative fictions constructs a semiotics of virtuality by showing how the central concepts of information and materiality can be mapped onto a multilayered semiotic square. The tutor texts for this analysis, which include *Snow Crash*, *Blood Music*, *Galatea 2.2*, and *Terminal Games*, indicate the range of what counts as the posthuman in the age of virtuality, from neural nets to hackers, biologically modified humans, and entities who live only in computer simulations.[44] In following the construction of the posthuman in these texts, I will argue that older ideas are reinscribed as well as contested. As was the case for the scientific models, change occurs in a seriated pattern of overlapping innovation and replication.

I hope that this book will demonstrate, once again, how crucial it is to recognize interrelations between different kinds of cultural productions, specifically literature and science. The stories I tell here—how information lost its body, how the cyborg was created as a cultural icon and technological artifact, and how humans became posthumans—and the waves of historical change I chart would not have the same resonance or breadth if they had been pursued only through literary texts or only through scientific discourses. The scientific texts often reveal, as literature cannot, the foundational assumptions that gave theoretical scope and artifactual efficacy to a particular approach. The literary texts often reveal, as scientific work cannot, the complex cultural, social, and representational issues tied up with conceptual shifts and technological innovations. From my point of view, literature and science as an area of specialization is more than a subset of cultural studies or a minor activity in a literature department. It is a way of understanding ourselves as embodied creatures living within and through embodied worlds and embodied words.

VIRTUAL BODIES
AND FLICKERING SIGNIFIERS

We might regard patterning or predictability as the very essence and raison d'être *of communication . . . communication is the creation of redundancy or patterning.*

Gregory Bateson, *Steps to an Ecology of Mind*

The development of information theory in the wake of World War II left as its legacy a conundrum: even though information provides the basis for much of contemporary U.S. society, it has been constructed never to be present in itself. In information theoretic terms, as we saw in chapter 1, information is conceptually distinct from the markers that embody it, for example newsprint or electromagnetic waves. It is a pattern rather than a presence, defined by the probability distribution of the coding elements composing the message. If information is pattern, then noninformation should be the absence of pattern, that is, randomness. This commonsense expectation ran into unexpected complications when certain developments within information theory implied that information could be equated with randomness as well as with pattern.[1] Identifying information with *both* pattern and randomness proved to be a powerful paradox, leading to the realization that in some instances, an infusion of noise into a system can cause it to reorganize at a higher level of complexity.[2] Within such a system, pattern and randomness are bound together in a complex dialectic that makes them not so much opposites as complements or supplements to one another. Each helps to define the other; each contributes to the flow of information through the system.

Were this dialectical relation only an aspect of the formal theory, its impact might well be limited to the problems of maximizing channel utility and minimizing noise that concern electrical engineers. Through the development of information technologies, however, the interplay between pattern and randomness became a feature of everyday life. As Friedrich Kittler has demonstrated in *Discourse Networks 1800/1900,* media come into existence when technologies of inscription intervene between the hand gripping the pen or the mouth framing the sounds and the production of the texts. In a literal sense, technologies of inscription are media when they are perceived as mediating, inserting themselves into the chain of textual production. Kittler identifies the innovative characteristics of the typewriter, originally designed for the blind, not with speed but rather with "spatially designated and discrete signs," along with a corresponding shift from the word as flowing *image* to the word "as a geometrical figure created by the spatial arrangements of the letter keys" (here Kittler quotes Richard Herbertz).[3] The emphasis on spatially fixed and geometrically arranged letters is significant, for it points to the physicality of the processes involved. Typewriter keys are directly proportionate to the script they produce. One keystroke yields one letter, and striking the key harder produces a darker letter. The system lends itself to a signification model that links signifier to signified in direct correspondence, for there is a one-to-one relation between the key and the letter it produces. Moreover, the signifier itself is spatially discrete, durably inscribed, and flat.

How does this experience change with electronic media? The relation between striking a key and producing text with a computer is very different from the relation achieved with a typewriter. Display brightness is unrelated to keystroke pressure, and striking a single key can effect massive changes in the entire text. The computer restores and heightens the sense of word as image—an image drawn in a medium as fluid and changeable as water.[4] Interacting with electronic images rather than with a materially resistant text, I absorb through my fingers as well as my mind a model of signification in which no simple one-to-one correspondence exists between signifier and signified. I know kinesthetically as well as conceptually that the text can be manipulated in ways that would be impossible if it existed as a material object rather than a visual display. As I work with the text-as-flickering-image, I instantiate within my body the habitual patterns of movement that make pattern and randomness more real, more relevant, and more powerful than presence and absence.

The technologies of virtual reality, with their potential for full-body mediation, further illustrate the kind of phenomena that foreground pattern and randomness and make presence and absence seem irrelevant. Already an industry worth hundreds of millions of dollars, virtual reality puts the user's sensory system into a direct feedback loop with a computer.[5] In one version, the user wears a stereovision helmet and a body glove with sensors at joint positions. The user's movements are reproduced by a simulacrum, called an avatar, on the computer screen. When the user turns his or her head, the computer display changes in a corresponding fashion. At the same time, audiophones create a three-dimensional sound field. Kinesthetic sensations, such as G-loads for flight simulators, can be supplied through more extensive and elaborate body coverings. The result is a multisensory interaction that creates the illusion that the user is *inside* the computer. From my experience with the virtual reality simulations at the Human Interface Technology Laboratory and elsewhere, I can attest to the disorienting, exhilarating effect of the feeling that subjectivity is dispersed throughout the cybernetic circuit. In these systems, the user learns, kinesthetically and proprioceptively, that the relevant boundaries for interaction are defined less by the skin than by the feedback loops connecting body and simulation in a technobio-integrated circuit.

Questions about presence and absence do not yield much leverage in this situation, for the avatar both is and is not present, just as the user both is and is not inside the screen. Instead, the focus shifts to questions about pattern and randomness. What transformations govern the connections between user and avatar? What parameters control the construction of the screen world? What patterns can the user discover through interaction with the system? Where do these patterns fade into randomness? What stimuli cannot be encoded within the system and therefore exist only as extraneous noise? When and how does this noise coalesce into pattern? Working from a different theoretical framework, Allucquère Roseanne Stone has proposed that one need not enter virtual reality to encounter these questions, although VR brings them vividly into the foreground. Merely communicating by email or participating in a text-based MUD (multi-user dungeon) already problematizes thinking of the body as a self-evident physicality.[6] In the face of such technologies, Stone proposes that we think of subjectivity as a multiple warranted by the body rather than contained within it. Sherry Turkle, in her fascinating work on people who spend serious time in MUDs, convincingly shows that virtual technologies, in a riptide of reverse influence, affect how real life is seen. "Reality is not my best window," one of her respondents remarks.[7]

In societies enmeshed within information networks, as the U.S. and other first world societies are, these examples can be multiplied a thousandfold. Money is increasingly experienced as informational patterns stored in computer banks rather than as the presence of cash; surrogacy and in vitro fertilization court cases offer examples of informational genetic patterns competing with physical presence for the right to determine the "legitimate" parent; automated factories are controlled by programs that constitute the physical realities of work assignments and production schedules as flows of information through the system;[8] criminals are tied to crime scenes through DNA patterns rather than through eyewitness accounts verifying their presence; access to computer networks rather than physical possession of data determines nine-tenths of computer law;[9] sexual relationships are pursued through the virtual spaces of computer networks rather than through meetings at which the participants are physically present.[10] The effect of these transformations is to create a highly heterogeneous and fissured space in which discursive formations based on pattern and randomness jostle and compete with formations based on presence and absence. Given the long tradition of dominance that presence and absence have enjoyed in the Western tradition, the surprise is not that formations based on them continue to exist but that these formations are being displaced so rapidly across a wide range of cultural sites.

These examples, taken from studies of information technologies, illustrate concerns that are also appropriate for literary texts. If the effects that the shift toward pattern/randomness has on literature are not widely recognized, perhaps it is because they are at once pervasive and elusive. A book produced by typesetting may look very similar to one generated by a computerized program, but the technological processes involved in this transformation are not neutral. Different technologies of text production suggest different models of signification; changes in signification are linked with shifts in consumption; shifting patterns of consumption initiate new experiences of embodiment; and embodied experience interacts with codes of representation to generate new kinds of textual worlds.[11] In fact, each category—production, signification, consumption, bodily experience, and representation—is in constant feedback and feedforward loops with the others.

As the emphasis shifts to pattern and randomness, characteristics of print texts that used to be transparent (because they were so pervasive) are becoming visible again through their differences from digital textuality. We lose the opportunity to understand the implications of these shifts if we mistake the dominance of pattern/randomness for the disappearance of the material world. In fact, it is precisely because material interfaces have changed that pattern and randomness can be perceived as dominant over presence and absence. The pattern/randomness dialectic does not erase the material world; information in fact derives its efficacy from the material infrastructures it appears to obscure. This illusion of erasure should be the *subject* of inquiry, not a presupposition that inquiry takes for granted.

To explore the importance of the medium's materiality, let us consider the book. Like the human body, the book is a form of information transmission and storage, and like the human body, the book incorporates its encodings in a durable material substrate. Once encoding in the material base has taken place, it cannot easily be changed. Print and proteins in this sense have more in common with each other than with magnetic encodings, which can be erased and rewritten simply by changing the polarities. (In chapter 8 we shall have an opportunity to see how a book's self-representations change when the book is linked with magnetic encodings.) The printing metaphors pervasive in the discourse of genetics are constituted through and by this similarity of corporeal encoding in books and bodies.

The entanglement of signal and materiality in bodies and books confers on them a parallel doubleness. As we have seen, the human body is understood in molecular biology simultaneously as an expression of genetic information and as a physical structure. Similarly, the literary corpus is at once a physical object and a space of representation, a body and a message. Because they have bodies, books and humans have something to lose if they are regarded solely as informational patterns, namely the resistant materiality that has traditionally marked the durable inscription of books no less than it has marked our experiences of living as embodied creatures. From this affinity emerge complex feedback loops between contemporary literature, the technologies that produce it, and the embodied readers who produce and are produced by books and technologies. Changes in bodies as they are represented within literary texts have deep connections with changes in textual bodies as they are encoded within information media, and both types of changes stand in complex relation to changes in the construction of human bodies as they interface with information technologies. The term I use to designate this network of relations is *informatics*. Following Donna Haraway, I take *informatics* to mean the technologies of information as well as the biological, social, linguistic, and cultural changes that initiate, accompany, and complicate their development.[12]

I am now in a position to state the thesis of this chapter explicitly. The contemporary pressure toward dematerialization, understood as an epistemic shift toward pattern/randomness and away from presence/absence, affects human and textual bodies on two levels at once, as a change in the body (the material substrate) and as a change in the message (the codes of representation). The connectivity between these changes is, as they say in the computer industry, massively parallel and highly interdigitated. My narrative will therefore weave back and forth between the represented worlds of contemporary fictions, models of signification implicit in word processing, embodied experience as it is constructed by interactions with information technologies, and the technologies themselves.

The compounding of signal with materiality suggests that new technologies will instantiate new models of signification. Information technologies do more than change modes of text production, storage, and dissemination. They fundamentally alter the relation of signified to signifier. Carrying the instabilities implicit in Lacanian floating signifiers one step further, information technologies create what I will call *flickering signifiers*, characterized by their tendency toward unexpected metamorphoses, attenuations, and dispersions. Flickering signifiers signal an important shift in the plate tectonics of language. Much of contemporary fiction is directly influenced by information technologies; cyberpunk, for example, takes informatics as its central theme. Even narratives without this focus can hardly avoid the rippling effects of informatics, however, for the changing modes of signification affect the *codes* as well as the subjects of representation.

Signifying the Processes of Production

"Language is not a code," Lacan asserted, because he wanted to deny one-to-one correspondence between the signifier and the signified.[13] In word processing, however, language *is* a code. The relation between machine and compiler languages is specified by a coding arrangement, as is the relation of the compiler language to the programming commands that the user manipulates. Through these multiple transformations, some quantity is conserved, but it is not the mechanical energy implicit in a system of levers or the molecular energy of a thermodynamical system. Rather it is the informational structure that emerges from the interplay between pattern and randomness. When a text presents itself as a constantly refreshed image rather than as a durable inscription, transformations can occur that would be unthinkable if matter or energy, rather than informational patterns, formed the primary basis for the systemic exchanges. This textual fluidity, which users learn in their bodies as they interact with the system, implies that signifiers flicker rather than float.

To explain what I mean by flickering signifiers, I will briefly review Lacan's notion of floating signifiers. Lacan, operating within a view of language that was primarily print-based rather than electronically mediated, not surprisingly focused on presence and absence as the dialectic of interest.[14] When he formulated the concept of floating signifiers, he drew on Saussure's idea that signifiers are defined by networks of relational differences between themselves rather than by their relation to signifieds. He complicated this picture by maintaining that signifieds do not exist in themselves, except insofar as they are produced by signifiers. He imagined them as an ungraspable flow floating beneath a network of signifiers, a network that itself is constituted through continual slippages and displacements.

Thus, for him, a doubly reinforced absence is at the core of signification—the absence of signifieds as things-in-themselves as well as the absence of stable correspondences between signifiers. The catastrophe in psycholinguistic development corresponding to this absence in signification is castration, the moment when the (male) subject symbolically confronts the realization that subjectivity, like language, is founded on absence.

How does this scenario change when floating signifiers give way to flickering signifiers? Foregrounding pattern and randomness, information technologies operate within a realm in which the signifier is opened to a rich internal play of difference. In informatics, the signifier can no longer be understood as a single marker, for example an ink mark on a page. Rather it exists as a flexible chain of markers bound together by the arbitrary relations specified by the relevant codes. As I write these words on my computer, I see the lights on the video screen, but for the computer, the relevant signifiers are electronic polarities on disks. Intervening between what I see and what the computer reads are the machine code that correlates alphanumeric symbols with binary digits, the compiler language that correlates these symbols with higher-level instructions determining how the symbols are to be manipulated, the processing program that mediates between these instructions and the commands I give the computer, and so forth. A signifier on one level becomes a signified on the next-higher level. Precisely because the relation between signifier and signified at each of these levels is arbitrary, it can be changed with a single global command. If I am producing ink marks by manipulating movable type, changing the font requires changing each line of type. By contrast, if I am producing flickering signifiers on a video screen, changing the font is as easy as giving the system a single command. The longer the chain of codes, the more radical the transformations that can be effected. Acting as linguistic transducers, the coding chains impart astonishing power to even very small changes. Such amplification is possible because the constant reproduced through multiple coding layers is a pattern rather than a presence.

Where does randomness enter this picture? Within information theory, information is identified with choices that reduce uncertainty, for example when I choose which book, out of eight on a reading list, my seminar will read for the first week of class. To get this information to the students, I need some way to transmit it. Information theory treats the communication situation as a system in which a sender encodes a message and sends it as a signal through a channel. At the other end is a receiver, who decodes the signal and reconstitutes the message. Suppose I write my students an email. The computer encodes the message in binary digits and sends a signal corresponding to these digits to the server, which then reconstitutes the message in a form the students can read. At many points along this route, noise can intervene. The message may be garbled by the computer system, so that it arrives looking like "°#e%^&s°°." Or I may have gotten distracted thinking about DeLillo halfway through the message, so that although I meant to assign Calvino for the first week, the message comes out, "If on a winter's night a white noise." These examples indicate that for real-life communication situations, pattern exists in dynamic tension with the random intrusions of noise.

Uncertainty enters in another sense as well. Although information is often defined as *reducing* uncertainty, it also *depends* on uncertainty. Suppose, for example, *Gravity's Rainbow* is the only text on the reading list. The probability that I would choose it is 1. If I send an email telling my students that the text for this week is *Gravity's Rainbow*, they will learn nothing they did not already know, and no information is communicated. The most surprising information I could send them would be a string of random letters. (Remember that information in the technical sense has nothing to do with meaning; the fact that such a message would be meaningless is thus paradoxically irrelevant to calculating the amount of information it contains.) These intuitions are confirmed by the mathematical theory of information.[15] For an individual message, the information increases as the probability that the event will occur diminishes; the more unlikely the event, the more information it conveys. Appropriately, this quantity is usually called the "surprisal." Let's say that nine of my reading assignments were on *Gravity's Rainbow*, and one was on *Vineland*. The students would gain more information from a message telling them that the assignment was *Vineland* than from a message stating that the assignment was *Gravity's Rainbow*—the more probable event and hence the more expected. Most of the time, however, electrical engineers are not interested in individual messages but in all the messages that can be produced from a given source. Thus they do not so much want to know the surprisal as the *average* amount of information coming from a source. This average reaches a maximum when it is equally likely that any symbol can appear in any position—which is to say, when there is no pattern or when the message is at the extreme of randomness. Thus Warren Weaver, in his interpretation of Shannon's theory of information, suggested that information should be understood as depending on both predictability and unpredictability, pattern and randomness.[16]

What happens in the case of mutation? Consider the example of the genetic code. Mutation normally occurs when some random event (for example, a burst of radiation or a coding error) disrupts an existing pattern and something else is put in its place instead. Although mutation disrupts pattern, it also presupposes a morphological standard against which it can be measured and understood as a mutation. If there were only randomness, as with the random movements of gas molecules, it would make no sense to speak of mutation. We have seen that in electronic textuality, the possibilities for mutation within the text are enhanced and heightened by long coding chains. We can now understand mutation in more fundamental terms. Mutation is crucial because it names the bifurcation point at which the interplay between pattern and randomness causes the system to evolve in a new direction. It reveals the productive potential of randomness that is also recognized within information theory when uncertainty is seen as both antagonistic and intrinsic to information.

We are now in a position to understand mutation as a decisive event in the psycholinguistics of information. Mutation is the catastrophe in the pattern/randomness dialectic analogous to castration in the presence/absence dialectic. It marks a rupture of pattern so extreme that the expectation of continuous replication can no longer be sustained. But as with castration, this only appears to be a disruption located at a specific moment. The randomness to which mutation testifies is implicit in the very idea of pattern, for only against the background of nonpattern can pattern emerge. Randomness is the contrasting term that allows pattern to be understood as such. The crisis named by mutation is as wide-ranging and pervasive in its import within the pattern/randomness dialectic as castration is within the tradition of presence/absence, for it is the visible mark that testifies to the continuing interplay of the dialectic between pattern and randomness, replication and variation, expectation and surprise.

Shifting the emphasis from presence/absence to pattern/randomness suggests different choices for tutor texts. Rather than studying Freud's discussion of "fort/da" (a short passage whose replication in hundreds of commentaries would no doubt astonish its creator), theorists interested in pattern and randomness might point to David Cronenberg's film *The Fly*. At a certain point, the protagonist's penis does fall off (quaintly, he puts it in his medicine chest as a memento of times past), but the loss scarcely registers in the larger mutation he is undergoing. The operative transition is not from male to female-as-castrated-male but from human to something radically other than human. Flickering signification brings together language with a psychodynamics based on the symbolic moment when the human confronts the posthuman.

As I indicated in chapter 1, I understand human and posthuman to be historically specific constructions that emerge from different configurations of embodiment, technology, and culture. My reference point for the human is the tradition of liberal humanism; the posthuman appears when computation rather than possessive individualism is taken as the ground of being, a move that allows the posthuman to be seamlessly articulated with intelligent machines. To see how technology interacts with these constructions, consider the picture that nineteenth-century U.S. and British anthropologists have drawn of "man" as a tool-user.[17] Using tools may shape the body (some anthropologists made this argument), but the tool nevertheless is envisioned as an object that is apart from the body, an object that can be picked up and put down at will. When the claim that man's unique nature was defined by tool use could not be sustained (because other animals were shown also to use tools), the focus shifted during the early twentieth century to man the tool-maker. Typical is Kenneth P. Oakley's 1949 *Man the Tool-Maker,* a magisterial work with the authority of the British Museum behind it. Oakley, in charge of the Anthropological Section of the museum's Natural History Division, wrote in his introduction, "Employment of tools appears to be [man's] chief biological characteristic, for considered functionally they are detachable extensions of the forelimb."[18] The kind of tool he envisioned was mechanical rather than informational; it goes *with* the hand, not *on* the head. Significantly, he imagined the tool to be at once "detachable" and an "extension," separate from yet partaking of the hand. If the placement and the kind of tool mark Oakley's affinity with the epoch of the human, the construction of the tool as a prosthesis points forward to the posthuman.

By the 1960s, Marshall McLuhan was speculating about the transformation that media, understood as technological prostheses, were effecting on human beings.[19] He argued that humans react to stress in their environments by withdrawing the locus of selfhood inward, in a numbing withdrawal from the world he called (following Hans Selye and Adolphe Jonas) "autoamputation." This withdrawal in turn facilitates and requires compensating technological extensions that project the body-as-prosthesis back out into the world. Whereas Oakley remains grounded in the human and looks only distantly toward the posthuman, McLuhan clearly sees that electronic media are capable of bringing about a reconfiguration so extensive as to change the nature of "man."

As we saw in chapter 1, similar shifts in orientation informed the Macy Conference discussions taking place during the same period (1946–53). Participants wavered between a vision of man as a homeostatic self-regulating mechanism whose boundaries were clearly delineated from the environment[20] and a more threatening, reflexive vision of a man spliced into an informational circuit that could change him in unpredictable ways.

By the 1960s, the consensus within cybernetics had shifted dramatically toward reflexivity. By the 1980s, the inertial pull of homeostasis as a constitutive concept had largely given way to self-organization theories implying that radical changes were possible within certain kinds of complex systems.[21] In the contemporary period, the posthuman future of humanity is increasingly evoked, ranging from Hans Moravec's argument for a "postbiological" future in which intelligent machines become the dominant life form on the planet, to the more sedate and in part already realized prospect of a symbiotic union between human and intelligent machine, a union that Howard Rheingold calls "intelligence augmentation."[22] Although these visions differ in the degree and kind of interfaces they imagine, they concur that the posthuman implies not only a coupling with intelligent machines but a coupling so intense and multifaceted that it is no longer possible to distinguish meaningfully between the biological organism and the informational circuits in which the organism is enmeshed. Accompanying this change is a corresponding shift in how signification is understood and corporeally experienced. In contrast to Lacanian psycholinguistics, derived from the generative coupling of linguistics and sexuality, flickering signification is the progeny of the fascinating and troubling coupling of language and machine.

Information Narratives and Bodies of Information

The shift from presence and absence to pattern and randomness is encoded into every aspect of contemporary literature, from the physical object that constitutes the text to such staples of literary interpretation as character, plot, author, and reader. The development is by no means even; some texts testify dramatically and explicitly to the shift, whereas others manifest this shift only indirectly. I will call those texts in which the displacement is most apparent *information narratives.* Information narratives show, in exaggerated form, changes that are more subtly present in other texts as well. Whether in information narratives or contemporary fiction generally, the dynamic of displacement is crucial. One could focus on pattern in any era, but the peculiarity of pattern in these texts is its interpenetration with randomness and its implicit challenge to physicality. *Pattern tends to overwhelm presence,* leading to a construction of immateriality that depends not on spirituality or even consciousness but only on information.

Consider William Gibson's *Neuromancer* (1984), the novel that—along with the companion volumes *Count Zero* (1986) and *Mona Lisa Overdrive*

(1988)—sparked the cyberpunk movement. The *Neuromancer* trilogy gave a local habitation and a name to the disparate spaces of computer simulations, networks, and hypertext windows that, before Gibson's intervention, had been discussed as separate phenomena. Gibson's novels acted like seed crystals thrown into a supersaturated solution; the time was ripe for the technology known as cyberspace to precipitate into public consciousness. In *Neuromancer* the narrator defines cyberspace as a "consensual illusion" accessed when a user "jacks into" a computer. Here the writer's imagination outstrips existing technologies, for Gibson imagines a direct neural link between the brain and the computer through electrodes. Another version of this link is a socket, implanted behind the ear, that accepts computer chips, allowing direct neural access to computer memory. Network users collaborate in creating the richly textured landscape of cyberspace, a "graphic representation of data abstracted from the banks of every computer in the human system. Unthinkable complexity. Lines of light ranged in the nonspace of the mind, clusters and constellations of data. Like city lights, receding."[23] Existing in the nonmaterial space of computer simulation, cyberspace defines a regime of representation within which pattern is the essential reality, presence an optical illusion.

Like the landscapes they negotiate, the subjectivities who operate within cyberspace also become patterns rather than physical entities. Case, the computer cowboy who is the protagonist of *Neuromancer,* still has a physical presence, although he regards his body as so much "meat" that exists primarily to sustain his consciousness until the next time he can enter cyberspace. Others have completed the transition that Case's values imply. Dixie Flatline, a cowboy who encountered something in cyberspace that flattened his EEG, ceased to exist as a physical body and lives now as a personality construct within the computer, defined by the magnetic patterns that store his identity.

The contrast between the body's limitations and cyberspace's power highlights the advantages of pattern over presence. As long as the pattern endures, one has attained a kind of immortality—an implication that Hans Moravec makes explicit in *Mind Children.* Such views are authorized by cultural conditions that make physicality seem a better state to be from than to inhabit. In a world despoiled by overdevelopment, overpopulation, and time-release environmental poisons, it is comforting to think that physical forms can recover their pristine purity by being reconstituted as informational patterns in a multidimensional computer space. A cyberspace body, like a cyberspace landscape, is immune to blight and corruption. It is no accident that the vaguely apocalyptic landscapes of films such as *Ter-*

minator, Blade Runner, and *Hardware* occur in narratives focusing on cybernetic life-forms. The sense that the world is rapidly becoming uninhabitable by human beings is part of the impetus for the displacement of presence by pattern.

These connections lie close to the surface in *Neuromancer.* "Get just wasted enough, find yourself in some desperate but strangely arbitrary kind of trouble, and it was possible to see Ninsei as a field of data, the way the matrix had once reminded him of proteins linking to distinguish cell specialities. Then you could throw yourself into a highspeed drift and skid, totally engaged but set apart from it all, and all around you the dance of biz, information interacting, data made flesh in the mazes of the black market."[24] The metaphoric slippages between urban sprawl, computer matrix, and biological protein culminate in the final elliptical phrase, "data made flesh." Information is the putative origin, physicality the derivative manifestation. Body parts sold in black-market clinics, body neurochemistry manipulated by synthetic drugs, body of the world overlaid by urban sprawl—all testify to the precariousness of physical existence. If flesh is data incarnate, why not go back to the source and leave the perils of physicality behind?

The reasoning presupposes that subjectivity and computer programs have a common arena in which to interact. Historically, that arena was first defined in cybernetics by the creation of a conceptual framework that constituted humans, animals, and machines as information-processing devices receiving and transmitting signals to effect goal-directed behavior.[25] Gibson matches this technical achievement with two literary innovations that allow subjectivity, with its connotations of consciousness and self-awareness, to be articulated together with abstract data. The first is a subtle modification in point of view, abbreviated in the text as "pov." More than an acronym, pov is a substantive noun that constitutes the character's subjectivity by serving as a positional marker substituting for his absent body.

In its usual Jamesian sense, point of view presumes the fiction of a person who observes the action from a particular angle and tells what he sees. In the preface to *The Portrait of a Lady,* James imagines a "house of fiction" with a "million windows" formed by "the need of the individual vision and by the pressure of the individual will." At each window "stands a figure with a pair of eyes, or at least with a field glass, which forms, again and again, for observation, a unique instrument, insuring to the person making use of it an impression distinct from every other."[26] For James, the observer is an embodied creature, and the specificity of his or her location determines what the observer can see when looking out on a scene that itself is physically specific. When an omniscient viewpoint is used, the limitations of the narrator's corporeality begin to fall away, but the suggestion of embodiment lingers in the idea of focus, the "scene" created by the eye's movement.

Even for James, vision is not unmediated technologically. Significantly, he hovers between eye and field glass as the receptor constituting vision. Cyberspace represents a quantum leap forward into the technological construction of vision. Instead of an embodied consciousness looking through the window at a scene, consciousness moves *through* the screen to become the pov, leaving behind the body as an unoccupied shell. In cyberspace, point of view does not emanate from the character; rather, the pov literally *is* the character. If a pov is annihilated, the character disappears with it, ceasing to exist as a consciousness in and out of cyberspace. The realistic fiction of a narrator who observes but does not create is thus unmasked in cyberspace. The effect is not primarily metafictional, however, but is in a literal sense metaphysical, above and beyond physicality. The crucial difference between the Jamesian point of view and the cyberspace pov is that the former implies physical presence, whereas the latter does not.

Gibson's technique recalls Alain Robbe-Grillet's novels, which were among the first information narratives to exploit the formal consequences of combining subjectivity with data. In Robbe-Grillet's work, however, the effect of interfacing narrative voice with objective description was paradoxically to heighten the narrator's subjectivity, for certain objects, like the jalousied windows or the centipede in *Jealousy*, are inventoried with obsessive interest, indicating a mindset that is anything but objective. In Gibson, the space in which subjectivity moves lacks this personalized stamp. Cyberspace is the domain of virtual collectivity, constituted as the resultant of millions of vectors representing the diverse and often conflicting interests of human and artificial intelligences linked together through computer networks.[27]

To make this space work as a level playing field on which humans and computers can meet on equal terms, Gibson introduces his second innovation. Cyberspace is created by transforming a data matrix into a landscape in which narratives can happen. In mathematics, "matrix" is a technical term denoting data that have been arranged into an n-dimensional array. Expressed in this form, data seem as far removed from the fascinations of story as random-number tables are from the *National Inquirer*. Because the array is already conceptualized in spatial terms, however, it is a small step to imagining the matrix as a three-dimensional landscape. Narrative becomes possible when this spatiality is given a temporal dimension by the pov's movement through it. The pov is *located* in space, but it *exists* in time.

Through the track it weaves, the desires, repressions, and obsessions of subjectivity can be expressed. The genius of *Neuromancer* lies in its explicit recognition that the categories Kant considered fundamental to human experience—space and time—can be used as a conjunction to join awareness with data. Reduced to a point, the pov is abstracted into a purely temporal entity with no spatial extension; metaphorized into an interactive space, the datascape is narrativized by the pov's movement through it. Data are thus humanized, and subjectivity is computerized, allowing them to join in a symbiotic union whose result is narrative.

Such innovations carry the implications of informatics beyond the textual surface into the signifying processes that constitute theme and character. I suspect that Gibson's novels have been so influential not only because they present a vision of the posthuman future that is already upon us—in this they are no more prescient than many other science fiction novels—but also because they embody within their techniques the assumptions expressed explicitly in the themes of the novels. This kind of move is possible when the cultural conditions authorizing the assumptions are pervasive enough that the posthuman is experienced as an everyday, lived reality as well as an intellectual proposition.

The shift of emphasis from ownership to access is another manifestation of the underlying transition from presence/absence to pattern/randomness. In *The Condition of Postmodernity,* David Harvey characterizes the economic aspects of the shift to an informatted society as a transition from a Fordist regime to a regime of flexible accumulation.[28] As Harvey and many others have pointed out, in late capitalism, durable goods yield pride of place to information.[29] A significant difference between information and durable goods is replicability. Information is not a conserved quantity. If I give you information, you have it and I do too. With information, the constraining factor separating the haves from the have-nots is not so much possession as access. Presence precedes and makes possible the idea of possession, for one can possess something only if it already exists. By contrast, access implies pattern recognition, whether the access is to a piece of land (recognized as such through the boundary pattern defining that land as different from adjoining parcels), confidential information (constituted as confidential through the comparison of its informational patterns with less-secure documents), or a bank vault (associated with knowing the correct pattern of tumbler combinations). In general, access differs from possession because the former tracks patterns rather than presences. When someone breaks into a computer system, it is not a physical presence that is detected but the informational traces that the entry has created.[30]

When the emphasis falls on access rather than ownership, the private/public distinction that was so important in the formation of the novel is radically reconfigured. Whereas possession implies the existence of private life based on physical exclusion or inclusion, access implies the existence of credentialing practices that use patterns rather than presences to distinguish between those who do and those who do not have the right to enter. Moreover, entering is itself constituted as access to data rather than as a change in physical location. In Don DeLillo's *White Noise* (1985), for example, the Gladneys' home, traditionally the private space of family life, is penetrated by noise and radiation of all wavelengths—microwave, radio, television.[31] The penetration signals that private spaces, and the private thoughts they engender and figure, are less a concern than the interplay between codes and the articulation of individual subjectivity with data. Jack Gladney's death is prefigured for him as a pattern of pulsing stars around a computerized data display, a striking image of how his corporeality has been penetrated by informational patterns that construct as well as predict his mortality.

Although the Gladney family still operates as a social unit (albeit with the geographical dispersion endemic to postmodern life), their conversations are punctuated by random bits of information emanating from the radio and TV. The punctuation points toward a mutation in subjectivity that comes from joining the focused attention of traditional novelistic consciousness with the digitized randomness of miscellaneous bits. The mutation reaches incarnation in Willie Mink, whose brain has become so addled by a designer drug that his consciousness is finally indistinguishable from the white noise that surrounds him. Through a route different from that used by Gibson, DeLillo arrives at a similar destination: a vision of subjectivity constituted through the interplay of pattern and randomness rather than presence and absence.

The bodies of texts are also implicated in these changes. The displacement of presence by pattern thins the tissue of textuality, making it a semipermeable membrane that allows awareness of the text as an informational pattern to infuse into the space of representation. When the fiction of presence gives way to the recognition of pattern, passages are opened between the text-as-object and those representations within the text that are characteristic of the condition of virtuality. Consider the play between text as physical object and as information flow in Italo Calvino's *If on a winter's night a traveler* (1979). The text's awareness of its own physicality is painfully apparent in the anxiety it manifests toward keeping the literary corpus intact. Within the space of representation, texts are subjected to birth defects, maimed and torn apart, lost and stolen. The text operates as if it knows it has a physical body and fears that its body is in jeopardy from a host of threats, from defective printing technologies and editors experiencing middle-age brain fade to nefarious political plots. Most of all, perhaps, the text fears losing its body to information.

When "you," the reader, are foiled in your pursuit of its story by the frailty of the text's physical corpus, the narrator imagines you hurling the book through a closed window, reducing the text's body to "photons, undulatory vibrations, polarized spectra." Not content with this pulverization, you throw it through the wall so that the text breaks up into "electrons, neutrons, neutrinos, elementary particles more and more minute." Still disgusted, in an act of ultimate dispersion, you send it through a computer line, causing the textual body to be "reduced to electronic impulses, into the flow of information." With the text "shaken by redundancies and noises," you "let it be degraded into a swirling entropy." Yet the very story you seek can be envisioned as a pattern, for that night you sleep and "fight with dreams as with formless and meaningless life, seeking a pattern, a route that must surely be there, as when you begin to read a book and you don't yet know in which direction it will carry you."[32]

Once the text's physical body is interfaced with information technologies, however, the pattern that is story stands in jeopardy of being disrupted by the randomness implicit in information. The disruptive power of randomness becomes manifest when you find yourself entangled with Lotaria, a reader who believes books are best read by scanning them into computers and letting the machine analyze word-frequency patterns. Seduced by Lotaria against your better judgment, you get tangled up with her and with rolls of printout covering the floor. The printouts contain part of the story that you desperately want to finish, which Lotaria has entered into the computer. Distracted by her multiple entanglements, Lotaria presses the wrong key, and the rest of the story is "erased in an instant demagnetization of the circuits. The multicolored wires now grind out the dust of dissolved words: the the, of of of of, from from from from, that that that that, in columns according to their respective frequency. The book has been crumbled, dissolved, can no longer be recomposed, like a sand dune blown away by the wind."[33] Now you can never achieve satiation, never reach the point of satisfied completion that comes with finishing a book. Your anxiety about *reading interruptus* is intensified by what might be called *print interruptus*, a print book's fear that once it has been digitized, the computer will garble its body, breaking it apart and reassembling it into the nonstory of a data matrix rather than an entangled and entangling narrative.

This anxiety is transmitted to readers within the text, who keep pursuing parts of textual bodies only to lose them, as well as to readers outside the text, who must try to make sense of the radically discontinuous narrative. Only when the chapter titles are perceived to form a sentence is the literary corpus reconstituted as a unity. Significantly, the recuperation is syntactical rather than physical. It does not arise from or imply an intact physical body. Rather, it emerges from the patterns—metaphorical, grammatical, narrative, thematic, and textual—that the parts together make. As the climactic scene in the library suggests, the reconstituted corpus is a body of information, emerging from the discourse community among whom information circulates. The textual body may be dismembered or ground into digital word dust, the narrative implies, but as long as there are readers who care passionately about stories and want to pursue them, narrative itself can be recuperated. Through such textual strategies, *If on a winter's night* testifies vividly to the impact of information technologies on bodies of books.

Human bodies are similarly affected. The correspondence between human and textual bodies can be seen as early as William Burroughs's *Naked Lunch,* written in 1959, in the decade that saw the institutionalization of cybernetics and the construction of the first large-scale electronic digital computer.[34] The narrative metamorphizes nearly as often as bodies within it, suggesting by its cut-up method a textual corpus that is as artificial, heterogeneous, and cybernetic as they are.[35] Since the fissures that mark the text always fall *within* the units that compose the textual body—within chapters, paragraphs, sentences, and even words—it becomes increasingly clear that they do not function to delineate the textual corpus. Rather, the body of the text is produced precisely by these fissures, which are not so much ruptures as productive dialectics that bring the narrative as a syntactic and chronological sequence into being.

Bodies within the text follow the same logic. Under the pressure of sex and addiction, bodies explode or mutate, protoplasm is sucked out of cocks or nostrils, plots are hatched to take over the planet or nearest life-form. Burroughs anticipates Fredric Jameson's claim that an information society is the purest form of capitalism. When bodies are constituted as information, they can be not only sold but fundamentally reconstituted in response to market pressures. Junk instantiates the dynamics of informatics and makes clear the relation of junk-as-information to late capitalism. Junk is the "ideal product" because the "junk merchant does not sell his product to the consumer, he sells the consumer to his product. He does not improve and simplify his merchandise. He degrades and simplifies the client."[36] The junkie's body is a harbinger of the postmodern mutant, for it demonstrates how presence yields to assembly and disassembly patterns created by the flow of junk-as-information through points of amplification and resistance.

The characteristics of information narratives include, then, an emphasis on mutation and transformation as a central thematic for bodies within the text as well as for the bodies of texts. Subjectivity, already joined with information technologies through cybernetic circuits, is further integrated into the circuit by novelistic techniques that combine it with data. Access vies with possession as a structuring element, and data are narrativized to accommodate their integration with subjectivity. In general, materiality and immateriality are joined in a complex tension that is a source of exultation and strong anxiety.

Information technologies leave their mark on books in the realization that sooner or later, the body of print will be interfaced with other media. All but a handful of books printed in the United States and Europe in 1998 will be digitized during some phase of their existence. Print texts such as *If on a winter's night a traveler* bear the imprint of this digitalization in their narratives, as if the text remembers the moment when it was nothing but electronic polarities on a disk. At moments of crisis, the repressed memory erupts onto the textual surface in the form of an acute fear that randomness will so interpenetrate its patterns that story will be lost and the textual corpus will be reduced to a body of meaningless data. These eruptions are vivid testimony that even print texts cannot escape being affected by information technologies.

To understand more about the effects of informatics on contemporary fictions, let us turn now to consider the relation between text and subjectivity, specifically how information narratives constitute both the voice speaking the narrative and the reader.

Functionalities of Narrative

The very word *narrator* implies a voice speaking, and a speaking voice implies a sense of presence. Jacques Derrida, announcing the advent of grammatology, focused on the gap that separates speaking from writing. Such a change transforms the narrator from speaker to scribe or, more precisely, someone who is absent from the scene but toward whom the inscriptions point.[37] Informatics pushes this transformation further. As writing yields to flickering signifiers underwritten by binary digits, the narrator becomes not so such a scribe as a cyborg authorized to access the relevant codes. The progression suggests that the dialectic between absence and presence came clearly into focus with the advent of deconstruction because it was already being displaced as a cultural presupposition by randomness and pattern. Presence and absence were forced into visibility, so to speak, because they were already losing their constitutive power to form the ground for discourse, becoming instead the subject of discourse. In this sense, deconstruction is the child of an information age, formulating its theories from strata pushed upward by the emerging substrata beneath.

To see how the function of the narrator changes as we progress deeper into virtuality, consider the seduction scene from "I Was an Infinitely Hot and Dense White Dot," one of the stories in Mark Leyner's *My Cousin, My Gastroenterologist*. The narrator, "high on Sinutab" and driving "isotropically," so that any destination is equally probable, finds himself at a "squalid little dive."

> I don't know . . . but there she is. I can't tell if she's a human or a fifth-generation gynemorphic android and I don't care. I crack open an ampoule of mating pheromone and let it waft across the bar, as I sip my drink, a methyl isocyanate on the rocks—methyl isocyanate is the substance which killed more than 2,000 people when it leaked in Bhopal, India, but thanks to my weight training, aerobic workouts, and a low-fat fiber-rich diet, the stuff has no effect on me. Sure enough she strolls over and occupies the stool next to mine. . . . My lips are now one angstrom unit from her lips . . . I begin to kiss her but she turns her head away. . . . I can't kiss you, we're monozygotic replicants—we share 100% of our genetic material. My head spins. You are the beautiful day, I exclaim, your breath is a zephyr of eucalyptus that does a pas de bourre across the Sea of Galilee. Thanks, she says, but we can't go back to my house and make love because monozygotic incest is forbidden by the elders. What if I said I could change all that. . . . What if I said that I had a miniature shotgun that blasts gene fragments into the cells of living organisms, altering their genetic matrices so that a monozygotic replicant would no longer be a monozygotic replicant and she could then make love to a muscleman without transgressing the incest taboo, I say, opening my shirt and exposing the device which I had stuck in the waistband of my black jeans. How'd you get that thing? she gasps, ogling its thick fiber-reinforced plastic barrel and the Uzi-Biotech logo embossed on the magazine which held two cartridges of gelated recombinant DNA. I got it for Christmas. . . . Do you have any last words before I scramble your chromosomes, I say, taking aim. Yes, she says, you first.[38]

Much of the wit in this passage comes from the juxtaposition of folk wisdom and seduction clichés with high-tech language and ideas. The narrator sips a chemical that killed thousands when it leaked into the environment, but he is immune to damage because he eats a low-fat diet. The narrator leans close to the woman-android to kiss her, but he has not yet made contact when he is an angstrom away, considerably less than the diameter of a hydrogen atom. The characters cannot make love because they are barred by incest taboos, being replicants from the same monozygote, which would make them identical twins, but this does not seem to prevent them from being opposite sexes. They are governed by kinship rules enforced by tribal elders, but they have access to genetic technologies that intervene in and disrupt evolutionary modes of descent. They think their problem can be solved by an Uzi-Biotech weapon that will scramble their chromosomes, but the narrator, at least, seems to expect their identities to survive intact.

Even within the confines of a short story no more than five pages long, this encounter is not preceded or followed by events that relate directly to it. Rather, the narrative leaps from scene to scene, all of them linked by only the most tenuous and arbitrary threads. The incongruities make the narrative a kind of textual android created through patterns of assembly and disassembly. There is no natural body to this text, any more than there are natural bodies within the text. As the title intimates, identity merges with typography ("I was a . . . dot") and is further conflated with such high-tech reconstructions as computer simulations of gravitational collapse ("I was an infinitely hot and dense white dot"). Signifiers collapse like stellar bodies into an explosive materiality that approaches the critical point of nova, ready to blast outward into dissipating waves of flickering signification.

The explosive tensions between cultural codes that familiarize the action and neologistic splices that dislocate traditional expectations do more than structure the narrative. They also constitute the narrator, who exists less as a speaking voice endowed with a plausible psychology than as a series of fissures and dislocations that push toward a new kind of subjectivity. To understand the nature of this subjectivity, let us imagine a trajectory that arcs from storyteller to professional to some destination beyond. Walter Benjamin's shared community of values and presence—the community that he had in mind when he evoked the traditional storyteller whose words are woven into the rhythms of work—echoes faintly in allusions to the Song of Songs and tribal elders.[39] Overlaid on this is the professionalization that Jean-François Lyotard wrote about in *The Postmodern Condition,* in which the authority to tell the story is constituted by possessing the appropriate credentials that qualify one as a member of a physically dispersed, electronically bound professional community.[40] This phase of the trajectory is signified in a number of ways. The narrator is driving "isotropically," indicating that physical location is no longer necessary or relevant to the

production of the story. His authority derives not from his physical participation in a community but from his possession of a high-tech language that
includes pheromones, methyl isocyanate, and gelated recombinant DNA,
not to mention the Uzi-Biotech phallus. This authority too is displaced
even as it is created, for the incongruities reveal that the narrative and
therefore the narrator are radically unstable, about to mutate into a
scarcely conceivable form, signified in the story by the high-tech, identity-
transforming orgasmic blast that never quite comes.

What is this form? Its physical manifestations vary, but the ability to manipulate complex codes is a constant. The looming transformation, already
enacted through the language of the passage, is into a subjectivity who derives his authority from possessing the correct codes. Popular literature
and culture contain countless scenarios in which someone fools a computer
into thinking that he or she is an "authorized" person because the person
possesses or stumbles upon the codes that the computer recognizes as constituting authorization. Usually these scenarios imply that the person exists
unchanged, taking on a spurious identity that allows him or her to move unrecognized within an informational system. There is, however, another way
to read these narratives. Constituting identity through authorization codes,
the person using the codes is changed into another kind of subjectivity, precisely one who exists and is recognized because of knowing the codes. The
surface deception is underlaid by a deeper truth. We become the codes we
punch. The narrator is not a storyteller and not a professional authority, although these functions linger in the narrative as anachronistic allusions and
wrenched referentiality. Rather, the narrator is a keyboarder, a hacker, a
manipulator of codes.[41] Assuming that the text was digitized at some phase
in its existence, in a literal sense he (it?) *is* these codes.

The construction of the narrator as a manipulator of codes obviously has
important implications for the construction of the reader. The reader is
similarly constituted through a layered archaeology that moves from listener to reader to decoder. Drawing on a context that included information
technologies, Roland Barthes in *S/Z* brilliantly demonstrated the possibility of reading a text as a production of diverse codes.[42] Information narratives make that possibility an inevitability, for they often cannot be
understood, even on a literal level, without referring to codes and the informatics that produce and are produced by these codes. Flickering signification extends the productive force of codes beyond the text to include the
signifying processes by which the technologies produce texts, as well as the
interfaces that enmesh humans into integrated circuits. As the circuits
connecting technology, text, and human expand and intensify, the point

where quantitative increments shade into qualitative transformation draws closer.

Because codes can be sent over fiber optics essentially instantaneously, there is no longer a shared, stable context that helps to anchor meaning and guide interpretation. Like reading, decoding takes place in a location arbitrarily far removed in space and time from the source text. In contrast to the fixity of print, decoding implies that there is no original text—no first editions, no fair copies, no holographic manuscripts. There are only the flickering signifiers, whose transient patterns evoke and embody what G. W. S. Trow has called the context of no context, the suspicion that all contexts, like all texts, are electronically mediated constructions.[43] What binds the decoder to the system is not the stability of being a member of an interpretive community or the intense pleasure of physically possessing the book, a pleasure that all bibliophiles know. Rather, it is the decoder's construction as a cyborg, the impression that his or her physicality is also data made flesh, another flickering signifier in a chain of signification that extends through many levels, from the DNA that in-formats the decoder's body to the binary code that is the computer's first language.

Against this dream or nightmare of the body as information, what alternatives exist? We can see beyond this dream, I have argued, by attending to the material interfaces and technologies that make disembodiment such a powerful illusion. By adopting a double vision that looks *simultaneously* at the power of simulation and at the materialities that produce it, we can better understand the implications of articulating posthuman constructions together with embodied actualities. One way to think about these materialities is through functionality. "Functionality" is a term used by virtual reality technologists to describe the communication modes that are active in a computer-human interface. If the user wears a data glove, for example, hand motions constitute one functionality. If the computer can respond to voice-activated commands, voice is another functionality. If the computer can sense body position, spatial location is yet another functionality. Functionalities work in both directions; that is, they describe the computer's capabilities and also indicate how the user's sensory-motor apparatus is being trained to accommodate the computer's responses. Working with a VR simulation, the user learns to move his or her hand in stylized gestures that the computer can accommodate. In the process, the neural configuration of the user's brain experiences changes, some of which can be long-lasting. The computer molds the human even as the human builds the computer.

When narrative functionalities change, a new kind of reader is produced by the text. The material effects of flickering signification ripple outward because readers are trained to read through different functionalities, which can affect how they interpret any text, including texts written before computers were invented. The impatience that some readers now feel with print texts, for example, no doubt has a physiological as well as a psychological basis. They miss pushing the keys and seeing the cursor blinking at them. Conversely, other readers (or perhaps the same readers in different moods) go back to print with a renewed appreciation for its durability, its sturdiness, and its ease of use. I began to appreciate certain qualities of print only after I had experience with computers. When I open a book, it almost always works, and it can maintain backward compatibility for hundreds of years. I also appreciate that on some occasions—when I am revising a piece of writing, for example—there isn't a cursor blinking at me, as if demanding a response. With print I can take as long as I want, and the pages never disappear or shut themselves down. As these examples illustrate, changes in narrative functionalities are deeper than the structural or thematic characteristics of a specific genre, for they shift the embodied responses and expectations that different kinds of textualities evoke. Arguing from a different historical context, Friedrich Kittler made a similar point when he wrote about medial ecology.[44] When new media are introduced, the changes transform the environment as a whole. This transformation affects the niches that older media have carved for themselves, so they change also, even if they are not directly involved with the new media. Books will not remain unaffected by the emergence of new media.

If my assessment—that the emphasis on information technologies foregrounds pattern/randomness and pushes presence/absence into the background—is correct, the implications extend beyond narrative into many cultural arenas. As I indicated in chapter 1, one of the most serious of these implications is a *systematic devaluation of materiality and embodiment.* I find this trend ironic, for changes in material conditions and embodied experience are precisely what give the shift its deep roots in everyday experience. Implicit in nearly everything I have written here is the assumption that presence and pattern are opposites existing in antagonistic relation. The more emphasis that falls on one, the less the other is noticed and valued. Entirely different readings emerge when one entertains the possibility that pattern and presence are mutually enhancing and supportive. Paul Virilio has observed that one cannot ask whether information technologies should continue to be developed.[45] Given market forces already at work, it is virtually (if I may use the word) certain that we will increasingly live, work, and play in environments that construct us as embodied virtualities.[46] I believe that our best hope to intervene constructively in this de-

velopment is to put an interpretive spin on it—one that opens up the possibilities of seeing pattern and presence as complementary rather than antagonistic. Information, like humanity, cannot exist apart from the
embodiment that brings it into being as a material entity in the world; and
embodiment is always instantiated, local, and specific. Embodiment can be
destroyed, but it cannot be replicated. Once the specific form constituting
it is gone, no amount of massaging data will bring it back. This observation
is as true of the planet as it is of an individual life-form. As we rush to explore
the new vistas that cyberspace has made available for colonization, let us remember the fragility of a material world that cannot be replaced.

CONTESTING FOR THE BODY
OF INFORMATION:
THE MACY CONFERENCES
ON CYBERNETICS

When and where did information get constructed as a disembodied medium? How were researchers convinced that humans and machines are brothers under the skin? Although the Macy Conferences on Cybernetics were not the only forum grappling with these questions, they were particularly important because they acted as a crossroads for the traffic in cybernetic models and artifacts. This chapter charts the arguments that made information seem more important than materiality within this research community. Broadly speaking, the arguments were deployed along three fronts. The first was concerned with the construction of information as a theoretical entity; the second, with the construction of (human) neural structures so that they were seen as flows of information; the third, with the construction of artifacts that translated information flows into observable operations, thereby making the flows "real."

Yet at each of these fronts, there was also significant resistance to the reification of information. Alternate models were proposed; important qualifications were voiced; objections were raised to the disparity between simple artifacts and the complex problems they addressed. Reification was triumphant not because it had no opposition but because scientifically and culturally situated debates made it seem a better choice than the alternatives. Recovering the complexities of these debates helps to demystify the assumption that information is more essential than matter or energy. Followed back to moments before it became a black box, this conclusion seems less like an inevitability and more like the result of negotiations specific to the circumstances of the U.S. techno-scientific culture during and immediately following World War II.

The Macy Conferences were unusual in that participants did not present finished papers. Rather, speakers were invited to sketch out a few main ideas to initiate discussion. The discussions, rather than the presentations, were the center of interest. Designed to be intellectual free-for-alls, the conferences were radically interdisciplinary. The transcripts show that researchers from a wide variety of fields—neurophysiology, electrical engineering, philosophy, semantics, literature, and psychology, among others—struggled to understand one another and make connections between others' ideas and their own areas of expertise. In the process, a concept that may have begun as a model of a particular physical system came to have broader significance, acting simultaneously as mechanism and metaphor.

The dynamics of the conferences facilitated this mixing. Researchers might not have been able to identify in their own work the mechanism discussed by a fellow participant, but they could understand it metaphorically and then associate the metaphor with something applicable to their own field. The process appears repeatedly throughout the transcripts. When Claude Shannon used the word "information," for example, he employed it as a technical term having to do with message probabilities. When Gregory Bateson appropriated the same word to talk about initiation rituals, he interpreted it metaphorically as a "difference that makes a difference" and associated it with feedback loops between contesting social groups. As mechanism and metaphor were compounded, concepts that began with narrow definitions spread out into networks of broader significance. Earlier I called these networks "constellations," suggesting that during the Macy period, the emphasis was on homeostasis. This chapter explores the elements that came together to form the homeostasis constellation; it also demonstrates the chain of associations that bound reflexivity together with subjectivity during the Macy period, which for many of the physical scientists was enough to relegate reflexivity to the category of "nonscience" rather than "science." Tracing the development of reflexive epistemologies after the Macy period ended, the chapter concludes by showing how reflexivity was modified so that it could count as producing scientific knowledge during the second wave of cybernetics.

The Meaning(lessness) of Information

The triumph of information over materiality was a major theme at the first Macy Conference. John von Neumann and Norbert Wiener led the way by making clear that the important entity in the man-machine equation was information, not energy. Although energy considerations are not entirely absent (von Neumann discussed at length the problems involved in dissipating the heat generated from vacuum tubes), the thermodynamics of heat was incidental. Central was how much information could flow through the system and how quickly it could move. Wiener, emphasizing the movement from energy to information, made the point explicitly: "The fundamental idea is the message ... and the fundamental element of the message is the decision."[1] Decisions are important not because they produce material goods but because they produce information. Control information, and power follows.

But what counts as information? We saw in chapter 1 that Claude Shannon defined information as a probability function with no dimensions, no materiality, and no necessary connection with meaning. Although a full exposition of information theory is beyond the scope of this book, the following explanation, adapted from an account by Wiener, will give an idea of the underlying reasoning.[2] Like Shannon, Wiener thought of information as representing a choice. More specifically, it represents a choice of one message from among a range of possible messages. Suppose there are thirty-two horses in a race, and we want to bet on Number 3. The bookie suspects the police have tapped his telephone, so he has arranged for his clients to use a code. He studied communication theory (perhaps he was in one of the summer-school classes on communication theory that Wiener taught at UCLA), and he knows that any message can be communicated through a binary code. When we call up, his voice program asks if the number falls in the range of 1 to 16. If it does, we punch the number "1"; if not, the number "0." We use this same code when the voice program asks if the number falls in the range of 1 to 8, then the range of 1 to 4, and next the range of 1 to 2. Now the program knows that the number must be either 3 or 4, so it says, "If 3, press 1; if 4, press 0," and a final tap communicates the number. Using these binary divisions, we need five responses to communicate our choice.

How does this simple decision process translate into information? First let us generalize our result. Probability theory states that the number of binary choices C necessary to uniquely identify an element from a set with n elements can be calculated as follows:

$$C = \log_2 n$$

In our case,

$$C = \log_2 32 = 5,$$

the five choices we made to convey our desired selection. (Hereafter, to simplify the notation, consider all logarithms taken to base 2). Working from this formula, Wiener defined information I as the log of the number n of elements in the message set.

$$I = \log n$$

This formula gives I when the elements are equally likely. Usually this is not the case; in English, for example, the letter "e" is far more likely to occur than "z." For the more general situation, when the elements $s_1, s_2, s_3, \ldots s_n$ are not equally likely, and $p(s)$ is the probability that the element s will be chosen,

$$I(s_i) = \log 1/p(s_i) = -\log p(s_i).$$

This is the general formula for information communicated by a specific event, in our case the call to the bookie. Because electrical engineers must design circuits to handle a variety of messages, they are less interested in specific events than they are in the average amount of information from a source, for example, the average of all the different messages that a client might communicate about the horse race. This more complex case is represented by the following formula:

$$I = -\Sigma \, p(s_i) \, [\log p(s_i)],$$

where $p(s_i)$ is the probability that the message element s_i will be selected from a message set with n elements (Σ indicates the sum of terms as i varies from 1 to n).[3]

We are now in a position to understand the deeper implications of information as it was theorized by Wiener and Shannon. Note that the theory is formulated entirely without reference to what information means. Only the probabilities of message elements enter into the equations. Why divorce information from meaning? Shannon and Wiener wanted information to have a stable value as it moved from one context to another. If it was tied to meaning, it would potentially have to change values every time it was embedded in a new context, because context affects meaning. Suppose, for example, you are in a windowless office and call to ask about the weather. "It's raining," I say. On the other hand, if we are both standing on a street corner, being drenched by a downpour, this same response would have a very different meaning. In the first case, I am telling you something you don't know; in the second, I am being ironic (or perhaps moronic). An information concept that ties information to meaning would have to yield two different values for the two circumstances, even though the message ("It's raining") is the same.

To cut through this Gordian knot, Shannon and Wiener defined information so that it would be calculated as the same value regardless of the contexts in which it was embedded, which is to say, they divorced it from meaning. *In context,* this was an appropriate and sensible decision. *Taken out of context,* the definition allowed information to be conceptualized as if it were an entity that can flow unchanged between different material substrates, as when Moravec envisions the information contained in a brain being downloaded into a computer. Ironically, this reification of information is enacted through the same kind of decontextualizing moves that the theory uses to define information as such. The theory decontextualizes information; Moravec decontextualizes the theory. Thus, a simplification necessitated by engineering considerations becomes an ideology in which a reified concept of information is treated as if it were fully commensurate with the complexities of human thought.[4]

Shannon himself was meticulously careful about how he applied information theory, repeatedly stressing that information theory concerned only the efficient transmission of messages through communication channels, not what those messages mean. Although others were quick to impute larger linguistic and social implications to the theory, he resisted these attempts. Responding to a presentation by Alex Bavelas on group communication at the eighth Macy Conference, he cautioned that he did not see "too close a connection between the notion of information as we use it in communication engineering and what you are doing here . . . the problem here is not so much finding the best encoding of symbols . . . but, rather, the determination of the semantic question of what to send and to whom to send it."[5] For Shannon, defining information as a probability function was a strategic choice that enabled him to bracket semantics. He did not want to get involved in having to consider the receiver's mindset as part of the communication system. He felt so strongly on this point that he suggested Bavelas distinguish between information in a channel and information in a human mind by characterizing the latter through "subjective probabilities," although how these were to be defined and calculated was by no means clear.

Not everyone agreed that it was a good idea to decontextualize information. At the same time that Shannon and Wiener were forging what information would mean in a U.S. context, Donald MacKay, a British researcher, was trying to formulate an information theory that would take meaning into account. At the seventh conference, he presented his ideas to the Macy group. The difference between his view and Shannon's can be seen in the way he bridled at Shannon's suggestion about "subjective probabilities." In the rhetoric of the Macy Conferences, "objective" was associated with being scientific, whereas "subjective" was a code word implying that one had fallen into a morass of unquantifiable feelings that might be magnificent but were certainly not science. MacKay's first move was to rescue information that affected the receiver's mindset from the "subjective" label. He proposed that both Shannon and Bavelas were concerned with what he called "selective information," that is, information calculated by considering the selection of message elements from a set. But selective information alone is not enough; also required is another kind of information that he called "structural." Structural information indicates how selective information is to be understood; it is a message about how to interpret a message—that is, it is a metacommunication.

To illustrate, say I launch into a joke and it falls flat. In that case, I may resort to telling my interlocutor, "That's a joke." The information content of this message, considered as selective information (measured in "metrons"), is calculated with probability functions similar to those used in the Shannon-Wiener theory. In addition, my metacomment also carries structural information (measured in "logons"), for it indicates that the preceding message has one kind of structure rather than another (a joke instead of a serious statement). In another image MacKay liked to use, he envisioned selective information as choosing among folders in a file drawer, whereas structural information increased the number of drawers (jokes in one drawer, academic treatises in another).

Since structural information indicates how a message should be interpreted, semantics necessarily enters the picture. In sharp contrast to message probabilities, which have no connection with meaning, structural information was to be calculated through changes brought about in the receiver's mind. "It's raining," heard by someone in a windowless office, would yield a value for the structural information different from the value that it would yield when heard by someone looking out a window at rain. To emphasize the correlation between structural information and changes in the receiver's mind, MacKay offered an analogy: "It is as if we had discovered how to talk quantitatively about size through discovering its effects on the measuring apparatus."[6] The analogy implies that representations created by the mind have a double valence. Seen from one perspective, they contain information about the world ("It's raining"). From another perspective, they are interactive phenomena that point back to the observer, for this information is quantified by measuring changes in the "measuring instrument," that is, in the mind itself. And how does one measure these changes? An observer looks at the mind of the person who received the message, which is to say that changes are made in the observer's mind, which in turn can also be observed and measured by someone else. The progression tends toward the infinite regress characteristic of reflexivity. Arguing for a strong correlation between the *nature* of a representation and its *effect,* MacKay's model recognized the mutual constitution of form and content, message and receiver. His model was fundamentally different from the Shannon-Wiener theory because it triangulated between reflexivity, information, and meaning. In the context of the Macy Conferences, his conclusion qualified as radical: subjectivity, far from being a morass to be avoided, is precisely what enables information and meaning to be connected.

The problem was how to quantify the model. To achieve quantification, a mathematical model was needed for the changes that a message triggered in the receiver's mind. The staggering problems this presented no doubt explain why MacKay's version of information theory was not widely accepted among the electrical engineers who would be writing, reading, and teaching the textbooks on information theory in the coming decades. Although MacKay's work continued to be foundational for the British school of information theory, in the United States the Shannon-Wiener definition of information, not MacKay's, became the industry standard.

Not everyone in the United States capitulated. As late as 1968, Nicolas S. Tzannes, an information theorist working for the U.S. government, sent Warren McCulloch a memorandum about his attempt to revise MacKay's theory so that it would be more workable.[7] He wanted to define information so that its meaning varied with context, and he looked to Kotelly's context algebra for a way to handle these changes quantitatively. In the process, he made an important observation. He pointed out that whereas Shannon and Wiener define information in terms of what it *is,* MacKay defines it in terms of what it *does.*[8] The formulation emphasizes the reification that information undergoes in the Shannon-Wiener theory. Stripped of context, it becomes a mathematical quantity weightless as sunshine, moving in a rarefied realm of pure probability, not tied down to bodies or material instantiations. The price it pays for this universality is its divorce from representation. When information is made representational, as in MacKay's model, it is conceptualized as an action rather than a thing. Verblike, it becomes a process that someone enacts, and thus it necessarily implies context and embodiment. The price it pays for embodiment is difficulty of quantification and loss of universality.

In the choice between what information is and what it does, we can see the rival constellations of homeostasis and reflexivity beginning to take shape. Making information a thing allies it with homeostasis, for so defined, it can be transported into any medium and maintain a stable quantitative value, reinforcing the stability that homeostasis implies. Making information an action links it with reflexivity, for then its effect on the receiver must be taken into account, and measuring this effect sets up the potential for a reflexive spiral through an infinite regress of observers. Homeostasis won in the first wave largely because it was more manageable quantitatively. Reflexivity lost because specifying and delimiting context quickly ballooned into an unmanageable project. At every point, these outcomes are tied to the historical contingencies of the situation—the definitions offered, the models proposed, the techniques available, the allies and resources mobilized by contending participants for their views. Conceptualizing information as a disembodied entity was not an arbitrary decision, but neither was it inevitable.

The tension between reified models and embodied complexities figures importantly in the next episode of our story. If humans are information-processing machines, then they must have biological equipment enabling them to process binary code. The model constructing the human in these terms was the McCulloch-Pitts neuron. The McCulloch-Pitts neuron was the primary model through which cybernetics was seen as having "a setting in the flesh," as Warren McCulloch put it. The problem was how to move from this stripped-down neural model to such complex issues as universals in thought, gestalts in perception, and representations of what a system cannot represent. Here the slippage between mechanism and model becomes important, for even among researchers dedicated to a hard-science approach, such as McCulloch, the tendency was to use the model metaphorically to forge connections between relatively simple neural circuits and the complexities of embodied experience. In the process, the disembodied logical form of the circuit was rhetorically transformed from being an *effect* of the model to a *cause* of the model's efficacy. This move, familiar to us as the Platonic backhand, made embodied reality into a blurred and messy instantiation of the clean abstractions of logical forms. Unlike others who make this move, however, McCulloch never relinquished his commitment to embodiment. The tension between logical form and embodiment in his work displays how the construction of a weightless information was complicated when cybernetics moved into the intimate context of the body's own neural functioning.

Neural Nets as Logical Operators

Warren McCulloch figured large in the Macy Conferences. He chaired the meetings and, according to all accounts, was a strong leader who exercised considerable control over who was allowed to speak and who was not. He had studied philosophy under F. S. C. Northrop and was familiar with Rudolf Carnap's propositional logic. When he turned to neurophysiology, he was driven by two questions as much philosophical as scientific. "What is a number, that a man may know it, and a man, that he may know a number?"[9] He sought the answers in a model of a neuron that he envisioned as having two aspects—one physical, the other symbolic. The McCulloch-Pitts neuron, as it came to be called, was enormously influential. Although it has now been modified in significant ways, for a generation of researchers it provided the standard model of neural functioning. In its day, it represented a triumph of experimental work and theoretical reasoning. As Steve Heims points out, it was not easy to extrapolate from amorphous pink tissue on the laboratory table to the clean abstractions of the model.[10] Before complicating our story by looking at the interplay between logical form and complex embodiment, let us first consider the model on its own terms.

The McCulloch-Pitts neuron has inputs that can be either excitatory or inhibitory. A threshold determines how much excitation is needed for it to fire. A neuron fires only if the excitation of its inputs exceeds the inhibition by at least the amount of the threshold. Neurons are connected into nets. Each net has a set of inputs (signals coming in to neurons in the net), an output set (signals leading out from neurons in the net), and a set of internal states (determined by input, output, and signals from neurons that operate inside the net but are not connected to incoming or outgoing neurons). McCulloch's central insight was that neurons connected in this way are capable of signifying logical propositions. For example, if neurons A and B are connected to C and both are necessary for C to fire, this situation corresponds to the proposition, "If A and B are both true, then C is true." If either A or B can cause C to fire, the signified proposition is "If A or B is true, then C is true." If B is inhibitory and C will fire on input from A only if B does not fire, the signified proposition is "C is true only if A is true and B is not true." This much McCulloch had formulated by 1941 when he met Walter Pitts, a brilliant and eccentric seventeen-year-old who was to become his most important collaborator.[11] Pitts worked out the mathematics proving several important theorems about neural nets. In particular, he showed that a neural net can calculate any number (that is, any proposition) that can be calculated by a Turing machine.[12] The proof was important because it joined a model of human neural functioning with automata theory. Demonstrating that the operations of a McCulloch-Pitts neural net and a Turing machine formally converge confirmed McCulloch's insight "that brains do not secrete thought as the liver secretes bile but . . . they compute thought the way electronic computers calculate numbers."[13]

Although McCulloch knew as well as anyone that the McCulloch-Pitts neuron was a simplified schematic of an actual neuron's complexity, not to mention the brain's complexity, he pushed toward connecting the operations of a neural net directly with human thought. In his view, when a neuron receives an input related to a sensory stimulus, its firing is a direct consequence of something that happened in the external world. When he says a proposition calculated by a neural net is "true," he means that the event to which the firing refers really happened. How did McCulloch account for hallucinations and such phenomena as causalgia, an amputee's burning sensation that refers to a limb no longer present? He proposed that neural nets can set up reverberating loops that, once started, continue firing even though no new signals are incoming. To distinguish between firings signifying an external event and those caused by past history, he called the former "signals" and the latter "signs." A signal "always implies its occasion," but a sign is an "enduring affair which has lost its essential temporal reference."[14] The multiple meanings that McCulloch and his colleagues attached to reverberating loops indicate how quickly speculation leaped from the simplified model to highly complex phenomena. Lawrence Kubie linked reverberating loops with the repetitive and obsessive qualities of neuroses; numerous Macy participants suggested that the loops could account for gestalt perception; and McCulloch himself connected them not only with physical sensations but also with universals in philosophical thought.[15]

The gap between the relatively simple model and the complex phenomena it was supposed to explain is the subject of an exchange of letters between McCulloch and Hans-Lukas Teuber, a young psychologist who joined the Macy group on the fourth meeting and later became a coeditor of the published transcripts. Here, in correspondence with a junior colleague, McCulloch lays bare the assumptions that make embodied reality derivative from logical form. In a letter dated November 10, 1947, Teuber argues that similarity in outcome between different cybernetic systems does not necessarily imply similarity in structure or process. "Your robot may become capable of doing innumerable tricks the nervous system is able to do; it is still unlikely that the nervous system uses the same methods as the robot in arriving at what might look like identical results. Your models remain models—unless some platonic demon mediate between the investigators of organic structure and the diagram-making mathematicians." Only the psychologist, he claims, can give the neurophysiologist information on what "the most relevant aspects of the recipient structures [in sensory function] might be."[16] Cybernetic mechanisms do not signify unless they are connected with how perception actually takes place in human observers.

In his response on December 10, 1947, McCulloch explained his position. "I look to mathematics, including symbolic logic, for a statement of a theory in terms so general that the creations of God and man must exemplify the processes prescribed by that theory. Just because the theory is so general as to fit robot and man, it lacks the specificity required to indicate mechanism in man to be the same as mechanism in robot." In this argument, universality is achieved by bracketing or "black-boxing" the specific mechanisms. It emerges by erasing particularity and looking for general forms. Rhetorically, however, McCulloch presents the theory as though it *preexisted* specific mechanisms and then was later imperfectly instantiated in them. This backhanded swing invests the theory with a coercive power that cannot be ignored, for it expresses "a law so general" that "every circuit built by God or man must exemplify it in some form."[17]

In actuality, the theorem to which McCulloch refers is proved only in relation to the simplified model of a McCulloch-Pitts neural net. It therefore can have the coercive power he claims for it only if the assumptions made for the model also hold for embodied actuality, a congruence that can be exact only if the model is as complex and noisy as reality itself. Building such a model would, of course, defeat the purpose of model-making, as Lewis Carroll (and later Jorge Luis Borges) playfully points out when he imagines a king's mad cartographer who is satisfied only when he creates a map that covers the entire kingdom, reflecting its every detail in a scale of 1:1.[18] Teuber points to a gap when he ironically asks if some "platonic demon" is mediating between organic structure and abstract diagrams, a gap that has not been closed despite McCulloch's backhand volley.

In a feminist critique of the history of logic, Andrea Nye traces similar Platonic backhands that were made to develop a logic coercive in its lawlike power.[19] Nye points out that such moves are always made in specific political and historical contexts in which they have important social implications—implications that are masked by being presenting as preexisting laws of nature.[20] Like the logicians, McCulloch stripped away context to expose (or create) a universal form. But unlike the logicians, McCulloch in 1947 does not want to leave embodied reality behind. He is searching for an "empirical epistemology," a way of combining embodied actuality with the force of logical propositions. Teuber's objections hit a nerve (or neuron) because he insisted that the abstraction is not the actuality.

Dedicated to an empirical epistemology, McCulloch cannot rest content with interpreting logical form as a universal command that embodied flesh must obey. A suture is needed to bind the flesh more tightly to the model. The suture appears in his invocation of mechanisms that had previously been black-boxed in his appeal to universality. He recounts two instances when circuits he had sketched out for pattern-recognition in robots were identified by colleagues as accurate representations of the auditory and visual portions of the cortex—*in humans*. Now McCulloch—like a knight that, moved from the diagonal to attack the queen, exposes the queen to the bishop's attack as well—has caught Teuber in a two-pronged attack. In the first approach, humans and robots are judged alike because they obey the same universal law, whatever their mechanisms. In the second approach, humans and robots are judged alike because they use the same mechanisms. This double attack is also invoked, as we shall see in the next chapter, by Norbert Wiener and his collaborators when a young upstart philosopher took issue with their cybernetic manifesto. It tends to appear when cybernetic arguments are challenged because it allows a defense on two fronts simultaneously. If mechanisms are black-boxed so that only behavior counts, humans and robots look the same because they (can be made to) behave the same. If the black boxes are opened up (and viewed from carefully controlled perspectives), the mechanisms inside the boxes look the same, again demonstrating the equivalence.

How can the queen be saved? By recognizing that the abstractions here are multilayered. When McCulloch goes down a level, away from what information is toward what it does, he still ends up several layers away from embodied complexity. Consider his claim that pattern-recognition circuits in a robot mechanism and in a human cortex are the same. These circuits are diagrams that have been abstracted from two different kinds of embodiments, neural tissue for the human and vacuum tubes or silicon chips for the robot. Although there may be a level of abstraction at which similarities can be made to appear, there is also a level of specificity at which differences create a significant gap. It depends on how the perspective is constructed. Controlling the context, particularly the movement from instantiated specificity to abstraction, was crucial to constructing the pathways through which the McCulloch-Pitts neuron was made to stand simultaneously for a computer code and for human thought. Transforming the body into a flow of binary code pulsing through neurons was an essential step in seeing human being as an informational pattern. *In context*, this transformation can be seen as a necessary simplification that made an important contribution to neurophysiology. *Taken out of context*, it is extrapolated to the unwarranted conclusion that there is no essential difference between thought and code.

I admire McCulloch because he made the audacious leap from amorphous tissue to logical model; I admire him even more because he resisted the leap. Although he emphasized the ability of his neurons to formulate propositions, he never saw them as disembodied. He was aware that information moves only through signals and that signals have existence only if they are embodied. "By definition, a signal is a proposition embodied in a physical process," he asserted in a speech, entitled "How Nervous Structures Have Ideas," to the American Neurological Association in 1949.[21] In the context of his writing as a whole, a commitment to embodiment exists in dynamic tension with an equally strong proclivity to see embodiment as the instantiation of abstract propositions.

This tension can be seen in the manuscript version of "What's in the Brain That Ink May Character?" dated August 28, 1964. McCulloch recounts about a recent trip to Ravello: "I was told that an automaton or a nerve net, like me, was a mapping of a free monoid onto a semigroup with the possible addition of identity." The parenthetical "like me" points up the incongruity between a highly abstract mathematical model involving monoids and semigroups and the embodied creature who pens these lines. "This is the same sort of nonsense one finds in the writings of those who never understood [abstract form] as an embodiment," he continues. "It is like mistaking a Chomsky language for a real language. You will find no such categorical confusion in the original Pitts and McCulloch of 1943. There the temporal propositional expressions are events occurring in time and space in a physically real net. The postulated neurons, for all their oversimplifications, are still physical neurons as truly as the chemist's atoms are physical atoms."[22] Here, in the slippages between abstract propositions, models of neurons, and "physically real" nets, we can see McCulloch trying to keep three balls in the air at once. Although the neurons are only "postulated" and are admittedly "oversimplifications," McCulloch fiercely wants to insist they are still physical. If he does not entirely succeed in creating an "empirical epistemology," he nevertheless achieves no small feat in insisting that none of the balls can be dropped without sacrificing the complexities of embodied thought.

The McCulloch-Pitts neuron is a liminal object, part abstraction and part embodied actuality, but other models were more firmly in the material realm. Part of what made cybernetics convincing to Macy participants and others were the electromechanical devices that showed cybernetic principles in action. Cybernetics was powerful because it *worked*. If you don't believe, watch William Grey Walter's robot tortoise returning to its cage for an electrochemical nip when its batteries are running low, or see Wiener's

Moth turning to follow the light and his Bedbug scuttling under a chair to avoid it. These devices were simple mechanisms by contemporary standards. Nevertheless, they served an important function because they acted as material instantiations of the momentous conclusion that humans and robots are siblings under the skin. Particularly important for the Macy Conferences were Shannon's electronic rat, a goal-seeking machine that modeled a rat learning a maze, and Ross Ashby's homeostat, a device that sought to return to a steady state when disturbed. These artifacts functioned as exchangers that brought man and machine into equivalence; they shaped the kinds of stories that participants would tell about the meaning of this equivalence. In conjunction with the formal theories, they helped to construct the human as cyborg.

The Rat and the Homeostat: Looping between Concept and Artifact

There are moments of clarity when participants came close to explicitly articulating the presuppositions informing the deep structure of the discussion. At the seventh conference, John Stroud, of the U.S. Naval Electronic Laboratory in San Diego, pointed to the far-reaching implications of Shannon's construction of information through the binary distinction between signal and noise. "Mr. Shannon is perfectly justified in being as arbitrary as he wishes," Stroud observed. "We who listen to him must always keep in mind that he has done so. Nothing that comes out of rigorous argument will be uncontaminated by the particular set of decisions that were made by him at the beginning, and it is rather dangerous at times to generalize. If we at any time relax our awareness of the way in which we originally defined the signal, we thereby automatically call all of the remainder of the received message the 'not' signal or noise."[23] As Stroud realized, Shannon's distinction between signal and noise had a conservative bias that privileges stasis over change. Noise interferes with the message's exact replication, which is presumed to be the desired result. The structure of the theory implied that change was deviation and that deviation should be corrected. By contrast, MacKay's theory had as its generative distinction the difference in the state of the receiver's mind before and after the message arrived. In his model, information was not opposed to change; it was change.

Applied to goal-seeking behavior, the two theories pointed in different directions. Privileging signal over noise, Shannon's theory implied that the goal was a preexisting state toward which the mechanism would move by making a series of distinctions between correct and incorrect choices. The goal was stable, and the mechanism would achieve stability when it reached the goal. This construction easily led to the implication that the goal, formulated in general and abstract terms, was less a specific site than stability itself. Thus the construction of information as a signal/noise distinction and the privileging of homeostasis produced and were produced by each other. By contrast, MacKay's theory implied that the goal was not a fixed point but was a changing series of values that varied with context. In his model, setting a goal temporarily marked a state that itself would become enfolded into a reflexive spiral of change. In the same way that signal/noise and homeostasis went together, so did reflexivity and information as a signifying difference.

These correlations imply that before Shannon's electronic rat ever set marker in maze, it was constituted through assumptions that affected how it would be interpreted. Although Shannon called his device a maze-solving machine, the Macy group quickly dubbed it a rat.[24] The machine consisted of a five-by-five square grid, through which a sensing finger moved. An electric jack that could be plugged into any of the twenty-five squares marked the goal, and the machine's task was to move through the squares by orderly search procedures until it reached the jack. The machine could remember previous search patterns and either repeat them or not, depending on whether they had been successful. Although Heinz von Foerster, Margaret Mead, and Hans Teuber—in their introduction to the eighth conference volume—highlighted the electronic rat's significance, they also acknowledged its limitations. "We all know that we ought to study the organism, and not the computers, if we wish to understand the organism. Differences in levels of organization may be more than quantitative." They go on to argue, however, that "the computing robot provides us with analogs that are helpful as far as they seem to hold, and no less helpful whenever they break down. To find out in what ways a nervous system (or a social group) differs from our man-made analogs requires experiment. These experiments would not have been considered if the analog had not been proposed."[25]

There is another way to understand this linkage. By suggesting certain kinds of experiments, the analogs between intelligent machines and humans *construct the human in terms of the machine.* Even when the experiment fails, the basic terms of the comparison operate to constitute the signifying difference. If I say a chicken is not like a tractor, I have characterized the chicken in terms of the tractor, no less than when I assert that the two are alike. In the same way, whether they are understood as like or unlike, ranging human intelligence alongside an intelligent machine puts the two into a relay system that constitutes the human as a special kind of information machine and the information machine as a special kind of human.[26] Although some characteristics of the analogy may be explicitly denied, the basic linkages it embodies cannot be denied, for they are intrinsic to being able to think the model. Presuppositions embodied in the electronic rat include the idea that both humans and cybernetic machines are goal-seeking mechanisms that learn, through corrective feedback, to reach a stable state. Both are information processors that tend toward homeostasis when they are functioning correctly.

Given these assumptions, it was perhaps predictable that reflexivity should be constructed as neurosis in this model. Shannon, demonstrating how his electronic rat could get caught in a reflexive loop that would keep it circling endlessly around, remarked, "It has established a vicious circle, or a singing condition."[27] "Singing condition" is a phrase that Warren McCulloch and Warren Pitts had used, in an earlier presentation, to describe neuroses modeled through cybernetic neural nets. If machines are like humans in having neuroses, humans are like machines in having neuroses that can be modeled mechanically. Linking humans and machines in a common circuit, the analogy constructs both of them as steady state systems that become pathological when they fall into reflexivity. This kind of mutually constitutive interaction belies the implication, inscribed in the volume's introduction, that such analogs are neutral heuristic devices. More accurately, they are relay systems that transport assumptions from one arena to the next.[28]

The assumptions traveling across the relay system set up by homeostasis are perhaps most visible in the discussion of W. Ross Ashby's homeostat.[29] The homeostat was an electrical device constructed with transducers and variable resistors. When it received an input changing its state, it searched for the configuration of variables that would return it to its initial condition. Ashby explained that the homeostat was meant to model an organism which must keep essential variables within preset limits to survive. He emphasized that the cost of exceeding those limits is death. If homeostasis equals safety ("Your life would be safe," Ashby responded when demonstrating how the machine could return to homeostasis), departure from homeostasis threatens death (p. 79). One of his examples concerns an engineer sitting at the control panel of a ship. The engineer functions like a homeostat, striving to keep the dials within certain limits to prevent catastrophe. Human and machine are alike in needing stable interior environments. The human keeps the ship's interior stable, and this stability preserves the homeostasis of the human's interior, in turn allowing the human to continue to ensure the ship's homeostasis. Arguing that homeostasis is a requirement "uniform among the inanimate and the animate," Ashby privileged it as a universally desirable state (p. 73).

The postwar context for the Macy Conferences played an important role in formulating what counted as homeostasis. Given the cataclysm of the war, it seemed self-evident that homeostasis was meaningful only if it included the environment as part of the picture. Thus Ashby conceived of the homeostat as a device that included both the organism and the environment. "Our question is how the organism is going to struggle with its environment," he remarked, "and if that question is to be treated adequately, we must assume some specific environment" (pp. 73–74). This specificity was expressed through the homeostat's four units, which could be arranged in various configurations to simulate organism-plus-environment. For example, one unit could be designated "organism" and the remaining three the "environment"; in another arrangement, three of the units might be the "organism," with the remaining one the "environment." Formulated in general terms, the problem the homeostat addressed was this: given some function of the environment E, can the organism find an inverse function E^{-1} such that the product of the two will result in a steady state? When Ashby asked Macy participants whether such a solution could be found for highly nonlinear systems, Julian Bigelow correctly answered, "In general, no" (p. 75). Yet, as Walter Pitts observed, the fact that an organism continues to live means that a solution does exist. More precisely, the problem was whether a solution could be articulated within the mathematical conventions and technologies of representation available to express it. These limits in turn were constituted through the model's specificities that translated between the question in the abstract and the particular question posed by that experiment. Thus the emphasis shifted from finding a solution to stating the problem.

This dynamic appears repeatedly throughout the Macy discussions. Participants increasingly understood the ability to specify exactly what was wanted as the limiting factor for building machines that could perform human functions. Von Neumann stated the thesis at the first conference, and Walter Pitts restated it near the end of the meetings, at the ninth conference. "At the very beginning of these meetings," Pitts recalled, "the question was frequently under discussion of whether a machine could be built which would do a particular thing, and, of course, the answer, which everybody has realized by now, is that as long as you definitely specify what you want the machine to do, you can, in principle, build a machine to do it" (p. 107). After the conferences were over, McCulloch repeated this dynamic in *Embodiments of Mind*. Echoing across two decades, the assertion has important implications for language.

If what is exactly stated can be done by a machine, the residue of the uniquely human becomes coextensive with the linguistic qualities that interfere with precise specification—ambiguity, metaphoric play, multiple encoding, and allusive exchanges between one symbol system and another. The uniqueness of human behavior thus becomes assimilated to the ineffability of language, and the common ground that humans and machines share is identified with the univocality of an instrumental language that has banished ambiguity from its lexicon. Through such "chunking" processes, the constellations of homeostasis and reflexivity assimilated other elements into themselves. On the side of homeostasis was instrumental language, whereas ambiguity, allusion, and metaphor stood with reflexivity.

By today's standards, Ashby's homeostat was a simple machine, but it had encoded within it a complex network of assumptions. Paradoxically, the model's simplicity facilitated the overlay of assumptions onto the artifact, for its very lack of complicating detail meant that the model stood for much more than it physically enacted. During discussion, Ashby acknowledged that the homeostat was a simple model and asserted that he "would like to get on to the more difficult case of the clever animal that has a lot of nervous system and is, nevertheless, trying to get itself stable" (p. 97). The slippage between the simplicity of the model and the complexity of the phenomena did not go unremarked. J. Z. Young, from the Anatomy Department at University College, London, sharply responded: "Actually that is experimentally rather dangerous. You are all talking about the cortex and you have it very much in mind. Simpler systems have only a limited number of possibilities" (p. 100). Yet the "simpler systems" helped to reinforce several ideas: humans are mechanisms that respond to their environments by trying to maintain homeostasis; the function of scientific language is exact specification; the bottleneck for creating intelligent machines lies in formulating problems exactly; and an information concept that privileges exactness over meaning is therefore more suitable to model construction than one that does not. Ashby's homeostat, Shannon's information theory, and the electronic rat were collaborators in constructing an interconnected network of assumptions about language, teleology, and human behavior.[30]

These assumptions did not go uncontested. The concept that most clearly brought them into question was reflexivity. As we have seen, during the Macy Conferences reflexivity was a nebulous cluster that was not explicitly named as such. To give the flavor of the discussions that both invoked the possibility of reflexivity and failed to coalesce into coherent theory about it, we can consider the image of the man-in-the-middle. The image was given currency by World War II engineering technologies that aimed to improve human performance by splicing humans into feedback loops with machines. The image takes center stage in the sixth conference during John Stroud's analysis of an operator sandwiched between a radar-tracking device on one side and an antiaircraft gun on the other. The gun operator, Stroud observed, is "surrounded on both sides by very precisely known mechanisms and the question comes up, 'What kind of a machine have we put in the middle?'"[31] The image as Stroud used it constructs the man as an input/output device. Information comes in from the radar, travels through the man, and goes out through the gun. The man is significantly placed in the *middle* of the circuit, where his output and his input are already spliced into an existing loop. Were he at the end, it might be necessary to consider more complex factors, such as how he was interacting with an open-ended and unpredictable environment. The focus in Stroud's presentation was on how information is transformed as it moves through the man-in-the-middle. As with the electronic rat and the homeostat, the emphasis was on predictability and homeostatic stability.

Countering this view was Frank Fremont-Smith's insistence on the observer's role in constructing the image of the man-in-the-middle. "Probably man is never only between the two machines," he pointed out. "Certainly he is never only in between two machines when you are studying him because you are the other man who is making an input into the man. You are studying and changing his relation to the machines by virtue of the fact that you are studying him." Fremont-Smith's introduction of the observer was addressed by Stroud in a revealing image that sought to convert the observer into a man-in-the-middle. "The human being is the most marvelous set of instruments," Stroud observed, "but like all portable instrument sets the human observer is noisy and erratic in operation. However, if these are all the instruments you have, you have to work with them until something better comes along."[32] In Stroud's remark, the man is converted from an open-ended system into a portable instrument set. The instrument may not be physically connected to two mechanistic terminals, the image implied, but this lack of tight connection only makes the splice invisible. It does not negate the suture that constructs the human as an information-processing machine spliced into a closed circuit that ideally should be homeostatic in its operation, however noisy it is in practice.

Fremont-Smith responded: "You cannot possibly, Dr. Stroud, eliminate the human being. Therefore what I am saying and trying to emphasize is that, with all their limitations, it might be pertinent for those scientific investigators at the general level, who find to their horror that we have to work with human beings, to make as much use as possible of the insights available as to what human beings are like and how they operate."[33] As his switch to formal address indicates, Fremont-Smith was upset at the recuperation of his comment back into the ideology of objectivism. His comment cuts to the heart of the objection against reflexivity. Just as with MacKay's model of structural information, reflexivity opens the man-in-the-middle to psychological complexity, so that he can no longer be constructed as a black box functioning as an input/output device. The fear is that under these conditions, reliable quantification becomes elusive or impossible and science slips into subjectivity, which to many conferees meant that it was not real science at all. Confirming traditional ideas of how science should be done in a postwar atmosphere that was already clouded by the hysteria of McCarthyism, homeostasis implied a return to normalcy in more than one sense.

The thrust of Fremont-Smith's observations was, of course, to intimate that psychological complexity was unavoidable. The responses of other participants reveal that this implication was precisely what they were most concerned to deny. They especially disliked reflexive considerations that took the personal form of suggesting that their statements were not assertions about the world but were revelations of their own internal states. The primary spokesperson for this disconcerting possibility was Lawrence Kubie, a psychoanalyst from the Yale University Psychiatric Clinic. In correspondence, Kubie enraged other participants by interpreting their criticisms of his theories as evidence of their subconscious resistances rather than as matters for scientific debate. In his presentations he was more tactful, but the reflexive thrust of his arguments remained clear. His presentations occupy more space in the published transcripts than those of any other participant, composing about one-sixth of the total. Although he met with repeated skepticism among the physical scientists, he continued to defend his position. At the center of his explanation was the multiply encoded nature of language, which operated at once as an instrument that the speaker could use to communicate and as a reflexive mirror that revealed more than the speaker knew. Like MacKay's theory of information, Kubie's psychoanalytic approach built reflexivity into the model. Also like MacKay's theory, Kubie's argument met the greatest (conscious?) resistance in the demand for reliable quantification.

Kubie's ideas will serve as a springboard for looking at the role that reflexivity played in the Macy Conferences and in the lives of some participants after the conferences ended, particularly the lives of Margaret Mead and Gregory Bateson and their daughter, Mary Catherine Bateson. Contrasting the Macy Conferences with Catherine Bateson's account of a similar conference held in 1968 will illustrate why the full implications of reflexivity could scarcely have been admitted during the Macy period. Once the observer is made a part of the picture, cracks in the frame radiate outward until the perspectives that controlled context are fractured as irretrievably as a safety-glass windshield hit by a large rock. The Macy participants were right to feel wary about reflexivity. Its potential was every bit as explosive as they suspected.

Kubie's Last Stand

Lawrence Kubie had been trained as a neurophysiologist. He won McCulloch's admiration for his 1930 paper suggesting that neuroses were caused by reverberating loops similar to those McCulloch later modeled in neural nets.[34] In midcareer Kubie converted to psychoanalysis. By the time of the Macy Conferences, he was affiliated with the hard-line Freudianism of the New York Psychoanalytic Institute. In his presentation at the sixth conference, he laid out the fundamentals of his position. Neurotic processes are dominated by unconscious motivations. As goal-seeking behavior, these processes are ineffective because the unconscious pursues its goals in symbolic form. A man wants to feel secure, and money symbolizes this security for him. But when he acquires money, he still does not feel secure. He has acquired the symbol but lacks what the symbol represents. With the gap between desire and reality yawning as widely as ever, he may actually feel more rather than less anxious as he approaches his putative goal.

Although McCulloch thought of Kubie as an experimentalist, from the beginning of the conferences Kubie resisted the reductive approach that was characteristic of McCulloch's work. At the first conference, Kubie expressed uneasiness over reducing complex psychological phenomena to mechanistic models equating humans and automata. At the sixth conference he was still resisting. In "Neurotic Potential and Human Adaptation," he explained why he had not addressed feedback mechanisms: "I wanted to make clear the complexity and subtlety of the neurotic process as it is encountered clinically. Without this we are constantly in danger of oversimplifying the problem so as to scale it down for mathematical treatment."[35] Instead of mechanistic models, his formulations emphasized the reflexivity of psychological processes. At the seventh conference, in "The Relation of Symbolic Function in Language Formation and in Neurosis," he insisted on "the fact that the human organism has two symbolic functions and not one. One is language. The other is neurosis." Moreover, the two functions converge into the same utterance. Fremont-Smith drove the point home.

"What Dr. Kubie is really trying to say is that language is a double coding: both a statement about the outside and a statement about the inside. It is that doubleness which gives this conscious/unconscious quality to it."[36]

In this view, a statement intended as an observation of the external world is pierced by reflections of the speaker's interior state, including neurotic processes of which the speaker is not conscious. If a scientist denies this is the case, insisting that he or she speaks solely about external reality, these objections themselves can be taken as evidence of unconscious motivations. For experimentalists like McCulloch, concerned to give an objective account of mental processes, psychoanalysis was the devil's plaything because it collapsed the distance between speaker and language, turning what should be scientific debate into a tar baby that clung to them the more they tried to push it away.

The damage that this view of reflexive utterance could do to scientific objectivity was dramatically laid out by McCulloch in a 1953 address to the Chicago Literary Club. Entitled "The Past of a Delusion," the speech was a fiery denunciation of Freudian psychoanalysis.[37] If all scientific utterance is tinged with subjectivity, McCulloch felt, then scientific theory must inextricably be tied to the foibles and frailties of humans as subjective beings. To show the disastrous effects that this close coupling could have on science, McCulloch took as his case study Freudian psychoanalysis, a theory that in his view both promoted the idea of close coupling and itself insidiously instantiated it. McCulloch ripped into Freud, suggesting that Freud had turned to psychoanalysis because he had wanted to make more money than he would have as a Jewish medical doctor. McCulloch recounted Freud's sex life, intimating that Freud put sexuality at the heart of his theory because he was sexually frustrated himself. McCulloch denounced psychoanalysts as charlatans who, motivated by greed, kept treating their patients as long as those patients had money to pay. He sneered at the empirical evidence used by Freud and other psychoanalysts. In his ironic conclusion, McCulloch cautioned his audience not to try to argue with psychoanalysts. All they would get for their pains, he predicted, were psychoanalytic interpretations of their objections as evidence of their own unconscious hostilities.

Kubie learned of this speech from a colleague who had been in the audience.[38] Although McCulloch went out of his way to exempt Kubie from his general scorn for psychoanalysis (in a 1950 letter to Fremont-Smith, he had written, "Of all the psychoanalysts I know, [Kubie] has the clearest head for theory"),[39] the attack was too stinging not to draw a rejoinder. As pat as McCulloch would have wished, Kubie interpreted the speech as a sign of

McCulloch's own psychological distress. Speaking to a colleague, Kubie noted that McCulloch's "vitriole may be due to an accumulation of personal frustrations of his own displaced onto analysis."[40] Later, when he heard about McCulloch's erratic behavior during a presentation at Yale, he wrote to McCulloch's host, sending a copy of the letter to Fremont-Smith: "I am distressed by this news about Warren . . . in him the boundary between sickness and health has always been narrow" (p. 137). Kubie even tried to arrange for psychoanalysts in the Boston area to meet with McCulloch "on a social pretext if necessary," with a view to getting him the "help" that Kubie thought he needed (p. 138). As Steve Heims observes in his account of these incidents, McCulloch would have been enraged had he known about Kubie's attempts at intervention.

McCulloch's "The Past of a Delusion" is vivid evidence that Fremont-Smith's attempts at reconciliation between psychoanalysts and physical scientists did not succeed. Kubie was well aware of the experimentalists' attitudes. After repeated attempts to win them over, he delivered his final presentation at the ninth conference in what sounds like a state of controlled rage. He likened the supposed "troublemaker" psychiatrist to "a naturalist, reporting on the facts of human nature as observed by him." By contrast, he noted, the physical scientists ignore complex psychological phenomena in favor of the simplifications of an abstract model. "The experimentalist and mathematician then offer their explanation, whereupon, the naturalist presents additional observations which confront the experimentalist and the mathematician with an even more complex version of natural phenomena." As the cycle continues, "these new complexities are accepted with increasing reluctance and skepticism."[41] In these remarks Kubie presented his version of his presentations at the Macy Conferences. He merely reported on the facts, whereas the others offered inadequate mechanistic explanations for them. This characterization ignores, of course, the Freudian framework he used to interpret his colleagues' behavior, a framework at least as theory-laden in its observations as anything McCulloch proposed.[42]

I think of this presentation, loaded with controlled anger as if in point/counterpoint to McCulloch's extravagant display of anger in his speech of the following year, as Kubie's last stand. The resistance it describes and inscribes went in both directions, from the psychoanalyst to the experimentalist and from the experimentalist to psychoanalyst. For the experimentalists, psychoanalysis strengthened the chain of association that bound reflexivity together with subjectivity, for it added to the already daunting problems of quantification the unfalsifiable notion of the unconscious. It is no wonder that reflexivity came to seem, for many of the participants, a dead end for legitimate scientific inquiry.

Even as one version of reflexivity fizzled out, other versions were being constructed in terms that made them more productive, in part because these versions avoided associating reflexivity with the unconscious. Temple Burling, reading the published transcripts in 1954, wrote to McCulloch: "I was surprised at the jamb that the group got into at this late date over the question of 'the unconscious.' It seems to me that is putting the cart before the horse. It isn't unconscious neuro activity that is puzzling but conscious. Consciousness is the great mystery."[43] Burling's comments point to another way into reflexivity, a way taken by a handful of participants, including Heinz von Foerster, Margaret Mead, and Gregory Bateson. Though they were not necessarily opposed to psychoanalytic interpretation, it was not the focus of their attention. The scale on which they wanted to play their tunes did not run up and down the conscious/unconscious keyboard. Rather, they wanted to create models that would take into account the observer's role in constructing the system. The important dichotomy for them was observer/system, and the important problems were how to locate the observer inside the system and the system inside the observer.

Circling the Observer

In 1969, near the end of his career, Fremont-Smith wrote (or rather, had his secretary write) to participants of the various Macy Conferences that he had organized over three decades, asking for their evaluation of the interdisciplinary programs and the discussion formats. The inquiry was clearly a career-closing move; he was looking for affirmation of what he considered his lifework. Some of the replies were disarmingly frank. Jimmie Savage wrote about how it felt to be a young man allowed to "hobnob with such a diverse group of illustrious and brilliant people." He recalled that he had frequently found himself thinking that the emperor had no clothes but wondering if he could trust his own feelings. He confessed, "Cybernetics itself seemed to me to be mostly baloney."[44] R. W. Gerard expressed similar dissatisfactions, recalling being "intensely frustrated by the perpetual tangents to tangents that developed during a meeting and the rare satisfaction of intellectual closure and completion of any line of thought or argument." He added, "You may recall that this frustration was sufficient so that I did not wish to attend later meetings."[45] These responses are interesting not only because they throw light on the conferences but also because they talk frankly about feelings. "Affect ran high," Savage recalled. In the transcripts, by contrast, emotions enter the discussion only as objects for scientific modeling. Almost never are they articulated as something the participants are experiencing. The contrast between the letters and the transcripts illuminates the scientific ethos that ruled at the meetings. Emotions were considered out of bounds for several reasons, all of which perhaps came down to the same reason. The framework of scientific inquiry had been constructed so as to ignore the observer.

Heinz von Foerster, in his letter to Fremont-Smith, saw the inclusion of the observer as the central issue of cybernetics.[46] He noted that at the beginning of the century, with the advent of relativity theory and the Uncertainty Principle, "a most enigmatic object was discovered which until then was carefully excluded from all scientific discourse: the 'observer.' 'Who is he?' was the question, indignantly asked by those who subscribed to a sour grape strategy, and seriously asked by those who felt that any science worth its name must include the subject that makes the observations at the first place." There were no precedents for this inclusion, he continued. "The whole methodology of a science that includes the observer had to be developed from scratch." He generously credited Fremont-Smith with the idea of bringing together people rather than disciplines and thus placing relationships at the center of the discussions (although the transcripts rarely acknowledge these relationships). He also commented that Fremont-Smith understood that including the observer would have to be an interdisciplinary task. In establishing the focus as "problems of communication," Fremont-Smith hoped the Macy group would see that the topic required an "intensive and comprehensive study of man." Thus the sciences were to be unified by an overarching framework that could simultaneously explain "man" and the people who studied "man." Cybernetics was to provide that framework.

In March 1976, two decades after the conferences had ended, Margaret Mead and Gregory Bateson were sitting with Stewart Brand at Bateson's kitchen table in a rare joint interview. Brand asked them about the Macy Conferences. They agreed that including the observer was one of the central problems raised by the cybernetic paradigm. Reaching for a scrap of paper, Bateson sketched a diagram (which Brand included in the published interview) of the communication system as it was envisioned before cybernetics. The drawing shows a black box with input, output, and feedback loops within the box. The space labeled "Engineer" remains outside the box. A second drawing represents Bateson's later understanding of cybernetics. Here the first black box, along with the names "Wiener, Bateson, Mead," is encapsulated within a larger box. In this drawing, the observers are included *within* the system rather than looking at it from the outside. The interview turned to a discussion of the dynamics that had prevailed at the Macy Conferences. Mead commented, "Kubie was a very important person at that point." She added: "McCulloch had a grand design in his mind. He got people into that conference, who he then kept from talking." Bateson continued, "Yes, he had a design for how the shape of the conversation would run over five years—what had to be said before what else could be said." When Brand asked what that design was, Bateson answered, "Who knows?" But Mead thought it was "more or less what happened."[47]

Brand wanted to know why cybernetics had run out of steam. "What happened?" he asked repeatedly. His sense of the situation is confirmed by correspondence exchanged between the transcript editors—Heinz von Foerster, Margaret Mead, and Hans Teuber—after the tenth conference in 1953. Fremont-Smith and McCulloch wanted the transcripts published, just as the transcripts for the previous four conferences had been published. But Teuber disagreed, noting that the discussions were too rambling and unfocused; if published, he said, they would be an embarrassment. Although he was the junior member of the editorial board, he stood his ground. He wrote to Fremont-Smith, sending a copy of the letter to McCulloch, that if the others decided to publish over his objections, he wanted his name removed from the list of editors.[48] As the junior member, he had the most to lose; the others already had established reputations. McCulloch must have written a stiff note in reply, for Teuber answered defensively. He insisted that the issue was not his reputation but the quality of the transcripts. "From your note, it is obvious that I sound stuffy to you and Walter. Do tell him that I wanted to get off the list of editors, not because I am worried about reputations, but simply because I can't do enough for this transcript to get it into any sort of shape. The transactions of this last meeting simply do not add to the earlier ones—they detract. Granted, there are a few sparks, but there is not enough of the old fire. I owed it to you and Frank Fremont-Smith to speak my mind on this matter."[49] Mead worked out a compromise. The three speakers would publish their talks as formal papers, and McCulloch's summary of all the conferences would be used as an introduction. No one thought of suggesting more conferences or more transcripts. It was the end of an era.

But not the end of reflexivity. Although a reflexive view of cybernetics failed to coalesce into a coherent theory during the Macy Conferences, Bateson did not want to let the idea go. He determined to go ahead on his own. He organized a conference in July 1968 to explore how the reflexive implications of cybernetics could provide the basis for a new epistemology, and he invited a group of scientists, social scientists, and humanists. Included were Warren McCulloch and Gordon Pask, both central players in cybernetics, along with Mary Catherine Bateson, known as Catherine (to her father as "Cap"), an anthropologist specializing in comparative religions.

Out of this week-long conference came Catherine's 1972 book, *Our Own Metaphor*.[50] Her account of this conference, in some ways a reflection of the Macy Conferences, contrasts sharply with the Macy transcripts. The best explanation for this difference, I think, is epistemological. Catherine assumes that *of course* the observer affects what is seen, so she takes care to tell her readers about her state of mind and situation at the time. She recounts, for example, finding out that she was pregnant in the months preceding the conference; how awed she felt by the life that, whether she consciously attended to it or not, continued to grow within her; and her devastation when the baby was born prematurely, lived for an afternoon, and died. Her grief was still fresh when she attended the conference, and it naturally colored, she feels, how she interpreted what she learned there.

The difference between her account and the Macy transcripts does not lie in the fact that one is technical and the other anecdotal. It is obviously important to Catherine to understand, as clearly as possible, what each presenter is saying, and she skillfully guides her reader through presentations fully as complex, technical, and detailed as any in the Macy transcripts. Rather, the difference lies in her attitude toward her material and her determination to include as much of the context as she can. She takes care to tell her readers not only what ideas were exchanged but also how the people looked and her interpretation of how they were feeling. In addition to the words exchanged, she includes appearance, body language, and emotional atmosphere. At the Macy Conferences, her mother, Margaret Mead, had repeatedly cautioned that the transcripts were a purely *verbal* record and therefore represented only a fraction of the communication taking place. Mead wanted a much fuller record that would include "posture, gesture, and intonation."[51] Two decades later, Catherine fulfilled that desire in her precisely crafted descriptions.

Here is Catherine's account of Warren McCulloch: "Warren had bright, fierce eyes and held his head dropped low between thin shoulders. He had white hair and a white beard and curious blend of glee and grief, of belligerence and gentleness" (*OOM*, pp. 23–24). When he gave a presentation, Catherine strained to follow his ideas and found it odd that he was not more responsive to the needs and situations of those who were listening. "More than anyone else present, Warren tended to use an uncompromisingly technical vocabulary, referring to scientists I knew nothing of and calling on unfamiliar mathematics and neurophysiology. As I listened I kept checking to see whether I was sorting out what each example was about, what kind of thing he was trying to say in this interdisciplinary context where not more than two or three people could follow the substance of most of his examples" (*OOM*, p. 65). In her contextualized account, McCulloch's fierce commitment to an "empirical epistemology" carries with it an obvious price—a tendency toward decontextualization that made him less than effective in communicating with this audience.

Catherine Bateson included in her prologue Gregory Bateson's document that set the agenda for the conference and laid out the problems it would explore. The influence of cybernetics as it had evolved during the Macy Conferences is apparent throughout. Equally clear are Gregory's revisions, critiques, and transformations of those concepts. He indicated that he wanted participants to consider "three cybernetic or homeostatic systems": the individual, the society, and the larger global ecosystem in which both are embedded. Although consciousness would be considered as "an important component in the *coupling* of these systems" (*OOM*, p. 13), epistemologically its role was limited. From an "enormously great plethora of mental events," it chooses a few on which to focus (*OOM*, p. 16). An important factor guiding this choice, he hypothesized, is "purpose." Problems arise when this purposeful selection is taken as the whole. "If consciousness has feedback upon the remainder of mind and if consciousness deals only with a skewed sample of the events of the total mind, then there must exist a systematic (i.e., non-random) difference between the conscious views of self and the world and the true nature of self and the world" (*OOM*, p. 16). Thus the emphasis on "purpose" so central to the Macy Conferences became here not an assumed orientation but a lens that consciousness wears and that distorts what it sees. Specifically, this lens obscures "the cybernetic nature of self and the world," an obfuscation that "*tends to be imperceptible to consciousness*" (*OOM*, p. 16).

Nowhere is the transformation that Gregory worked on the Macy Conferences clearer than in what he considers the "cybernetic nature" of world and self. For him, cybernetics is no longer the homeostatic model of the Macy Conferences (although echoes of this language still linger). Rather, it has become the reflexivity of the larger box that he would sketch a decade later at his kitchen table. Equally striking is the changed significance of separating a system from its surrounding context. For Bateson, decontextualization is not a necessary scientific move but a systematic distortion.

The inclination of the conscious mind toward purpose makes it focus on an arc of causally related events leading to a perceived goal. Obliterated or forgotten is the matrix in which these arcs are embedded. A truly cybernetic approach, for Bateson, concentrates on the couplings that bind the parts into interactive wholes.

The revisionist thrust of Gregory's view of cybernetics is apparent in a letter he wrote to Catherine in June 1977, a year after his interview with Stewart Brand. The letter begins with Gregory remarking on how rereading *Our Own Metaphor* vividly brought the conference back to his mind. Then Gregory lays out the gist of his new "cybernetic" epistemology. He starts from the premise that we never know the world as such. We know only what our sensory perceptions construct for us. In this sense, we know nothing about the world. But we know something, and what we know is the end result of the internal processes we use to construct our inner world. Thus we know ourselves as complex beings, including processes that extend below consciousness and beyond ourselves out into the world, through the inner world available to a consciousness that exists only because of those processes. "We are our epistemology" is Gregory's formulation.[52] Catherine's phrasing is similar: "Each person is his own central metaphor" (*OOM*, p. 285). In this view, the dualism between subject and object disappears, for the object as a thing in itself cannot exist for us. There is only the subjective, inner world. The world, as this "cybernetics" constructs it, is a monism. Nevertheless, it is not solipsistic, for Gregory believes that the microcosm of the inner world is functional within the larger ecosystem only because it is an appropriate metaphor for the macrocosm. In her concluding chapter, Catherine amplifies on this view by supposing that we can understand the complexity of the outer world only because our codes for constructing the inner world are similarly diverse and complex. In this sense, we are a metaphor not only for ourselves but also for the larger system in which we are embedded. This leads her into a subtly nuanced analysis of couplings between inner world and outer world, including the insight that because the worlds are coupled, they must in the last analysis be regarded as a single system.

For Gregory, McCulloch represents a Moses-like figure who could lead others to the brink of this new epistemology but was unable to enter into it himself. "His last speech makes a special sort of sense if you read it as spoken in that context," Gregory suggests.[53] Catherine uses McCulloch's speech to end her account of the conference, and the speech is worth quoting in detail. "I am by nature a warrior, and wars don't make sense anymore," McCulloch begins (*OOM*, p. 311). The recognition rings true. I think of the statement in his summary of the Macy Conferences: "Our most notable agreement is that we have learned to know one another a bit better, and to fight fair in our shirt sleeves."[54] For him, scientific debate was a form of agonistic conflict. He continues in his speech by recalling the nitty-gritty details of his experimental work, its difficulties and funny moments. Then his thoughts turn to human mortality. He is an old man; although he cannot know it now, within a year he will die. Earlier in the conference, he "snapped" (says Catherine): "I don't particularly like people. Never have. Man to my mind is about the nastiest, most destructive of all the animals. I don't see any reason, if he can evolve machines that can have more fun than he himself can, why they shouldn't take over, enslave us, quite happily. They might have a lot more fun. Invent better games than we ever did" (*OOM*, p. 226).

Now, at the penultimate moment of the conference, of Catherine's book that she will dedicate to him, and of his life, he confesses to mortal feelings. "'The difficulty is that we, who are not single-cell organisms, cannot simply divide and pass on our programs. We have to couple and there is behind this a second requirement.' Warren began to weep. 'We learn . . . that there's a utility in death because . . . the world goes on changing and we can't keep up with it. If I have any disciples, you can say this of every one of them, they think for themselves'" (*OOM*, p. 311).

If Gregory Bateson thought of himself as McCulloch's disciple, the epitaph that McCulloch wanted for himself is certainly true in Bateson's case, for he both learned from his mentor and went beyond him. Taking the cybernetic paradigm of McCulloch's "empirical epistemology" and making it into "our own metaphor," Bateson reintroduced the reflexive dimension that McCulloch had fought so hard to exorcise when it was associated with psychoanalysis. Yet Bateson's reinterpretation succeeded in articulating a version of reflexivity that did not depend on a psychoanalytic entanglement of conscious and unconscious meanings in scientific statements. Moreover, his epistemology gave an important role to objective constraints, for it insisted that only those constructions that were compatible with reality were conducive to long-term survival. And survival was very much the name of the game for Catherine and Gregory Bateson. The larger issues they wanted their conference to address included the increasing degradation of the environment. In looking for an epistemology that would proceed from a sense of the world's complexity, they did not give up the idea that some constructions are better than others.

Let me now anticipate connections between the path the Batesons followed and those paths traced in subsequent chapters. In breaking new conceptual ground, Gregory Bateson drew on a famous article on the frog's visual cortex. The article had been coauthored by several people from the Macy Conferences, including Warren McCulloch, Walter Pitts, and Jerome Lettvin; also listed as coauthor was a newcomer who did not attend the Macy Conferences, Humberto Maturana.[55] In using this article to develop "our own metaphor," Bateson went where no experimentalist could easily follow, for he made speculative leaps that would take decades of experimental work to confirm. He went into the inner world and turned it inside out, so to speak, so that the inner world became a metaphor for the outer world. Maturana was to follow a similar yet different path. He went into the inner world and insisted that it can't be turned inside out, that it is a metaphor for nothing other than its own creation of itself as a system. This is the theory of autopoiesis, which we will discuss in chapter 6. Maturana did not identify with cybernetics as much as Bateson did, and he did not generally use that term to describe his work. Nevertheless, his theory took up certain problems that were left hanging after the Macy Conferences ended. Like Bateson, Maturana found reflexivity more promising than homeostasis. Also like Bateson, he both appropriated concepts from the Macy context and changed them profoundly.

Janet Freud/Freed

Like Bateson, Mead, and Brand sitting at a kitchen table on that March morning in 1976, I am sitting at my kitchen table in March 1996. I'm looking at the pages on which their interview is published. I'm particularly intrigued by a photograph that Brand included, one evidently given to him by Mead or Bateson. It's a large picture, too large to include in one frame, so it stretches across two pages. The caption identifies the setting as the 1952 Macy Conference—the ninth, the conference with the last real Macy transcript, for the tenth volume (as noted above) was not a transcript but was instead formal papers. This was the conference of Kubie's last stand. The photograph shows a large group of men and one woman—Margaret Mead —sitting around cloth-covered tables pulled into a U-shape. A speaker stands at the mouth of the U; the caption identifies him as Yehoshua Bar-Hillel. But wait. That must mean the date is incorrect, since Bar-Hillel spoke at the tenth conference. He wasn't present at the ninth. So this photograph must have been taken in 1953, at the conference in which the conversation was so meandering and dilatory that it couldn't be published. I wonder where the caption came from. I imagine Bateson digging out the photograph and giving it to Brand while he and Mead clue Brand in on who was who as Brand scribbles down the names, probably while they are all still sitting at the kitchen table.

Now I notice that Mead isn't the only woman in the picture. Another woman sits with her back to the photographer, her arms extended, hands reaching out to a machine I can't quite see. The caption identifies her as "Janet Freud," but I know this can't be right either. She must be Janet Freed, listed in the published transcripts as "assistant to the conference program." I have seen her name in the typed transcripts of the editorial meetings that followed the later conferences, and I know more or less what she did.

She was responsible for turning these men's (and a couple of women's) words into type. She was the one who listened to the tape-recordings of the early conferences and strained to catch inaudible strange words. When she sent McCulloch the typed transcript of the second Macy conference, she plaintively wrote that she knew there were "many, many blank spaces" but that Dr. Fremont-Smith had ordered her and her staff to listen to the recordings only twice and to type what they heard.[56] Evidently, transcribing the tape-recordings was taking too much staff time, and Fremont-Smith did not want to waste his resources that way—his resources, her time.

The quirk of memory or handwriting that made Brand call her "Janet Freud" seems eerily appropriate, for this was the woman who, like Freud's patients, had no voice in the transcripts, although the transcripts have a voice that we can read only because of her. She was the one who presided over the physical transformations of signifiers as they went from tape-recording to transcript to revised copy to galley to book. Others—the editors Teuber, Mead, and von Foerster, the organizer Fremont-Smith, and the chairman McCulloch—worried about content—but her focus was the materiality of the processes that make sounds into words, marks into books. She did the best she could, but the transcription took much time and she had many other things to do. When she was told not to take time, the transcript had more ellipses than words, and she felt bad. What to do? She suggested to Fremont-Smith that he and McCulloch insist the speakers deliver drafts of their talks ahead of time.[57] Then she wouldn't have to strain to listen to tape-recordings that were noisy beyond endurance by today's standards. She wouldn't have to guess at unfamiliar words (the manuscripts of the transcripts are peppered with misspellings). She learned stenotypography (or perhaps arranged to hire someone else who knew it) so that the words could be transcribed directly into the machine. This, combined with the drafts of the presentations, allowed her to come up with reasonable transcripts of both presentation and discussion without driving herself crazy. At an editorial meeting, when others suggested that it was too much work to pressure the speakers to get their drafts into the office ahead of time, she spoke up. The drafts were essential. She defended the other woman who was lower on the totem pole than she was—her staff—and said that this woman could be expected to do only so much. She didn't say so, but surely she had herself in mind as well.

Janet Freed's role in the Macy group is teasingly hinted at in the transcripts to the 1949 Editors' Meeting. Fremont-Smith depended on her to keep him on track. He decided to make up a little booklet for the Macy Conference chairmen to supply them with guidelines, commenting, "It occurred to us, in fact, it was Miss Freed's suggestion . . . " Elsewhere, when he realized that he had "jumped around a good deal" and gotten off track, he referred to the list of topics that Freed had made up for him to follow.[58] When one of the men remarked that there were now thirteen Macy groups and wondered if his office was going "to be able to do it," Fremont-Smith must have looked at Freed, for he uttered a comment that, in this professional and overwhelmingly male meeting, comes across as almost shocking in its personal nature. "You write and get a lovely smile. Do you have anything else you want to say at this point?" "No," she replies, not elaborating. Nowhere else in the Macy transcripts, to my knowledge, does someone simply answer, "No." Perhaps she was embarrassed, or perhaps she simply felt her position made it inappropriate for her to say more.

Fremont-Smith's remark, faithfully preserved by the transcription technologies that Janet Freed oversaw, has a slightly odd phrasing, and I puzzle over it. She writes and gets a smile, as if she had to go somewhere to fetch it, as if it were produced elsewhere and transported back to her face. I feel I don't know where the smile comes from because Janet Freed effaces herself. Rarely do we see her directly; we glimpse her largely through her reflections in the speech of others. More than anyone else, she qualifies as the outside observer who watches a system that she constructs through the marks she makes on paper, although the system itself has a great deal of trouble including her within the names of those people who are authorized to speak and make meaning.

What are we to make of Janet F., this sign of the repressed, this Freudian slip of a female who, with a flick of a "u" (the U-shaped table at which she sits?), goes from Freed to Freud, Freud to Freed? Thinking of her, I am reminded of Dorothy Smith's suggestion that men of a certain class are prone to decontextualization and reification because they are in a position to command the labors of others.[59] "Take a letter, Miss Freed," he says. Miss

Freed comes in. She gets a lovely smile. The man speaks, and she writes on her stenography pad (or perhaps on her stenography typewriter). The man leaves. He has a plane to catch, a meeting to attend. When he returns, the letter is on his desk, awaiting his signature. From his point of view, what has happened? He speaks, giving commands or dictating words, and things happen. A woman comes in, marks are inscribed onto paper, letters appear, conferences are arranged, books are published. *Taken out of context,* his words fly, by themselves, into books. The full burden of the labor that makes these things happen is for him only an abstraction, a resource diverted from other possible uses, because he is not the one performing the labor.

Miss Freed has no such illusions. *Embedded in context,* she knows that words never make things happen by themselves—or rather, that the only things they can make happen are other abstractions, like getting married or opening meetings. They can't put marks onto paper. They can't get letters in the mail. They can't bring twenty-five people together at the right time and in the right place, at the Beekman Hotel in New York City, where white tablecloths and black chalkboards await them. For that, material and embodied processes must be used—processes that exist never in isolation but always *in contexts* where the relevant boundaries are permeable, negotiable, instantiated.

On a level beyond words, beyond theories and equations, in her body and her arms and her fingers and her aching back, Janet Freed knows that information is never disembodied, that messages don't flow by themselves, and that epistemology isn't a word floating through the thin, thin air until it is connected up with incorporating practices.

LIBERAL SUBJECTIVITY IMPERILED:

NORBERT WIENER AND CYBERNETIC ANXIETY

Of all the implications that first-wave cybernetics conveyed, perhaps none was more disturbing and potentially revolutionary than the idea that the boundaries of the human subject are constructed rather than given. Conceptualizing control, communication, and information as an integrated system, cybernetics radically changed how boundaries were conceived. Gregory Bateson brought the point home when he puzzled his graduate students with a question koanlike in its simplicity, asking if a blind man's cane is part of the man.[1] The question aimed to spark a mind-shift. Most of his students thought that human boundaries are naturally defined by epidermal surfaces. Seen from the cybernetic perspective coalescing into awareness during and after World War II, however, cybernetic systems are constituted by flows of information. In this viewpoint, cane and man join in a single system, for the cane funnels to the man essential information about his environment. The same is true of a hearing aid for a deaf person, a voice synthesizer for someone with impaired speech, and a helmet with a voice-activated firing control for a fighter pilot.

This list is meant to be seductive, for over the space of a comma, it moves from modifications intended to compensate for deficiencies to interventions designed to enhance normal functioning. Once this splice is passed, establishing conceptual limits to the process becomes difficult. In "A Manifesto for Cyborgs," Donna Haraway wrote about the potential of the cyborg to disrupt traditional categories.[2] Fusing cybernetic device and biological organism, the cyborg violates the human/machine distinction; replacing cognition with neural feedback, it challenges the human-animal difference; explaining the behavior of thermostats and people through theories of feedback, hierarchical structure, and control, it erases the animate/inanimate distinction. In addition to arousing anxiety, the cyborg can also spark erotic fascination: witness the female cyborg in *Blade Runner.* The flip side of the cyborg's violation of boundaries is what Haraway calls its "pleasurably tight coupling" between parts that are not supposed to touch. Mingling erotically charged violations with potent new fusions, the cyborg becomes the stage on which are performed contestations about the body boundaries that have often marked class, ethnic, and cultural differences. Especially when it operates in the realm of the Imaginary rather than through actual physical operations (which act as a reality check on fantasies about cyborgism), cybernetics intimates that body boundaries are up for grabs.

As George Lakoff and Mark Johnson have shown in their study of embodied metaphors, our images of our bodies, their limitations and possibilities, openings and self-containments, inform how we envision the intellectual territories we stake out and occupy.[3] When the body is revealed as a construct, subject to radical change and redefinition, bodies of knowledge are similarly apt to be seen as constructs, no more inevitable than the organic form that images them. At the same time that cybernetics was reconfiguring the body as an informational system, it was also presenting itself as a science of information that would remap intellectual terrains. Branching out into disciplines as different as biology, psychology, and electrical engineering, it claimed to be a universal solvent that would dissolve traditional disciplinary boundaries.[4] Norbert Wiener, the father of cybernetics, could be supposed to endorse this imperialist ambition. Yet, contemplating the penetration of cybernetics into social and humanistic fields, he found himself confronted with some disturbing questions. Where should the cybernetic dissolution of boundaries stop? At what point does the anxiety provoked by dissolution overcome the ecstasy? His writings testify to both the exhilaration and the uneasiness that cybernetics generated when its boundary disruptions threatened to get out of hand. They illustrate the complex dynamics that marked the construction of the cyborg during the foundational period of the late 1940s and 1950s.

As this brief summary suggests, to engage Wiener's work is to be struck by contradiction. Envisioning powerful new ways to equate humans and machines, he also spoke up strongly for liberal humanist values. A talk given to an audience of physicians in 1954 illustrates the breadth of his concern and ambivalence.[5] He predicted the existence of the automatic factory, argued that electronic computers were thinking machines capable of taking over many human decision-making processes, and cautioned that humans must not let machines become their masters. As I indicated in chapter 1, the values of liberal humanism—a coherent, rational self, the right of that self to autonomy and freedom, and a sense of agency linked with a belief in enlightened self-interest—deeply inform Wiener's thinking. Often these values stand him in good stead, for example when he rejected the practice of lobotomy at a time when Lawrence Kubie, along with many others, was endorsing it. During World War II he frantically immersed himself in military-funded research, but after the war he announced his opposition to nuclear weapons and from then on refused to do military research.[6] The tension between Wiener's humanistic values and the cybernetic viewpoint is everywhere apparent in his writing. On the one hand, he used cybernetics to create more effective killing machines (as Peter Galison has noted),[7] applying cybernetics to self-correcting radar tuning, automated antiaircraft fire, torpedoes, and guided missiles. Yet he also struggled to envision the cybernetic machine in the image of a humanistic self. Placed alongside his human brother (sisters rarely enter this picture), the cybernetic machine was to be designed so that it did not threaten the autonomous, self-regulating subject of liberal humanism. On the contrary, it was to extend that self into the realm of the machine.

But the confluence of cybernetics with liberal humanism was not to run so smoothly. The parallel between self-regulating machinery and liberal humanism has a history that stretches back into the eighteenth century, as Otto Mayr demonstrates in *Authority, Liberty, and Automatic Machinery in Early Modern Europe.*[8] Mayr argues that ideas about self-regulation were instrumental in effecting a shift from the centralized authoritarian control that characterized European political philosophy during the sixteenth and seventeenth centuries (especially in England, France, and Germany) to the Enlightenment philosophies of democracy, decentralized control, and liberal self-regulation. Because systems were envisioned as self-regulating, they could be left to work on their own—from the Invisible Hand of Adam Smith's self-regulating market to the political philosophy of enlightened self-interest. These visions of self-regulating economic and political systems produced a complementary notion of the liberal self as an autonomous, self-regulating subject. By the mid-twentieth century, liberal humanism, self-regulating machinery, and possessive individualism had come together in an uneasy alliance that at once helped to create the cyborg and also undermined the foundations of liberal subjectivity. Philip K. Dick tapped into this potential instability when he used his fiction to pose a disturbing question: should a cybernetic machine, sufficiently powerful in its self-regulating processes to become fully conscious and rational, be allowed to own itself?[9] If owning oneself was a constitutive premise for liberal humanism, the cyborg complicated that premise by its figuring of a rational subject who is always already constituted by the forces of capitalist markets.

The inconsistencies in liberal philosophy that Dick's fiction exposes are also apparent in Wiener's texts. His writing indulges in many of the practices that have given liberalism a bad name among cultural critics: the tendency to use the plural to give voice to a privileged few while presuming to speak for everyone; the masking of deep structural inequalities by enfranchising some while others remain excluded; and the complicity of the speaker in capitalist imperialism, a complicity that his rhetorical practices are designed to veil or obscure. The closest that Wiener comes to a critique of these complicities is a rigid machine he constructs in opposition to the cybernetic machine. This alien and alienating machine is invested with qualities he wants to purge from cybernetics, including rigidity, oppression, militaristic regulation of thought and action, reduction of humans to antlike elements, manipulation, betrayal, and death. The scope of the critique is limited, for it distances the negative values away from his projects instead of recognizing his complicity with them. When he predicted the automatic factory, for example, he foresaw that it would result in large-scale economic displacements (with all the implications that this would have for working-class people as autonomous independent agents), but he offered no remedy other than the platitude that men must not let machines take over.[10]

Wiener was not unaware of the ironies through which cybernetics would imperil the very liberal humanist subject whose origins are enmeshed with self-regulating machinery. Throughout his mature writings, he struggled to reconcile the tradition of liberalism with the new cybernetic paradigm he was in the process of creating. When I think of him, I imagine him laboring mightily to construct the mirror of the cyborg. He stands proudly before this product of his reflection, urging us to look into it so that we can see ourselves as control-communication devices, differing in no substantial regard from our mechanical siblings. Then he happens to glance over his shoulder, sees himself as a cyborg, and makes a horrified withdrawal. What assumptions underlie this intense ambivalence? What threads bind them together into something we might call a worldview? How are the ambivalences negotiated, and when do they become so intense that the only way to resolve them is to withdraw? What can these complex negotiations tell us about the pleasures and dangers of the posthuman subjectivity that would soon displace the liberal humanist self?

To explore these questions, we will begin with Wiener's early work on probability. In his view, it is because the world is fundamentally probabilistic that control is needed, for the path of future events cannot be accurately predicted. By the same token, control cannot be static or centralized, for then it would not be able to cope with unexpected developments. The necessity for a flexible, self-regulating system of control based on feedback *from the system itself* starts with the system thumbing its nose at Newtonian predictability. From this, we will follow a web of sticky connections: a reinscription of homeostasis; an information construction that grows out of Wiener's deep belief in a probabilistic universe; an interpretation of noise linking noise with entropy, degradation, and death; and above all, an analogical mode of thinking that moves easily across boundaries to identify (or construct) pattern similarities between very different kinds of structures. As much as anything, it was these analogical moves that helped to construct the cyborg as Wiener envisioned it. All this from a man so uncomfortable with his own body that he could not throw horseshoes in even approximately the right direction and had to abandon a career in biology because he was too clumsy to do the lab work. These physical characteristics are not, I shall argue, entirely irrelevant to the cybernetic viewpoint that Wiener was instrumental in forging.

Of Molecules and Men: Cybernetics and Probability

Like Venus, cybernetics was born from the froth of chaos. Wiener's important early work was done on Brownian motion, the random motion that molecules make as they collide with each other, bounce off each other, and collide again, as if they were manic bumper cars.[11] Given this chaos, it is impossible to know the microstates in enough detail to predict from the laws of motion how individual molecules will behave. Therefore, probabilistic and statistical methods are required. (The Uncertainty Principle introduced additional complications of a profound nature by setting limits on how precisely positions and momenta can be known.) Probability calculations are facilitated if one assumes that the chaotic motion is homogeneous, that is, that it is the same regardless of how the system is sliced to analyze it. This leads to the famous ergodic hypothesis: "an ensemble of dynamic systems in some way traces in the course of time a distribution of parameters which is identical with the distribution of parameters of all systems at a given time."[12] Following George David Birkhoff, Wiener helped to make this hypothesis more limited, precise, and mathematically rigorous than had Willard Gibbs when he first conceived the idea.

Refining Gibbs's methods and ideas, Wiener saw Gibbs as a seminal figure not only for his own work but for all of twentieth-century science. "It is . . . Gibbs rather than Einstein or Heisenberg or Planck to whom we must attribute the first great revolution of twentieth century physics," Wiener wrote in *The Human Use of Human Beings*.[13] Gibbs deserved this honor, Wiener believed, because he realized the deeper implications of probability theory. One explanation for this uncertainty is the limit placed on knowledge by the Uncertainty Principle, mentioned above. In addition to reflecting our ignorance of microstates, uncertainty also stems from our finitude as human beings. Thirty years before this became an important element in chaos theory, Wiener shrewdly realized that initial conditions can never be known exactly because physical measurements are never completely precise. "What we have to say about a machine or other dynamic system really concerns not what we must expect when the initial positions and momenta are given with perfect accuracy (which never occurs), but what we are to expect when they are given with attainable accuracy" (*HU*, p. 8).

Related to these epistemological issues is the shift of orientation implicit in Gibbs's approach. Rather than use probabilistic methods to address large numbers of particles (like the bumper cars), Gibbs used probability to consider how different initial velocities and positions might cause a system to evolve in different ways. Thus, he considered not many sets within one world but many worlds generated from a single set or, in Wiener's phrase, "all the worlds which are possible answers to a limited set of questions concerning our environment." So important did Wiener consider this perspective that he argued, "It is with this point of view at its core that the new science of Cybernetics began its development" (*HU*, p. 12). To see why Wiener considered the innovation profound, we have only to compare it with Laplace's famous boast that given the initial conditions, a being with enough computing power would be able to predict a system's evolution for eternity. In this view, the universe is completely deterministic and knowable, as precise and predictable as a clock made by God—or, amounting to the same thing for Laplace, a clock governed by Newton's laws of motion. By contrast, the probabilistic world of Gibbs and Wiener operates like a baggy pair of pants, holding together all right but constantly rearranging itself every time one tries to sit down.

Already steeped in probability theory and inclined to view the world as one evolution realized from a range of possible worlds, Wiener thought about information in the same terms. Working more or less independently of Leon Brillouin and Claude Shannon, he came to similar conclusions.[14] As we saw in chapter 3, Wiener defined information as a function of probabilities representing a choice of one message from a range of possible messages that might be sent. In a sense, he took Gibbs's idea and substituted word for world. Instead of one world coming into being from among a galaxy of possible worlds, one message comes into being from a cacophony of possible messages. When the theory worked, Wiener took it as further confirmation that Gibbs's approach expressed something fundamental about reality; the word and the world are both essentially probabilistic in their natures. This interpretation, though fascinating as a window into Wiener's view of the relation between information and physical reality, seriously understates the constructive aspect of information theory. Far from being a passive confirmation, information theory was an active extension of a probabilistic worldview into the new and powerfully synthetic realm of communication theory. We can now understand on a deeper level Wiener's view of cybernetics as a universal theory of knowledge. Such a universal perspective would succeed, he thought, because it reflects the way that we—as finite, imperfect creatures—know the universe. Statistical and quantum mechanics deal with uncertainty on the microscale; communication reflects and embodies it on the macroscale. *Envisioning relations on the macroscale as acts of communication was thus tantamount to extending the reach of probability into the social world of agents and actors.*

For us, in the late age of information, it may seem obvious that communication should be understood as requiring control and that control should be construed as a form of communication. Underlying this construction, however, is a complex series of events, with its own seriated history of engineering problems, material forms, and bureaucratic structures—a history that James Beniger has written about so well in *The Control Revolution: Technological and Economic Origins of the Information Society.*[15] In broad outline, the forms of control moved from mechanical (a cam directing a mechanical rod to follow a certain path) to thermodynamic (a governor directing the action of a heat engine) to informational (cybernetic mechanisms of all kinds, from computers to the hypothalamus understood in cybernetic terms). In mechanical exchanges, determinism and predictability loom large. When the center of interest turns to the furnace, with its fiery enactments of Brownian motion, probability necessarily enters the picture.[16] When information comes to the fore, probability moves from being ignorance of microstates to becoming a fundamental attribute of the communication act. As each new form of exchange came to the fore, the older ones did not disappear. An automobile is essentially a heat engine, but it nevertheless continues to use levers and rods of the kind known since the classical era. Similarly, a computer is an information machine, but it also uses molecular processes governed by the laws of thermodynamics. The new forms are distinguished not by the disappearance of the old but rather by a shift in the nature of their control mechanisms, which in turn are determined by the kinds of exchanges the machine is understood to transact.

The move toward cybernetic control theory is itself driven by feedback loops between theory and artifact, research and researcher. Envisioning different kinds of exchanges demanded different kinds of control mechanisms, and constructing new control mechanisms facilitated the construction of more exchanges in that mode.[17] The circularity among experimenter, control mechanism, and system interface is part of the story I want to tell. This story includes not only the mechanisms of cybernetic systems but also the mindsets of those who constructed themselves and their machines in a cybernetic image. Wiener's assumptions, as we have seen, were rooted in a probabilistic worldview. He realized that one of the subtle implications of this view is that messages are constituted, measured, and communicated not as things-in-themselves but as relational differences between elements in a field. Communication is about relation, not essence.

Across the range of Wiener's writing, the rhetorical trope that figures most importantly is analogy. Understanding communication as relation suggests a deeper reading of this figure. Analogy is not merely an ornament of language but is a powerful conceptual mode that constitutes meaning through relation. Seen in this way, analogy is a crucial operator in everything from Wiener's passion for mathematics to his advocacy of "black box" engineering and behaviorist philosophy. Indeed, cybernetics as a discipline could not have been created without analogy. When analogy is used to constitute agents in cybernetic discourse, it makes an end run around questions of essence, for objects are constructed through their relations to other objects. Writing in the years immediately preceding and following World War II, Wiener anticipated some aspects of poststructuralist theories. He questioned whether humans, animals, and machines have any "essential" qualities that exist in themselves, apart from the web of relations that constituted them in discursive and communicative fields. "Whatever view we have of the 'realities' underlying our introspections and experiments and mathematical truths is quite secondary; any proposition which cannot be translated into a statement concerning the observable is nugatory," he wrote in 1936 in "The Role of the Observer."[18] Wiener also saw sense perception as working through analogy. In his most extreme pronouncement on the matter, he asserted, "Physics itself is merely a coherent way of describing the readings of physical instruments" (a statement deeply regretted by his mathematical biographer, Pesi Masani.)[19] Among the mappings in his view of the world-as-analogy were metaphors that overlaid mathematics onto emotion, sense perception onto communication, and machines onto biological organisms. These mappings throw a different light on his attempts to reconcile cybernetics with a liberal humanist subject. If meaning is constituted through relation, then juxtaposing men and machines goes beyond bringing two preexisting objects into harmonious relation. Rather, the analogical relation constitutes both terms through the process of articulating their relationship. To see this meaning-making in process, let us turn now to a consideration of analogy in Wiener's texts and practices.

Crossing Boundaries: Everything Is an Analogy, Including This Statement

In his autobiography *I Am a Mathematician,* Wiener tells of retreating to the family farm for a weekend after a row with a couple of influential Harvard mathematicians. Coming home cold and wet, he fell ill and slipped into delirium. "All through the pneumonia," he wrote, "my delirium assumed the form of a peculiar depression and worry [about the row and] . . . anxiety about the logical status of my mathematical work. It was impossible for me to distinguish among my pain and difficulty in breathing, the flapping of the window curtain, and certain as yet unresolved parts of the [mathematical] potential problem on which I was working." Retrospectively musing on how his pain merged with external stimuli and mental abstraction, he arrived at a key insight about his relation to mathematics. "I cannot say merely that the pain revealed itself as a mathematical tension, or that the mathematical tension symbolized itself as a pain: for the two were united too closely to make such a separation significant." He realized "the possibility that almost any experience may act as a temporary symbol for a mathematical situation which has not yet been organized and cleared up." Identifying an unsolved scientific problem with emotional conflict and physical pain, he became "more and more conscious" that for him, mathematics served to "reduce such a discord to semipermanent and recognizable terms." Once he solved the conceptual problem, its link with a personal conflict seemed to resolve that as well, allowing him to "release it and pass on to something else."[20] Mapping mathematics onto emotional conflicts is one way, then, that Wiener used analogy. No doubt on more complex grounds than Jacob Bronowski intended, he enthusiastically endorsed Bronowski's suggestion that all of mathematics is a metaphor. Mathematics, Wiener wrote in *The Human Use of Human Beings,* "which most of us see as the most factual of all sciences, constitutes the most colossal metaphor imaginable, and must be judged, aesthetically as well as intellectually, in terms of the success of this metaphor" (*HU*, p. 95).

His identification of personal conflicts with conceptual problems was so strong that he perceived it as "driving" him to mathematics, almost as if against his will.[21] The coercive imagery is significant. He was the son of a domineering father who consciously wanted to mold him into a prodigy. Once out from under his father's tutelage, he often found it difficult to motivate himself. Steve Heims, in his biography of Wiener, observes that Wiener apparently used the identification between emotional states and mathematical problems as a spur to goad himself onward.[22] While working on a difficult problem, he would fall into a depression, which he would deliberately exacerbate to make himself work harder. Relying on analogical equivalencies he set up between mathematics and emotion, he anticipated that solving the intellectual problem would allow him to regain psychological homeostasis.

The flip side of drawing analogies is constructing boundaries. Analogy as a figure draws its force from the boundaries it leapfrogs across. Without boundaries, the links created by analogy would cease to have revolutionary impact. For Wiener, analogy and boundary work went hand in hand. In both his professional and his private life, he saw boundaries playing important roles. He included in his first autobiography, *Ex-Prodigy: My Childhood and Youth*, an account of his mother's anti-Semitism and his feeling of being unwanted and alienated from her when he discovered, as a teenager, that his father's side of the family was Jewish.[23] Perhaps because of this formative experience, the construction of inside/outside markers characterized his response to many life situations. In his autobiographies, he frequently depicted himself as an outsider, standing apart from a privileged group whose boundaries did not include him. He made it a point to decline scientific prizes and to resign from prestigious professional groups in which he was offered membership if he did not agree with their goals.

Boundaries also played important roles in his scientific work (as they do in electrical engineering generally). The problem that engaged him when he fell ill and felt the flapping curtain woven into the mathematics was a boundary problem, having to do with what happens to an electrical field around a sharp physical discontinuity. In his later work on cybernetics, boundary formation and analogical linking collaborate to create a discursive field in which animals, humans, and machines can be treated as equivalent cybernetic systems. The central text displaying this interplay is the influential cybernetic manifesto that Wiener coauthored in 1943 with

Julian Bigelow and Arturo Rosenblueth, "Behavior, Purpose, and Teleology."[24] Offering an agenda for the nascent field of cybernetics, this work also created a discursive style that *produced* the objects of its analysis.

"Behavior, Purpose, and Teleology" begins by contrasting behaviorism with functionalism. Whereas functionalism (in the authors' definition) foregrounds internal structure and is relatively unconcerned with the organism's relation to the environment, behaviorism focuses on relations between the organism and environment and is relatively unconcerned with internal structure. In the laboratory, the behaviorist approach leads to "black box" engineering, in which one assumes that the organism is a "black box" whose contents are unknown. Producing equivalent behavior, then, counts as producing an equivalent system. The obvious justification is that even when little or nothing is known about internal structure, meaningful conclusions can still be drawn about behavior. Bracketing internal structure did more than this, however. It also produced the assertion that because humans and machines sometimes behave similarly, they are essentially alike. Note the slippage in this passage comparing living organisms and machines. "The methods of study for the two groups are at present similar. Whether they should always be the same may depend on whether or not there are one or more qualitatively distinct, unique characteristics present in one group and absent in the other. Such qualitative differences have not appeared so far" (p. 22). "Appeared" is an apt choice of verb, for the behaviorist viewpoint was constructed precisely to elide the very real differences existing between the internal structure of organisms and that of machines. The analogy is produced by how the focus of attention is constructed. The authors make a similar move when they perform successive cuts in the kinds of behavior they find interesting, focusing, for example, on purposeful rather than random behavior. This series of boundary formations, they contend, "reveals that a uniform behavioristic analysis is applicable to both machines and living organisms, regardless of the complexity of the behavior" (p. 22). What tends to drop from sight is the fact that the equation between organism and machine works because it is seen from a position formulated precisely so that it will work.

Another rhetorical move is a reinscription of two important terms: purpose and teleology. Each is carefully defined to fit the cybernetic situation. *Purpose* implies action directed toward a goal (p. 18); *teleology* implies a goal achieved through negative feedback. In terms of the offered definitions, teleological behavior means simply "behavior controlled by negative feedback" (p. 24). But keeping a loaded term like *teleology* in play is not an innocent reinscription. It carries with it the sense of moving toward a goal meaningful to the system pursuing that goal, thus implying that meaning can exist for machines. It also suggests that the behaviorist project has a cosmological dimension appropriate to the sweeping vistas of time and space that teleology is usually taken to imply.

The authors reinforce these implications when they point out that teleology fell into scientific disrepute because it posits a "final cause" that exists in time *after* the effects it is supposed to bring about. Their version of teleology circumvents this problem; it does not rely on Aristotelian causality of any kind but only on purposeful action toward a goal. They suggest that the opposite of teleology is not deterministic causality but is nonteleology, that is, random behavior that is not goal-directed. They thus shift onto new ground the centuries-old debate between Newtonian causality and Christian teleology. The important tension now is not between science and God but between purpose and randomness. Purpose achieved through negative feedback is the way that goal-seeking devices deal with a probabilistic universe. By implication, the proper cosmological backdrop for the workings of teleological mechanisms is neither the cosmos infused by divine purpose as imagined by Christians nor the world of infinite predictability as dreamed by Laplace but a Gibbsian universe of probabilistic relations and entropic decay. Through these reinscriptions and analogical links, cybernetics becomes philosophy by other means.

A young philosopher, Richard Taylor, took up the gauntlet thrown down in the cybernetic manifesto. In a critique published seven years later in the same journal, *Philosophy of Science,* he sought to show that either "purpose" had been stretched so far that it could apply to any behavior or else it had been used to smuggle in inferences that referred to a machine's behavior but that had properly originated within a human observer.[25] He intended to demonstrate that the rhetoric of "black box" engineering had covertly opened the boxes and put into them qualities produced by the very analysis that treated them as unopened black boxes.

In their rebuttal, Wiener and Rosenblueth make clear that they are appealing to a discourse community of scientists, whom they deem superior to philosophers. They constitute this community by distinguishing between verbal analysis, which they call "trivial and barren," and their analysis, which is motivated by "scientific" concerns.[26] The implicit contrast between the verbal ambiguities that might interest a philosopher and the weighty concerns of "science" is underscored by the contrast they draw between Taylor's "beliefs" and their repeated use of "science" and "scientific" to describe their project (eleven times in a short article). Taylor had used several examples to illustrate that "purpose," as they defined it, could be applied to nonteleological mechanisms (a clock that breaks down at midnight on New Year's Eve, a submarine that follows a boat to which it is attached by a cable). In riposte, Wiener and Rosenblueth contend that these examples are easily distinguished from true servomechanisms using negative feedback. To make the point, however, they are necessarily led into a discussion of the internal structures of the mechanisms—exactly the position they did not want to take in their original article arguing for a behaviorist approach. Their rebuttal is effective, then, only to the extent that it complements a strict behaviorist approach with an analysis that, *contra* behaviorist principles, uses differences in internal structures to sort behaviors into different categories.

This alternating focus on behavior and internal structure is similar to the rhetorical strategies that Geof Bowker analyzes in his article showing how cybernetics constituted itself as a universal science.[27] Bowker points out that cybernetics positioned itself both as a metascience and as a tool that any other science could use. It offered a transdisciplinary vocabulary that could be adapted for a variety of disciplinary purposes, presenting itself in this guise as content-free, and it simultaneously offered a content-rich practice in which cybernetic mechanisms were analyzed, modeled, and occasionally built. Operating on these two different levels, cybernetic discourse was able to penetrate into other disciplines while also maintaining its turf as a disciplinary paradigm. In Wiener and Rosenblueth's rebuttal to Taylor, the alternation between a structure-free and a structure-rich cybernetics produces a similar rhetorical effect. In its structure-free guise, cybernetics links men and machines by eliding internal structure; in its structure-rich form, it presents information flow and negative feedback as important structural elements. It is no accident that Warren McCulloch used a similar rhetorical strategy in his argument with Hans Teuber, as discussed in chapter 3. Just as the alternation between content-free and content-rich cybernetics allows a deeper penetration into disciplinary sites than would otherwise be possible, so the alternation between behavior and structure allowed the discourse simultaneously to assimilate biological organisms and machines into the same category and to distinguish them from plain-vanilla mechanical systems.

In his rejoinder, Taylor missed the opportunity to point out that the focus of Wiener and Rosenblueth's analysis alternated between behavior and structure.[28] Instead he chose to pursue a line of questioning similar to that in his original article, as he again tried to show that if one relies only on external observations of behavior, "purpose" cannot be reliably distinguished from chance or random events. In contesting for what counts as "purpose," he wanted to deny to the behaviorist approach a distinction crucial to generating their system (the difference between purposeful and random behavior). He sensed that *behavior* had been defined so as to allow intention and desire to be imputed to machines. But he let slip by the larger point that behaviorist assumptions were used selectively to accomplish a political agenda implicit in the way that categories were constructed. For Wiener, this agenda included constituting one category that encompassed cybernetic machines and humans, which were put together because they shared the ability to use probabilistic methods to control randomness, and another category for noncybernetic mechanical systems. These boundary markers implied larger assumptions about the nature of the universe (probabilistic rather than deterministic), about effective strategies for dealing with this universe (controlling randomness through negative feedback), and about a system hierarchy that had moral connotations as well as practical values (flexible systems using negative feedback were *better* than mechanical devices that did not use feedback). More than the definition of *purpose*, it was these larger inscriptions that made "Behavior, Purpose, and Teleology" the founding document for cybernetics.

One of the most frequent criticisms made of cybernetics during this period was that it was not really a new science but was merely an extended analogy (men are like machines). Wiener heard the charge often enough that he finally felt it was time to take the cybernetic bull by the horns. In "The Nature of Analogy," a manuscript fragment dated 1950, he offers a strong defense for analogy, moving the argument onto new and more compelling ground.[29] Its brevity notwithstanding, "The Nature of Analogy" is a wide-ranging meditation on what analogy means in science, mathematics, language, and perception. It argues that those who object to Wiener's analogical moves do so because they hold realist assumptions that do not stand up to rigorous scrutiny. Cybernetics as Wiener envisioned it is about relation, not essence. The analogical relations it constructs are therefore not merely rhetorical figures but are systems that generate the only kind of significance available to us as perceiving, finite beings with no access to unmediated reality.

Wiener begins by pointing out that language is always analogical, in the sense that it puts forth propositions that listeners must interpret from their own experiences, which are never identical to the speaker's. This observation anticipates Michael Arbib and Mary Hesse's argument that signification occurs through category constitution, not through the communication of an Aristotelian essence.[30] Like them, Wiener also denies that language communicates an Aristotelian essence. The convergence points to similarities between his definition of *information* and Ferdinand de Saussure's view of *la langue*, or language as a system. In both cases, communication proceeds through selection from a field of possible alternatives rather than through the direct articulation of inherent reference. Just as Saussurian linguistics is associated with deconstructive theories that reveal the indeterminacy of reference and that expose the inability of language to ground itself, so Wiener's cybernetics sees communication as a probabilistic act in a probabilistic universe, where initial conditions are never known exactly and where messages signify only through their relation to other messages that might have been sent. For Wiener no less than Saussure, signification is about relation, not about the world as a thing-in-itself.[31]

It is in this context that pattern, associated with information (as we saw in chapter 2), assumes paramount importance. Wiener's view of sense perception makes the point clear. Perception does not reflect reality directly but rather relies on transformations that preserve a pattern across multiple sensory modalities and neural interfaces. Representation emerges through the analogical relation of these transformations to the original stimulus. In this respect, sense perception is like mathematics and logic, for they too "deal preeminently with pattern apart from content."[32] The behaviorist approach is well suited to this relational epistemology because it concentrates on transmission of patterns rather than communication of essence. Consider the antiaircraft predictor that Wiener developed in collaboration with Julian Bigelow during World War II.[33] The prognosticator received tracking data as input (for example, radar following a plane) and gave, as output, predictions of where the plane would go. Statistical analysis was used to find patterns in these data, and the data themselves were understood as patterns analogically related to events in the world. Thus, perception, mathematics, and information all concentrate on pattern rather than content. As data move across various kinds of interfaces, analogical relationships are the links that allow pattern to be preserved from one modality to another. Analogy is thus constituted as a universal exchange system that allows data to move across boundaries. It is the *lingua franca* of a world (re)constructed through relation rather than grasped in essence.

Border crossings accomplished through analogy include the separation between flesh and world (sense perception), the transition between one discipline and another (for example, moving from the physiology of living organisms to the electrical engineering of a cybernetic machine), and the transformation of embodied experience, noisy with error, into the clean abstractions of mathematical pattern. Even the prostheses that Wiener designed can be understood as operating through analogy, for they transformed information from one modality into another.[34] The "hearing glove," for example, was an apparatus that converted sounds (auditory signals) into touch (tactile signals) by stimulating a deaf person's fingers with electromagnetic vibrators that were analogical transformations of sound frequencies. For Wiener, analogy was communication, and communication was analogy. Objecting that cybernetics is "merely an analogy" was for him akin to saying that cybernetics is "merely about how we know the world."

The problem with this approach lies not so much in the analogical relations that Wiener constructed between living and mechanical systems as in his tendency to erase from view the very real differences in embodied materiality, differences that the analogies did not express. Confronted with two situations, he was much more inclined to move easily and quickly to an abstract level, where similarities in patterns became evident, than to remain attentive to the particularities that made each situation unique. No doubt his own lack of involvement in the nitty-gritty work of the lab was a contributing factor in this elision of embodied materiality. He noted the impatience he felt with the exacting procedures of the biological laboratory. "This impatience was largely the result of my mental quickness and physical slowness. I could see the end to be accomplished long before I could labor through the manipulative stages that were to bring me there."[35] The problem was serious enough to force him to give up his hope of earning a Ph.D. in biology. In his later professional collaborations with Rosenblueth and others, he left the lab work to them. Colleagues recall how he would wander into Rosenblueth's laboratory when an experiment was under way, make a few notes and ask a few questions, and retreat to his office to work out the mathematical analogies expressing the physical situation. When Wiener and his collaborators wrote such phrases as "We cut the attachment of the muscle," the plural was purely honorary, as Masani points out in his excellent biography of Wiener.[36] Other colleagues suggested that his ineptitude in the lab made him less attentive to the particularities of actual neurophysiological structures. In a posthumous tribute to Wiener, Walter Rosenblith and Jerome Wiesner wrote, "In areas in which Wiener's intuition was less educated than in engineering, he was often impatient with experimental details; for example, he sometimes unwilling to learn that the brain did not behave the way he expected it to."[37] For Wiener, the emphasis on analogy went hand in hand with a certain estrangement from the flesh. In this respect, the contrast between him and McCulloch is clear. As a dedicated experimentalist, McCulloch was sensitive in a way that Wiener was not to the tension between the plenitude of embodiment and the sparseness of abstraction.

As we have seen, Wiener wanted to inscribe cybernetics into a larger drama that would reinforce the liberal humanist subject. Given his inclination toward a Gibbesian universe, that drama focused on probability. In addition to operating on the microscale of subatomic particles and the macroscale of cybernetic circuits, probability also operates on the cosmological level of universal dissipation and decay. Linking probability with information allowed Wiener to script the cybernetic subject into a cosmological drama of chaos and order. It is here, on this cosmological level, that he staged the moral distinctions between good cybernetic systems, which reinforce the autonomous liberal subject, and evil machines, which undermine or destroy the autonomy of the subject. An important player in this titanic struggle between good and evil machines is entropy, a protean concept with a richly complex history.

Entropy as Cultural Relay: From Heat Engines to Information

We can begin our investigation into entropy with the series of transformations that Mark Seltzer traces in *Bodies and Machines*. Seltzer, concentrating on the social formations of late-nineteenth-century naturalism, finds at the heart of naturalism a double and seemingly contradictory thrust: on the one hand, "the insistence on the *materiality* or *physicality* of persons, representations, and actions"; on the other hand, "the insistent *abstraction* of persons, bodies, and motions to models, numbers, maps, charts, and diagrammatic representations." Calling the ideology that resulted from this double thrust a "dematerialized materialism," Seltzer instances such phenomena as the emergence of statistical representations for human behavior and the renewed interest in the ergonomics of the human body.[38] One focuses on behavior abstracted into statistical ensembles of data, the other on the material processes of energy consumption and dissipation. They illustrate the construction of bodies both as material objects and as probability distributions.

The duality that Seltzer locates in nineteenth-century culture continued into the twentieth century with renewed force when statistical thermodynamics merged with information theory. One of the principal sites for this merger was cybernetics. The emphasis on pattern constructed bodies as immaterial flows of information; the alternating emphasis on structure recognized that these "black boxes" were heavy with materiality. Complex couplings between the two registers worked to set up a series of exchanges between biological organisms and machines. To see how these couplings evolved, let us start with the exchanges that thermodynamics set up and follow them forward into cybernetics.

The first law of thermodynamics, stating that energy is neither created nor destroyed, points to a world in which no energy is lost. The second law, stating that entropy always tends to increase in a closed system, forecasts a universe that is constantly winding down. This tension between the first and second laws, between stability and degradation, runs like a leitmotiv through turn-of-the-century cultural formations. According to Seltzer, the tension itself acts like a thermodynamic exchanger, allowing incompatible terms such as production and reproduction, machines and bodies, to be articulated together. The body is like a heat engine because it cycles energy into different forms and degrades it in the process; the body is not like a heat engine because it can use energy to repair itself and to reproduce. In one sense the comparison constructs the difference between body and machine; in another sense it acts as an exchanger that allows bodies and heat engines to be linked together. Through such comparisons, Seltzer argues, "what is gradually elaborated is a more or less efficient, more or less effective system of transformations and relays between 'opposed' and contradictory registers." These ambiguous linkages were reinforced because thermodynamics itself was perceived as operating in the two different registers of conservation and dissipation. Thus, he concludes that thermodynamics, wrapping both conservative stability and dissipative decay within the mantle of scientific authority, "provided a working model of a new mechanics and biomechanics of power."[39]

Already functioning as an exchange system within the culture, thermodynamics evolved into "dematerialized materialism" when Ludwig Boltzmann gave entropy a much more general formulation by defining it as a probability function. In this "dematerialized" construction, entropy was interpreted as a measure of randomness. The second law was then reformulated to state that closed systems tend to move from order to randomness. Encompassing the earlier definition of entropy, Boltzmann's formulation also added something new, for it allowed entropy to be linked with systems that had nothing to do with heat engines.

This dematerialization was carried further when entropy was connected with information. As early as 1929, the connection was made through Leo Szilard's interpretation of Maxwell's Demon,[40] a mythical being in a thought experiment proposed by James Clerk Maxwell in 1871. The Demon gained energy by sorting molecules. Szilard and Leon Brillouin, among others, pointed out that to sort molecules, the Demon has to have information about them.[41] The container in which the Demon sits is imagined as a "black body" (a technical term meaning that the radiation is uniformly dispersed) so that there is no way for the Demon to "see" the molecules. Brillouin calculated that the energy the Demon would have to expend to get information about the molecules is greater than what the Demon could gain by the sorting process. The immediate result was to rescue the second law, which in any case was too well-established to be seriously in doubt. The more important implication was to suggest that entropy and information are inversely related to each other. The more information there is, the less entropy; the more entropy is present, the less information. Brillouin therefore proposed that information be considered as negative entropy, or negentropy. Maxwell's Demon was one of the relay points through which a relationship was established between entropy and information.

Like Brillouin and many others of his generation, Wiener accepted the idea that entropy was the opposite of information. The inverse relation made sense to him because he thought of information as allied with structure and viewed entropy as associated with randomness, dissipation, and death. "As entropy increases," he wrote, "the universe, and all closed systems in the universe, tend naturally to deteriorate and lose their distinctiveness, to move from the least to the most probable state, from a state of organization and differentiation in which distinctions and forms exist, to a state of chaos and sameness. In Gibbs' universe order is least probable, chaos most probable." In this view, life is an island of negentropy amid a sea of disorder. "There are local enclaves whose direction seems opposed to that of the universe at large and in which there is a limited and temporary tendency for organization to increase. Life finds its home in some of these enclaves" (*HU,* p. 12). In a related metaphor, he envisioned a living organism as an informational system swimming upstream against the entropic tide.

This view of entropy makes sense when viewed in the context of nineteenth-century thermodynamics. But it is not a necessary implication of information as *information* is technically defined. Claude Shannon took the opposite view and *identified* information and entropy rather than opposed them.[42] Since the choice of sign was conventional, this formulation was also a possibility. Heuristically, Shannon's choice was explained by saying that the more unexpected (or random) a message is, the more information it conveys.[43] This change in sign did not affect the dematerialization that entropy had undergone, but it did reverse entropy's value in more than a mathematical sense. In retrospect, identifying entropy with information can be seen as a crucial crossing point, for this allowed entropy to be reconceptualized as the thermodynamic motor driving systems to self-organization rather than as the heat engine driving the world to universal heat death.

Space will not permit me to tell the story of this reversal here, and in any event, it has been chronicled elsewhere.[44] Suffice it to say that as a result, chaos went from being associated with dissipation in the Victorian sense of dissolute living and reckless waste to being associated with dissipation in a newly positive sense of increasing complexity and new life.

Wiener came close to making this crossing. In one of his astonishing analogical leaps, he saw a connection between the "light" that the Demon needs to sort the molecules and the lights that plants use in photosynthesis. He argued that in photosynthesis, plants act as if their leaves were studded with Maxwell's Demons, all sorting molecules to allow the plant to run uphill toward increasing complexity rather than downhill toward death.[45] But he did not go beyond this isolated insight to the larger realization that large entropy production could drive systems to increasing complexity. Finally, he remains on the negative side of this divide, seeing life and homeostasis as contrarian islands that, although they may hold out for a while, must eventually be swamped by the entropic tide.

So firmly rooted is Wiener in this perspective that he comes close on several occasions to saying that entropic decay is evil. Entropy becomes morally negative for Wiener when he sees it operating against the differential probability distributions on which the transfer of information depends. Recall that Gregory Bateson defined information as a difference that makes a difference; if there is no difference, there is no information. Since entropy tends always to increase, it will eventually result in a universe in which all distributions are in their most probable state and in which universal homogeneity prevails. Imagine Dr. Zhivago sitting at his desk in a cold, cold room, trying to telegraph a message to his beloved Laura, while in the background Laura's theme plays and entropy keeps relentlessly increasing. Icicles hanging from his fingers and the telegraph key; he tries to tap out "I love you," but he is having trouble. He not only is freezing from heat death but also is stymied by information death. No matter what he taps, the message always comes out the same: "eeeeeee" (or whatever letter is most common in the Russian alphabet). This whimsical scenario illustrates why Wiener associated entropy with oppression, rigidity, and death. Communication can be seen, he suggested, as a game that two humans (or machines) play against noise.[46] To be rigid is inevitably to lose the game, for rigidity consigns the players to the mechanical repetition of messages that can only erode over time as noise intervenes. Only if creative play is allowed, if the mechanism can adapt freely to changing messages, can homeostasis be maintained, even temporarily, in the face of constant entropic pressure toward degradation.

In the "dematerialized materialism" of the battlefield where life struggles against entropy and noise, the body ceases to be regarded primarily as a
material object and instead is seen as an informational pattern. The struggle,
then, is between strategists who try to preserve this pattern intact and noisy
opponents (or, rather, noise as an opponent) who try to disrupt it. During the
1940s and 1950s, Wiener was one of the important voices casting the cosmological drama between cybernetic mechanisms and noise in these terms.
In *The Human Use of Human Beings,* he suggests that human beings are not
so much bone and blood, nerve and synapse, as they are patterns of organization. He points out that over the course of a lifetime, the cells composing a
human being change many times over. Identity cannot therefore consist in
physical continuity. "Our tissues change as we live: the food we eat and the
air we breathe become flesh of our flesh and bone of our bone, and the mo-
mentary elements of our flesh and bone pass out of our body every day with
our excreta. We are but whirlpools in a river of ever-flowing water. We are
not stuff that abides, but patterns that perpetuate themselves" (*HU*, p. 96).
Consequently, to understand humans, one needs to understand how the
patterns of information they embody are created, organized, stored, and re-
trieved. Once these mechanisms are understood, they can be used to create
cybernetic machines. If memory in humans is the transfer of informational
patterns from the environment to the brain, machines can be built to effect
the same kind of transfer. Even emotions may be achievable for machines if
feelings are considered not as "merely a useless epiphenomenon of nervous
actions" (*HU*, p. 72) but as control mechanisms governing learning.[47] Considered as informational patterns, cybernetic machines and men can make
common cause against the disruptive forces of noise and entropy.

The picture that emerges from these conjectures shows the cybernetic
organism—human or mechanical—responding flexibly to changing situations, learning from the past, freely adapting its behavior to meet new circumstances, and succeeding in preserving homeostatic stability in the
midst of even radically altered environments. Nimbleness is an essential
weapon in this struggle, for to repeat mindlessly and mechanically is inevitably to let noise win. Noise has the best chance against rote repetition,
where it goes to work at once to introduce randomness. But a system that
already behaves unpredictably cannot be so easily subverted. If a Gibbesian universe implies eventual information death, it also implies a universe
in which the best shot for success lies in flexible and probabilistic behavior.
The Greek root for *cybernetics,* "steersman," aptly describes the cybernetic man-machine: light on its feet, sensitive to change, a being that both
is a flow and knows how to go with the flow.

Reinforcing the boundary work that assimilates the liberal humanist subject and the cybernetic machine into the same privileged space are the distinctions Wiener makes between good and bad machines. When machines are evil in *The Human Use of Human Beings,* it is usually because they have become rigid and inflexible. Whereas the cybernetic machine is ranged alongside man as his brother and peer, metaphors that cluster around the rigid machine depict it through tropes of domination and engulfment. The ultimate horror is for the rigid machine to absorb the human being, co-opting the flexibility that is the human birthright. "When human atoms are knit into an organization in which they are used, not in their full right as responsible human beings, but as cogs and levers and rods, it matters little that their raw material is flesh and blood. What is used as an element in a machine, is in fact an element in the machine" (*HU*, p. 185). Here the analogical mapping between humans and machines turns sinister, trapping humans within inflexible walls that rob them of their autonomy. The passage shows how important it is to Wiener to construct the boundaries of the cybernetic machine so that it reinforces rather than threatens the autonomous self. When the boundaries turn rigid or engulf humans so that they lose their agency, the machine ceases to be cybernetic and becomes simply and oppressively mechanical.

The cosmological stage upon which the struggle between oppressive machines and cybernetic systems unfolds is—no surprise—the Gibbesian universe in which probability reigns supreme. "The great weakness of the machine—the weakness that saves us so far from being dominated by it— is that it cannot yet take into account the vast range of probability that characterizes the human situation." Here the probability differentials that make communication possible are assimilated to humans and good machines, leaving bad machines to flounder around in probabilities too diverse for them to assess. The rules of the contest are laid down by the second law of thermodynamics, which allows a margin in which cybernetic men-machines can operate because it is still cranking up its death engine. "The dominance of the machine presupposes a society in the last stages of increasing entropy, where probability is negligible and where the statistical differences among individuals are nil. Fortunately we have not yet reached such a state" (*HU*, p. 181). When in the end the universe ceases to manifest diverse probabilities and becomes a uniform soup, control, communication, cybernetics—not to mention life—will expire. In the meantime, men and cybernetic machines stand shoulder to shoulder in building dikes that temporarily stave off the entropic tide.

The boundary work that links cybernetic machines and humans perhaps reaches its most complex articulation in the distinction that Wiener makes between Augustinian and Manichean opponents. At issue in this distinction is the difference between an opponent who plays "honorably," that is, abiding by rules that do not change, and one who tries to win by manipulation. For Wiener, the exemplar of an Augustinian opponent is nature. Nature—including noise—may sometimes frustrate the scientist's attempt to control it, but it does not consciously try to manipulate its opponent. The exemplar of the Manichean opponent is the chess player, including chess-playing machines. Unlike nature, the chess player acts deviously and, if possible, manipulatively. When the chess player is contrasted with the scientist, it is almost always to the chess player's detriment. In pointing out that nature does not try to outwit the scientist, Wiener observes that having an Augustinian opponent means that the scientist has time to reflect on and correct his or her strategy, because no one is trying to take advantage of the scientist's mistakes. Scientists are thus governed by their best moments, whereas chess players are governed by their worst (*HU*, p. 36).

Peter Galison, in "The Ontology of the Enemy: Norbert Wiener and the Cybernetic Vision," argues that cybernetics (along with game theory and operations research) should be called a "Manichean science."[48] In a fine-grained analysis of Wiener's collaboration with Julian Bigelow to develop an antiaircraft (AA) weapon during World War II, Galison brilliantly shows that Wiener's construction of "the enemy" was significantly different from that portrayed in war propaganda or even in other technical reports. Rather than seeing the enemy in conventionally human (or, in the case of propaganda, subhuman) terms, Wiener modeled the enemy—for example, a fighter pilot trying to evade AA fire—as a probabilistic system that could effectively be countered using cybernetic modeling. Unlike other fire systems, which had fixed rules derived from probabilistic modeling, Wiener's imagined firing machine could evolve new rules based on prior observation—that is, it could learn. Thus the firing system would evolve to become as Manichean as the enemy it faced. Galison argues that this strategy enabled a series of substitutions and identifications that mapped the enemy pilot onto the servo-controller and ultimately onto the allied war personnel behind the servo-controller. In a "Summary Report for Demonstration," Wiener and Bigelow wrote: "We realized that the 'randomness' or irregularity of an airplane's path is introduced by the pilot; that in attempting to force his dynamic craft to execute a useful manoeuvre . . . the pilot *behaves like a servo-mechanism*" (quoted in Galison, p. 236). Thus cybernetics, itself constituted through analogies, creates further analogies through theories and artifacts that splice man to machine, German to American. Through this relay system, the enemy becomes like us and we become like the enemy: enemy mine. If these analogical mappings kept the enemy pilot from being demonized, they also made the cybernetic machine (and, by extension, cybernetics itself) party to a bloody struggle in which Manichean tactics were used by both sides to kill as many humans as possible.

Partly in reaction to this co-optation of cybernetics by the military, Wiener half a decade after the war wrote the significantly entitled *The Human Use of Human Beings.*[49] Although Wiener had done everything in his power during the war to further cybernetics as a "Manichean science," his writings after the war show a deep aversion to the manipulation that a Manichean strategy implies. From his autobiographies, it is clear that he was hypersensitive to being manipulated, perhaps with good reason. When he was first beginning to establish himself as a mathematician, his father tried to get him to use his contacts to advance his father's philological ideas—an instance of manipulation that made Wiener increasingly wary of how others might try to use his talents and influence to further their own ends. It is no accident that he associated the manipulative chess-playing machine with the military projects that he resolutely turned away from after atomic bombs vaporized hundreds of thousands of Japanese civilians. Remarking on Claude Shannon's suggestion that chess-playing machines have military potential, he wrote, "When Mr. Shannon speaks of the development of military tactics, he is not talking moonshine, but is discussing a most imminent and dangerous contingency" (*HU,* p. 178). The problem, of course, was that cybernetics adapted all too readily to Manichean tactics, making it possible to play these deadly games even more effectively.

Wiener's war work, combined with his antimilitary stance after the war, illustrates with startling clarity how cybernetics functioned as a source of both intense pride and intense anxiety for him. This tension, often expressed as an anxious desire to limit the *scope* of cybernetics, takes a different but related form when he considers the question of body boundaries, always a highly charged issue. When the physical boundaries of the human form are secure, he celebrates the flow of information through the organism. All this changes, however, when the boundaries cease to define an autonomous self, either through manipulation or engulfment. In the next section, we will see how this anxiety erupts into his 1948 book *Cybernetics* at critical points, causing him to withdraw from the more subversive implications of the discipline he fathered. It is no accident that erotic metaphors are used to carry the thrust of the argument. Like cybernetics, eroticism is intensely concerned with the problematics of body boundaries. It is not for nothing that sexual orgasm is called "the little death" or that writers from Marquis de Sade to J. G. Ballard have obsessively associated eroticism with penetrating and opening the body. At stake in the erotically charged discourse in which Wiener considers the pleasures and dangers of coupling between parts that are not supposed to touch is how extensively the body of the subject may be penetrated or even dissolved by cybernetics as a body of knowledge. It is here, as much as anywhere else, that Wiener's concern to preserve the liberal subject comes into uneasy tension with his equally strong desire to advance the cause of cybernetics. As we shall see, resolution can be achieved only by withdrawal, pointing toward a future in which the cybernetic subject could not finally be contained within the assumptions of liberal humanism.

The Argument for Celibacy: Preserving the Boundaries of the Subject

In *Cybernetics,* the technical text from which *The Human Use of Human Beings* was adapted, Wiener looks into the mirror of the cyborg but then withdraws.[50] The scenarios he constructs to enact and justify this withdrawal suggestively point to the role that erotic anxiety plays in cybernetic narratives. In my analysis, I will focus on the chapter entitled "Information, Language, and Society." Here Wiener entertains the possibility that cybernetics has provided a way of thinking so fertile that it will allow the social and natural sciences to be synthesized into one great field of inquiry. Yet he finally demurs from this palpable object of desire. Given that he is as imperialistic as most other scientists who think they have invented a new paradigm, why does he prefer to maintain the intellectual celibacy of his discovery? I will argue that central to his decision is a fantasy scene that expresses and controls anxiety by reconstituting boundaries. This fantasy gives rise to a series of erotically encoded metaphors that appear whenever anxiety becomes acute. The metaphors also have literal meanings that reveal how intermingled the physical remains with the conceptual, the erotic with the cybernetic. As gestures of separation disconcertingly transform into couplings, the cybernetics of the subject and the subject of cybernetics interpenetrate.

Wiener works up to the fantasy by pointing out that there are many organizations whose parts are themselves small organizations. Hobbes's Leviathan is a Man-State made up of men; a Portuguese man-of-war is composed of polyps that mirror it in miniature; a man is an organism made up of cells that in some respects also function like organisms. This line of thought leads Wiener to ask how these "bodies politic" function. "Obviously, the secret is in the intercommunication of its members." The flow of information is thus introduced as a principle explaining how organization occurs across multiple hierarchical levels. To illustrate, he instances the "sexually attractive substances" that various species secrete to ensure that the sexes will be brought together (*HU*, p. 156). For example, the pheromones that guide insect reproduction are general and omnidirectional, acting in this respect like hormones secreted within the body. The analogy suggests that external hormones organize internal hormones, so that a human organism becomes, in effect, a sort of permeable membrane through which hormonal information flows. At this point we encounter his first demurral. "I do not care to pronounce an opinion on this matter," he announces rather pretentiously after introducing it, preferring to "leave it as an interesting idea" (*HU*, p. 157).

I think that the idea is left because it is disturbing as well as speculative. It implies that personal identity and autonomous will are merely illusions that mask the cybernetic reality. If our body surfaces are membranes through which information flows, who are we? Are we the cells that respond to the stimuli? Are we the larger collectives whose actions are the resultant of the individual members? Or are we the host organisms who, as Richard Dawkins later claimed using cybernetic arguments, engage in sex because we are controlled by selfish genes within?[51] The choice of examples foregrounds sexuality, but this is a kind of sex without sexuality. Implying the deconstruction of the autonomous self as a locus of erotic pleasure, it circumvents the assenting, demurring, intensifying, delaying, and consummating that constitute sexual play. When Wiener is confronted with this sexless sex, his first impulse is to withdraw: coitus interruptus.

His second impulse is to reconstruct himself as a liberal subject through a disguised erotic fantasy that allows him to control the flow of information rather than be controlled by it. Similar fantasies appear everywhere in American literature, from Natty Bumppo and Chingachgook to Ishmael and Queequeg. They are ubiquitous because they are about the American values of masculine autonomy and control, about deferred intimacy between men in a society that is homophobic, racist, and misogynist. What is this fantasy? What else but for the American male to imagine himself alone in the woods with an "intelligent savage," giving himself over to the pursuits that men follow when they are alone together (*HU*, p. 157)?

The fantasy's ostensible purpose is to show that Wiener and his savage companion could achieve intimacy even if they did not touch and shared no language. Wiener imagines himself "alert to those moments when [the savage] shows the signs of emotion or interest," noticing at these moments what he watches. After a time, the savage would learn to reciprocate by "pick[ing] out the moments of my special, active attention," thus creating between them "a language as varied in possibilities as the range of impressions that the two of us are able to encompass" (*HU*, p. 157). Alone together in the woods, the two men construct a world of objects through the interplay of their gazes. In the process they also reconstitute themselves as autonomous subjects who achieve intimacy through their voyeuristic participation in each other's emotion and "special, active attention." There remains, of course, a necessary difference between them. Wiener can move from this fantasy to the rest of his argument, whereas the "intelligent savage" reappears in his discourse only when Wiener finds it convenient to invoke the savage. The passage reveals in miniature how the use of the plural by the liberal humanist subject can appropriate the voices of subaltern others, who if they could speak for themselves might say something very different.

Having reassured himself of intimacy, autonomy, and control, Wiener returns to the problem of the "body politic," concentrating on its alarming lack of homeostasis. In contrast to the regulated, orderly exchanges between him and his savage friend, the body politic is dominated by exchanges between knaves and fools, with "betrayal, turncoatism, and deception" the order of the day (*HU*, p. 59). The economy of this society is clear-cut: the fools desire; the knaves manipulate their desires. The economy is reinforced by statisticians, sociologists, and economists who prostitute themselves by figuring out for the knaves exactly how the calculus of desire can be maximized. The only respite from this relentless manipulation is found in small, autonomous populations. There homeostasis can still work, whether in "highly literate communities . . . or villages of primitive savages" (*HU*, p. 160). The reappearance of the savage here is significant, for anxiety about the manipulation of desire is reaching its height. No doubt this reappearance has a soothing effect on Wiener's imagination, for it reminds him that he need not be manipulated after all.

We come now to the crux of the argument. The danger of cybernetics, from Wiener's point of view, is that it can potentially annihilate the liberal subject as the locus of control. On the microscale, the individual is merely the container for still smaller units within, units that dictate actions and desires; on the macroscale, these desires make the individual into a fool to be manipulated by knaves. Under a cybernetic paradigm, these two scales of organization would be joined to each other. What chance then for intimate communication alone with an intelligent savage in the woods? No, despite the "hopes which some . . . friends have built for the social efficacy of whatever new ways of thinking this book may contain," Wiener finds himself unable to attribute "too much value to this type of wishful thinking" (*HU*, p. 162). Ironically, expanded too far across the bodies of disciplines, the science of control might rob its progenitor of the very control that was no doubt for him one of its most attractive features.

Having reached this conclusion, Wiener reenacts the anxiety that gave rise to it. Through a series of interactive metaphors that connect his fantasy with his anxiety, he claims that it is a "misunderstanding of the nature of all scientific achievement" to suppose that "the physical and social sciences can be joined" (*HU*, p. 162). They must be kept apart, for they permit different degrees of coupling between the scientist and the object of his interest. The precise sciences "achieve a sufficiently loose coupling with the phenomena we are studying [to allow us] to give a massive total account of this coupling." Erotic interest is not altogether lacking, for "the coupling may not be loose enough for us to be able to ignore it altogether" (*HU*, p. 163). Nevertheless, the restrained science that Wiener practices is different from the social sciences, where the coupling is much tighter and more intense. The contrast shows how central the concept of the autonomous self is to cybernetics as Wiener envisioned it.

The savage makes one last appearance in Wiener's anxious consideration of how tightly the scientist can be coupled with his object without losing his objectivity. To illustrate the dangers of tight coupling, Wiener observes that primitive societies are very often changed by the anthropologists who observe them. He makes the point specifically in terms of language: "Many a missionary has fixed his own misunderstandings of a primitive language as law eternal in the process of reducing it to writing" (*HU*, p. 63). In implicit contrast to this violation is the pristine intimacy Wiener achieved with his savage, where no misunderstandings disrupted the perfect sympathy of their gazes.

Concluding that "we are too much in tune with the objects of our investigations to be good probes," Wiener counsels that cybernetics had best be left to the physical sciences, for to carry it into the human sciences would only build "exaggerated expectations" (*HU*, p. 164). Behind this conclusion is the prospect of an interpenetration so complete that it would link the little units within to the larger social units without, thereby reducing the individual to a connective membrane with no control over desires and with no ability to derive pleasure from them. Not only sex but the sex organs themselves disappear in this construction. Thus, Wiener decides that however tempting the prospect of penetrating the boundaries of other disciplines might be, cybernetics is better off remaining celibate.

The conjunction of erotic anxiety and intellectual speculation in Wiener's text implies that cybernetics cannot be adequately understood simply as a theoretic and technological extension of information theory. The analogies so important to his thought are constituted not only through similarities between abstract forms (such as probability ratios and statistical analysis) but also through the complex lifeworld of embedded physicality that natural language expresses and evokes through its metaphoric resonances. Natural language is not extraneous to understanding the full complexities of Wiener's thinking, as his mathematical biographer Pesi Masani implies when Masani contrasts the disembodied abstractions of mathematics with the "long-winded verbosity [of natural language], the hallmark of bureaucratic chicanery and fake labor."[52] On the contrary, the embodied metaphors of language are crucial to understanding the ways in which Wiener's construction of the cybernetic body and the body of cybernetics both privilege and imperil the autonomous humanistic subject.

Viewed in historical perspective, Wiener was not successful in containing cybernetics within the circle of liberal humanist assumptions. Only for a relatively brief period in the late 1940s and 1950s could the dynamic tension between cybernetics and the liberal subject be maintained—uneasy and anxious as that accommodation often was for Wiener. By the 1960s, the link between liberal humanism and self-regulation, a link forged in the eighteenth century, was already stretched thin; by the 1980s, it was largely broken. It is to Wiener's credit that he tried to craft a version of cybernetics that would enhance rather than subvert human freedom. But no person, even the father of a discipline, can single-handedly control what cybernetics signifies when it propagates through the culture by all manner of promiscuous couplings. Even as cybernetics lost the momentum of its drive to be a universal science, its enabling premises were mutating and reproducing at other sites. The voices that speak the cyborg do not speak as one, and the stories they tell are very different from the narratives that Wiener struggled to authorize.

FROM HYPHEN TO SPLICE:
CYBERNETIC SYNTAX IN *LIMBO*

In Bernard Wolfe's *Limbo*, the 1952 novel that has become an underground classic, anxiety about boundaries becomes acute. Like Norbert Wiener, by whom he was deeply influenced, Wolfe recognized the revolutionary potential of cybernetics to reconfigure bodies. Also like Wiener, he tried unsuccessfully to contain that potential, fearing that if it went too far it could threaten the autonomy of the (male) liberal subject. Abrasive, outrageous, transgressive, frustratingly misogynistic, and occasionally brilliant, *Limbo* rarely leaves its readers feeling neutral. David Samuelson ranks it with *Brave New World* and *1984* as one of the three great dystopian novels of the century.[1] At the other end of the spectrum are readers (including some of my students) who see it as remarkable mostly for its egregious sexism and tendentious argument. Whatever one's view of *Limbo's* literary value, it is clear that the text is powerfully marked by the turn to a post–World War II cybernetic economy of information and simulacra.

Limbo arrived at a pivotal moment in U.S. history, at a time when changes in speed and communication were forcing technologies of control into a reorganization that would result in the computer revolution and when the cold war loomed large in the national consciousness. It was in this climate that cybernetics was beginning to change what counted as "human." As we saw in preceding chapters, cybernetics constructed humans as information-processing systems whose boundaries are determined by the flow of information. Cybernetics problematized body boundaries at the same time that the culture was generally anxious about communist penetrations into the body politic. The time was right for a text that would overlay the cybernetic reconfiguration of the human body onto the U.S. geopolitical body and (given Wolfe's misogynistic views) onto the contested terrain of the gendered body. *Limbo* creates that imaginary geography and imbues it with the hypnagogic force of a nightmare. As a novel of *ideas,* it displays some of the passageways through which cybernetic notions began to circulate throughout U.S. culture and connect up with contemporary political anxieties. As a *novel* of ideas, it is an important literary document because it stages encounters between literary form and bodies represented within the text. The textual corpus, no less than the represented world, bears the imprint of the cybernetic paradigm upon its body.

War, acknowledged and covert, is the repressed trauma that threatens to erupt throughout *Limbo.* But this is war transfigured, so compounded with neocortical forays and cybernetic refashionings that the terrains on which it is fought include synapses and circuits as well as checkpoints and borders. Although the novel is set in 1990, Wolfe asserts in an afterword: "Anybody who 'paints a picture' of some coming year is kidding—he's only fancying up something in the present or past, not blueprinting the future. All such writing is essentially satiric (today-centered), not utopic (tomorrow-centered)."[2] His insistence on the novel's satiric intent is a useful reminder that *Limbo* refracts its cybernetics concerns through the hysterical denunciations and national delirium precipitated by the cold war. In *Pure War,* Paul Virilio argues that postmodern technologies, especially global information networks and supersonic transport, have changed how military organizations conceptualize the enemy.[3] Whereas a country's borders were previously presumed adequate to distinguish between citizen and alien, in the post–World War II period the distinction between inside and outside ceased to signify in the same way. The military no longer thought of its task as protecting the body politic against an exterior enemy. Rather military resources were deployed against a country's own population, as in Latin American death squads. Such military operations are not aberrations, Virilio contends, but harbingers of a deep shift from exo-colonization to endo-colonization throughout postmodern cultures. Although Virilio's thesis is overstated, it nevertheless provides useful insight into the McCarthy era in the United States. During McCarthyism, paranoia about the inability to distinguish between citizen and alien, "loyal American" and communist spy, was at its height. In a scenario that, following Virilio, I call endo-colonization, *Limbo* joins political and geographical remappings with the cybernetic implosion into the body's interior.

As Donna Haraway has pointed out, cyborgs are simultaneously entities and metaphors, living beings and narrative constructions.[4] The conjunction of technology and discourse is crucial.[5] Were the cyborg only a product of discourse, it could perhaps be relegated to science fiction, of interest to SF aficionados but not of vital concern to the culture. Were it only a technological practice, it could be confined to such technical fields as bionics, medical prostheses, and virtual reality. Manifesting itself as both technological object and discursive formation, it partakes of the power of the imagination as well as of the actuality of technology. Cyborgs actually exist. About 10 percent of the current U.S. population are estimated to be cyborgs in the technical sense, including people with electronic pacemakers, artificial joints, drug-implant systems, implanted corneal lenses, and artificial skin. A much higher percentage participates in occupations that make them into metaphoric cyborgs, including the computer keyboarder joined in a cybernetic circuit with the screen, the neurosurgeon guided by fiber-optic microscopy during an operation, and the adolescent game player in the local video-game arcade. "Terminal identity" Scott Bukatman has named this condition, calling it an "unmistakably doubled articulation" that signals the end of traditional concepts of identity even as it points toward the cybernetic loop that generates a new kind of subjectivity.[6]

Limbo edges uneasily toward this subjectivity and then only with significant reservations. Instead of a circuit, it envisions polarities joined by a hyphen: human-machine, male-female, text-marginalia. The difference between hyphen and circuit lies in the tightness of the coupling (recall Wiener's argument about the virtues of loose coupling) and in the degree to which the hyphenated subject is transfigured after becoming a cybernetic entity. Whereas the hyphen joins opposites in a metonymic tension that can be seen as maintaining the identity of each, the circuit implies a more reflexive and transformative union. When the body is integrated into a cybernetic circuit, modification of the circuit will necessarily modify consciousness as well. Connected by multiple feedback loops to the objects it designs, the mind is also an object of design. In *Limbo* the ideology of the hyphen is threatened by the more radical implications of the cybernetic splice. Like Norbert Wiener, the patron saint of *Limbo,* Wolfe responds to this threat with anxiety. To see how this anxiety both generates the text and fails to contain the subversive implications of cybernetics, let us turn now to a consideration of this phantasmic narrative.

Limbo presents itself as the notebooks of Dr. Martine, a neurosurgeon who defiantly left his medical post in World War III and fled to an uncharted Pacific island. He finds the islanders, the Mandunji tribe, practicing a primitive form of lobotomy to quiet the "tonus" in antisocial people.[7] Thus the text reinscribes the privileged status of homeostasis during the Macy period and also glances toward Wiener's devastating criticism of lobotomy in the 1948 *Cybernetics* and the 1950 *The Human Use of Human Beings.*[8] Wiener's interest in lobotomy is played out in a short story he wrote entitled "The Brain," with which Wolfe may have been familiar. Published in a 1950 science fiction anthology under the transparent pseudonym "W. Norbert," the story was explicitly attributed by the editor to Norbert Wiener.

In the story, a mental patient's attending physician brings the patient as a guest to an intellectual dinner club for "a small group of scientists."[9] The dinner conversation is reminiscent of the Macy discussions. During dinner the patient, a victim of amnesia, faints. When he comes to with the help of drugs, he begins to recall the trauma that caused his amnesia. He remembers that he himself was a physician and that his wife was fatally injured and his child made into a vegetable in a hit-and-run accident caused by a fiendishly clever gangster called "The Brain." Later, fate delivered the gangster into the doctor's hands when he was called to perform emergency surgery on the gangster, who had received a bullet wound to the head. During the operation, the doctor quietly performs a lobotomy. Later the gangster is caught because he has become stupid.

Like the protagonist of "The Brain," Dr. Martine in *Limbo* performs lobotomies for the social good, rationalizing that it is better to do the surgery properly than to let people die from infections and botched jobs. He uses the operations to do neuroresearch on brain-function mapping. He discovers that no matter how deeply he cuts, certain characteristics appear to be twinned. One twin cannot be excised without sacrificing the other. When aggression is cut out, eroticism goes too; when violence yields to the surgeon's knife, creativity also disappears. Martine expands his observations into a theory of human nature. Humans are essentially hyphenated creatures, he asserts, creative-destructive, peaceful-aggressive. The appearance on the island of "queer limbs," men who have had their arms and legs amputated and replaced by atomic-powered plastic prostheses, brings Martine's philosophy of the hyphen into juxtaposition with the splice, the neologistic cutting, rejoining, and recircuiting that makes a cyb/ernetic org/anism into a cyborg. On the level of plot, the intrusion of the cyborgs gives Martine an excuse to leave his island family and find out how the world has shaped up in the aftermath of the war.

The island/mainland dichotomy is the first of a proliferating series of divisions. Their production follows a characteristic pattern. First the narrative presents what appears to be a unity (the island locale; the human psyche), which nevertheless cleaves in two (mainlanders come to the island; twin impulses are located within the psyche). The cleavage arouses anxiety, and textual representations try to achieve unity again by undergoing metamorphosis, usually truncation or amputation (Martine and the narrative leave the island behind and concentrate on the mainland, which posits itself as a unity; the islanders undergo lobotomies to make them "whole" citizens again). The logic implies that truncation is necessary if the part is to reconfigure itself as a whole. Better to formalize the split and render it irreversible so that life can proceed according to a new definition of what constitutes wholeness. Without truncation, however painful it may be, the part is doomed to exist as a remainder. But amputation always proves futile in the end because the truncated part splits in two again and the relentless progression continues.

Through delirious and savage puns, the text works out the permutations of this geography of the Imaginary. America has been bombed back to the Inland Strip, its coastal areas now virtually uninhabited wastelands. The image of a truncated country, its outer extremities blasted away, proves prophetic, for the ruling political ideology is Immob. Immob espouses such slogans as "No Demobilization without Immobilization" and "Pacifism means Passivity." Citing Napoleon, Paul Virilio wrote, *"The capacity for war is the capacity for movement."*[10] Immob reinscribes that proposition and reverses its import, reasoning that the only way to end war is to remove the capacity for motion. True believers become "vol-amps," men who have undergone voluntary amputations of their limbs. Social mobility paradoxically translates into physical immobility. Upwardly mobile executives have the complete treatment to become quadroamps; janitors are content to be uniamps; women and blacks are relegated to the limbo of unmodified bodies. But like the constructions that preceded it, Immob ideology also splits in two. The majority party, discovering that its adherents are restless lying around with nothing to do, approves the replacement of missing limbs with powerful prostheses (or "pros"), which bestow enhanced mobility and enable those who wear them to perform athletic feats impossible for unaltered bodies. These cyborgs are called (in a twinning pun that tries to encompass the cyborg under the sign of the hyphen) "Pro-pros." The logic of the hyphen dictates that Pro-pros be mirrored by Anti-pros, who believe that cyborgism is a perversion of Immob philosophy. Anti-pros spend their days proselytizing for voluntary amputation, using microphones hooked up to the baby baskets that are just the right size to accommodate their limbless human torsos, a detail that later becomes significant.

Unity, cleavage, truncation, and further cleavage—these are the counters through which geopolitical and cybernetic endo-colonization are represented in *Limbo*. Amputations, undertaken in an effort to stop the proliferation of doubleness, only drive the plot toward the next phase of the cycle, for they are nostalgic attempts to recover a unity that never was.

This much Wolfe sees clearly. Less clear is the increasingly urgent issue of how the parts should be reassembled: through a hyphen or through a circuit? I suggested earlier that the cyborg subverts Martine's (and Wolfe's) theory of the hyphen, for it implies that the hyphenated polarities will not be able to maintain their identity unchanged. This possibility, although not explicitly recognized by the narrator, is already encoded into the text, for the amputations intended to ensure that pacifism is irrevocable have instead ensured that the interface between human and machine is irrevocable. Although the Pro-pros justify the use of prostheses by pointing out that the pros can be detached, many of the changes (such as permanently installed bio-sockets into which the pros are snapped) have become integral parts of the organism. In a larger sense, the conversions have worked such far-reaching changes in social and economic infrastructures that a return to a precybernetic state is not possible. Whether functioning as an amputee or a prosthetic athlete, the citizen of *Limbo*'s world is spliced into cybernetic circuits that irreversibly connect his body to the truncated, military-industrial limbo that the world has become. In the circuit of metaphoric exchanges that the cyborg sets up, the narrator finds it increasingly difficult to maintain the hyphenated separations that allow Wolfe to criticize capitalist society while maintaining intact his own sexist and technological assumptions. Breakdown occurs when the hyphen is no longer sufficient to keep body, gender, and political categories separate from one another.

In exploring this breakdown, I will go further into Wolfe's background and his relation to cybernetics. Not one to disguise his sources, he adds an afterword in which he lists the books that have influenced him. In case anyone missed his frequent allusions to Norbert Wiener, the afterword makes clear that Wiener is a seminal figure. The title Wolfe cites is Wiener's 1948 *Cybernetics.* I noted earlier that the cyborg is both a technological entity and a discursive construction. The chapters of Wiener's book illustrate how discourse collaborates with technology to create cyborgs. The transformations that Wiener envisions are for far simpler mechanisms than human beings, but his explanations work as rhetorical software (Richard Doyle's phrase)[11] to extend his conclusions to complex human behaviors as well. We saw the same kind of slippage during the Macy discussions. Here is how it characteristically occurs in Wiener's text. First a behavior is noted—an intention tremor, a muscle contraction, a phobic or philic reaction to a stimulus. Next an electronic or mathematical model that can produce the same behavior is proposed. Sometimes the model is used to construct a cybernetic mechanism that can be tested experimentally. Whether actual construction takes place or the idea remains a thought experiment, the claim is made that the human mechanism, although unknown, might plausibly be the same as the mechanism embodied in the model. The laboratory "white box" is thus discursively equated with the human "black box," with the result that the human is now also a "white box," that is, a servo-mechanism whose workings are known. Once the correlation is made, cybernetics can be used not only to correct dysfunction but also to improve normal functioning. As a result, the cyborg signifies something more than a retrofitted human. It points toward an improved hybrid species that has the capacity to be humanity's evolutionary successor. As we saw in chapter 4, the problem that Wiener encountered was how to restrain this revolutionary potential of cybernetics so that it would not threaten the liberal humanism that so deeply informed his thinking.

In "Self Portrait," a short story published a few months before *Limbo* and concerned with similar themes, Wolfe shows that he understands the limitations as well as the potential of Wiener's method. "Cybernetics is simply the science of building machines that will duplicate and improve on the organs and functions of the animal, based on what we know about the systems of communication and control in the animal," the narrator says. But he acknowledges that "everything depends on just how *many* of the functions you want to duplicate, just how *much* of the total organ you want to replace."[12] In charge of a cybernetics laboratory, he decides to separate kinesthetic and neural functioning. He can be reasonably sure of creating an artificial limb that moves like a real one, but connecting it to the body's sensory-neural circuits is another matter.[13] His hesitation points up how speculative many of Wiener's claims were. More than a technology, they functioned as an ideology. Without mentioning Wolfe, Douglas D. Noble, in "Mental Materiel: The Militarization of Learning and Intelligence in U.S. Education," argues that the cybernetic paradigm has in fact brought about massive transformations in U.S. social, economic, and educational infrastructures, as Wolfe predicted it would.[14] In his view, these transformations have been driven primarily by the U.S. military. The cyborg, Noble insists, is no science fiction fantasy but an accurate image of the modern American soldier, including pilots wired into "intelligent cockpits," artillery gunners connected to computerized guidance systems, and infantry soldiers whose ground attacks are instantaneously broadcast on global television. His analysis, consistent with arguments by military strategists for "neocortical warfare" and with the picture that Chris Gray draws of the military's interest in the cyborg,[15] indicates that Wiener's antimilitary stance was not sufficient to prevent the marriage of war and cybernetics, a union that he both feared and helped to initiate.

Limbo takes the leap that "Self Portrait" resists, imagining that under the stimulus of war, the machine component, no longer limited to mimicking an organic limb, is hardwired into the human nervous system to form an integrated cybernetic circuit. This movement toward the splice is figured in *Limbo* through tropes of motion. Here Wolfe follows Wiener's lead, for most of Wiener's examples concentrate on dysfunctions of movement. The intention tremor provided Wiener with one of his first experimental successes. Through a mechanism that duplicated the behavior of an intention tremor, Wiener diagnosed the problem as an inappropriate positive amplification of feedback and showed how it could be cured. Other kinds of movement dysfunctions are similarly diagnosed in the 1948 *Cybernetics.* Even phenomena not obviously associated with motor skills are figured as various kinds of motion. Thinking, for example, is figured as movement across neural synapses, and schizophrenia is represented as a feedback problem in the cognitive-neural loop. Wiener's emphasis on movement implies that curing dysfunctions of movement can cure the patient of whatever ails him, whether muscular, neural, or psychological. Given this context, what could be more cybernetic than to construct war as a dysfunction of movement? In this sense, *Limbo* follows the line of thought that Wiener mapped out in *Cybernetics,* down to particular phrasings that Wolfe appropriates. Because in many respects Wolfe follows Wiener so closely, the departure he makes in insisting on the typographic hyphen rather than the cybernetic splice is even more significant. In the end, however, his resistance to the splice fails to restrain the scarier implications of cybernetics, much as Wiener's resistance to the cybernetic penetration of boundaries failed to prevent the dissolution of the liberal humanist subject.

The breakdown of Wolfe's "hyphenation" theory occurs, perhaps predictably, when the hyphen is no longer sufficient to contain the repressed violence that the cyborg unleashes (uncannily so, for the text operates as if Wolfe were unconsciously reenacting, from Wiener's war work on antiaircraft devices, the mapping of enemy onto self). In the world of *Limbo*, warfare has been replaced by a Superpower Olympics between the capitalist Inland Strip and the communist East Union, a competition designed to sublimate lethal violence into healthy competition. But in the 1990 Olympics, as if in recognition of Wiener's failure to prevent the promiscuous coupling of cybernetics with military research after World War II, cyborg competition neologistically slides into warfare rather than metonymically substitutes for it. Athletes from both sides are vol-amps, and they owe their victories as much to the technicians who design the prostheses as they do to their athletic abilities. Traditionally the Inland

Strip, with its superior technology, has dominated the Olympics. Vishinu, leader of the East Union, announces that this year it will be different. His people are tired of the imperialist smugness of the Inland Strip and will demonstrate that they are no second-rate colonials but are superior cyberneticians. The East Union cyborgs proceed to sweep the competition, winning every category.

For weeks before the event, Vishinu has darkly hinted at the growing schism between the two countries. The rare metal columbium is needed to make the prostheses on which both sides depend, and the East Union alleges that the Inland Strip has been trying to hoard the world's columbium supply. At the final ceremony, instead of confirming that East Union cyberneticians will share their technology with the Inland Strip (as custom dictates), Vishinu signals the East Union athletes to unveil their newest prosthetic innovation: artificial arms that terminate in guns. According to the West's own logic, Vishinu satirically argues, the East Union's triumph in cybernetics means that it has won the right to all the world's columbium. While Martine watches incredulously on his television at a remote mountain retreat, the East Union cyborgs open fire on the reviewing stand where Inland Strip officials are seated. The apparatus of war has imploded inward to join with flesh and bone. As a result of this cybernetic splice, war radiates from body zones outward.

In the last war, when the EMSIAC computer mindlessly tried to return Martine's plane to base—which would have returned him to almost certain death—Martine ripped out the circuit cables and destroyed the communication-control box. But now endocolonization has proceeded far enough into the human and political body so that he can no longer disable the circuit simply by ripping out cables. Instead of fleeing to the margins, he rushes toward the center, returning to the capital and demanding an audience with Helder, Vishinu's western counterpart. He uses as his calling card cryptic allusions to an incident that only he and Helder know about—an incident that hints at the network of anxieties that have been activated through cyborg circuitry. The return of these repressed anxieties takes the form of a corpse that, refusing to stay buried, haunts the narrative. Throughout Martine's notebook, references to it have surfaced in puns and half-remembered flashes. Finally, with the outbreak of war, the repressed memories erupt into full articulation. The corpse's name is Rosemary, a nurse that Helder took to a college peace rally at which he delivered a fiery speech. He returned with her to her apartment, tried to have sex with her, and when she refused, brutally raped her. After he left, she committed suicide by slashing her wrists. Martine's part in the affair was to provide a

reluctant alibi for his roommate Helder, allowing Helder to escape prose-
cution for the rape-manslaughter. The placement of Martine's recollection
of the incident and his use of it to address the political crisis hint that the
body politic and the politics of the body, like prostheses and trunk, are
spliced together in an integrated circuit.

Throughout the text, the narrator—and behind him, the author—has
exhibited profound ambivalence toward women. This ambivalence, like so
much else in Wolfe's cybernetic novel, is figured through tropes of motion.
Shortly after his arrival at the Inland Strip, Martine looks down onto the
balcony of the apartment below and sees a quadroamp lying on a lounge
reading a book. A young woman tries to arouse him sexually and begins to
remove his prostheses. Uninterested, the young man pushes her away and
resumes reading. The incident illustrates how sexual politics work under
Immob. Prohibited from becoming vol-amps, women have taken the initiative in sexual encounters. They refuse to have sex with men wearing
prostheses, for the interface between organism and mechanism is not perfect and at moments of stress or tension the limbs are apt to careen out of
control, smashing whatever is in the vicinity. Partnered with truncated, im-
mobilized men, women have perfected techniques that are performed in
the female superior position and that give them satisfaction while requiring
no motion from the men. Martine gets a firsthand demonstration when
Neen, an artist visiting from the East Union, seduces him. To Martine, the
idea that men would be immobile during sex is obscene, for he believes that
the only normal sexual experience for women is a "vaginal" orgasm reached
using the male superior position. Like his Victorian antecedents, Martine
atavistically polices what kinds of movements are proper for women during
sexual intercourse, enforcing them with violence when necessary. To revenge himself on Neen and assure himself that he has not been emascu-
lated after her "clitoral" orgasm, he rapes her and forces her to have a
"vaginal" orgasm, which the text assures us she enjoys in spite of herself.
Here the rape occurs in a context where Wolfe is in control of the dynamics,
for it reflects his own deeply misogynistic views. Nevertheless, the narrative keeps moving toward the moment when another rape will be recalled
and when the cyborg circuitry in which the narrator is enmeshed will make
authorial control much less certain.

On a structural level, the text strives to maintain the ideological purity of
male identity by constructing categorical and hierarchical differences between men and women. The man has a real penis, the woman a shadowy
surrogate that the narrator calls a "phantom penis"; the man is active, the
woman passive; the man has a single orgasm of undoubted authenticity,

whereas the woman's orgasms are duplicitous as well as double. The man responds to sexual aggression, but (the narrator insists) it is the woman who initiates sexual violence, even when she is raped. So far the novel reads like a devil's dictionary of sexist beliefs that are Neolithic even by the standards of the 1950s. Yet at the same time, the text also edges toward a realization that it cannot unequivocally articulate. Like man and machine, male and female are spliced together in a feedback circuit that makes them mutually determine each other. No less than geopolitical ideology, sexual ideology is subverted and reconfigured by the cybernetic paradigm.

Wolfe's outrageously sexist views echo those of his psychoanalyst Edmund Bergler, by whom he was deeply influenced.[16] Bergler acknowledged that it could be difficult for some women to reach orgasm in the male superior position, but he nevertheless insisted that only this position and only "vaginal" orgasms were normal for women. The view is inscribed in *Limbo*, where the usual (as distinct from "normal") state for women is frigidity. Martine applies the label liberally, using it to describe every woman with whom he is intimate except one, his island wife, Ooda. Frigidity applies both to women who are too aggressive (like Neen) and to women who find sex with Martine unsatisfying (like his first wife, Irene, whose connection with Neen is signified by the rhyme connecting their names). With astonishing blindness, he never considers that the dysfunction might lie in him or his view of women. The text strives to endorse the narrator's blindness. Yet it also engenders ambiguities beyond the narrator's control and perhaps beyond Wolfe's.

The kind of cyborg that Wolfe envisions locates the cybernetic splice at the joining of appendage to trunk. As the placement of the splice suggests, the novel's sexual politics revolve around fear of symbolic and actual castration, manifested as extreme anxiety about issues of control and domination. Wolfe, described by his biographer as a small man with a large mustache and fat cigar, creates in Immob a fantasy about technological extensions of the male body that endow it with supernatural power.[17] During the sex act, however, the extensions are laid aside, and only a truncated body remains. If the artificial limbs swell to an unnatural potency, the hidden price is the withering of the limb called, in U.S. slang, the third leg or the short arm. The connection becomes explicit when Martine discovers that his son, Tom, whom he has not seen in twenty years, has become an activist in the Anti-pro cause. Tom is a quadroamp, spending his days spreading the word from the baby basket that accommodates his limbless trunk. When war breaks out, his already truncated body is mutilated by exploding glass shards. Martine finds him in the street, lifts the blanket that covers his trunk, and sees the mark of castration as well as the wounded torso. He then shoots Tom, ostensibly to put him out of his misery and perhaps also to exorcise the specter of castration he represents.

In more than one sense, *Limbo* is a masculine fantasy that relates to women through mechanisms of projection. It is, moreover, a fantasy fixated in male adolescence. Wavering between infantile dependence and adult potency, Immob re-creates, every time a man takes off his prostheses to have sex, the dynamic typical of male adolescence. With the pros on, a vol-amp is capable of feats that even pros like O. J. Simpson and Mike Tyson would envy. With the pros off, he is reduced to infantile dependence on women. The unity sought in becoming a vol-amp is given the lie by the split he experiences within himself as a superman and a symbolically castrated infant. The woman is constructed in correspondingly ambivalent ways—as a willing victim to male violence, a nurturing mother who infantilizes her son, and a domineering sex partner all too willing to find pleasure in the man's symbolic castration. The instabilities in her subject position are consistent with the ambiguities characteristic of male adolescence. The narrative's overwritten prose, penchant for puns, and hostility toward women all recall a perpetually adolescent male who has learned to use what Martine calls a "screen of words" to compete with other men and to insulate himself from emotional involvements with women.

Were this all *Limbo* was, the novel would be merely frustrating rather than frustrating and brilliant. What makes it compelling is its ability to represent and comment on its own limitations. Consider the explanation that Martine gives for why Immob has been so successful. The author drops a broad hint in the baby baskets that Immob devotees adopt. In a theory adapted from Bergler's book on narcissism, Wolfe has his narrator suggest that the narcissistic wound from which the amputations derive is the male infant's separation from the mother and his outraged discovery that his body is not coextensive with the world.[18] Amputation allows the man to return to his pre-Oedipal state, where he will have his needs cared for by attentive and nurturing females. In locating the moment of trauma before the Oedipal triangle, Wolfe reenacts the same kind of move that Lacan makes in his revision of Freud. Whereas Freud identified the male child's fear of castration with the moment when he sees female genitalia and constructs them as lack, Wolfe (following Bergler) places the anxious moment considerably earlier, in the series of "splittings" and separations that the infant experiences from his primary love object, the mother.[19] Given this scenario, the catalyst for anxiety is not the woman's lack but the ambiguity of boundaries between infant and mother. The mother is the object of projected anger for two contradictory but paradoxically reinforcing reasons. When she withdraws from the infant, she traumatizes him; when she does not withdraw, she engulfs him. The question of who is responsible for the narcissistic wound and its aftermath is a matter for anxious consideration in *Limbo*—a query presupposing that the violation of boundaries is central to the formation of male subjectivity. In its stagings of traumatic moments of "splitting," the text vacillates on its answers to the question. At times it seems that the woman is appropriating the male infant into her body; at other times it seems that the amputated men are willfully forcing women into nurturing roles they would rather escape. In fact, once male and female are plugged into a cybernetic circuit, the question of origin becomes irrelevant. Each constitutes the other. In approaching this realization, the text goes beyond the presuppositions that underlie its sexual politics as it gropes, however tentatively, toward cyborg subjectivity.

Crucial to this process are transformations in the textual body, transformations that reenact and re-present the textual dynamics of Immob. The textual body begins by figuring itself as Martine's notebook "mark ii," written in the narrative present. In this notebook, Martine notices that Immob slogans have a disturbingly familiar ring, particularly the icon of a man getting run over by a steamroller (intended to symbolize technology before Immob, although for the narrator and reader it precisely characterizes Immob). Only when war erupts does Martine realize why the steamroller image is eerily familiar. In a notebook that he wrote two decades earlier and that he entitled "mark i," he used the steamroller as an ironic emblem for the war machine. In the same notebook and in a similarly ironic vein, he wrote a satiric fantasy of a society in which people pre-empt the atrocities of war by voluntarily cutting off their own limbs. After Martine deserted and rerouted his plane to the island, Helder found the notebook among Martine's gear and decided to use the satire as a blueprint for an actual postwar society. Surrounding Martine's bitter jokes with his own flat-footed, self-serving commentary, he ventriloquized Martine's words, making them speak the message he wanted, not what Martine intended. Martine's notebook thus functions like a child whom he abandoned (just as he abandoned his son, Tom, when he fled) and who then was turned into the very thing Martine dreaded most (as was Tom). The present narrative is recorded in the "mark ii" notebook. The revelation that the Immob bible is actually Martine's (mis)appropriated "mark i" notebook demonstrates that the body of the text is subject to the same kind of cleavages, truncations, and further cleavages that mark the bodies represented within the text.

Although Martine tries to heal the split narrative by renouncing the first notebook and destroying the second, the narrative continues to fragment. The form this fragmentation takes is significant, for it follows the geography of Immob. The text splits into a trunk, consisting of the main narrative, and prosthetic extensions constituted through drawings that punctuate the text and lines that scrawl down the page where the trunk ends. Prosthesis and trunk are connected through puns that act like cyborg circuitry, splicing the organic body of the writing together with the prosthetic extensions that operate in a subvocal margin. Pros are thus punningly and cunningly linked not only with the hyphenated Pro-pros but also with the more dangerous and circuitous cyborg "pros/e," the truncated/spliced noun that speaks the name of the text's body (prose) as well as the name of the prostheses (pros) attached to it and represented within it.

Wolfe is not the only writer to link writing and prosthesis. In *Prosthesis*, David Wills explores connections between his father's wooden leg and the language that the son adopted as his prosthesis of choice. Trunk and prosthesis, body and writing, are alike in having limits and in having relations with something beyond those limits. "Prosthesis is the writing of my self as a limit to writing," Wills explains as he interrogates the boundaries and splicings between the body of his prose and his (father's) body in prose. "There is no simple name for a discourse that articulates with, rather than issuing from, the body, while at the same time realizing that there is no other discourse—in the sense of no other translation, transfer, or relation—no other conception of it except as it is a balancing act performed by the body, a shift or transfer between the body and its exteriority."[20] The conflation that Wills addresses in this difficult and subtle passage is the superimposition of a body of prose with bodies constituted within prose. The passage points toward a double entanglement of the textual corpus and the physical body. Writing is a way to extend the author's body into the exterior world; in this sense, it functions as a technological aid so intimately bound up with his thinking and neural circuits that it acts like a prosthesis. At the same time, the writing within itself is trying to come to terms with what it means to have a prosthesis, particularly with whether the prosthesis should be incorporated into the subject's identity (in which case he becomes a cyborg) or should remain outside (in which case the prosthesis is necessarily alien from the self and so not something one can use with the "natural" dexterity). For Wolfe, the choice cannot be made in a clear-cut or unambiguous way. He can neither embrace the transformations that becoming a textual cyborg would imply nor remain content with an amputated text that has a limited range of motion. So he simultaneously crafts prosthetic extensions for the text and forbids the text to recognize them as itself. Just as pros/e destabilizes the concept of the natural human body, so it also destabilizes the notion of a text contained and embodied solely within its typographic markers. Pros/e implies a text spliced into a cybernetic circuit that reaches beyond the typography of the printed book into a variety of graphic and semiotic prostheses that it both authorizes and denies.

It is no surprise to find, then, that the pros/e of *Limbo*'s corpus implies a dispersed subjectivity. Whereas the voice that speaks (from) the text's trunk is clearly characterized as Martine, the subject that produces and is produced by the prosthetic marginalia is more difficult to identify. The question of which voice speaks from what textual body was a complicated issue in Wolfe's career. To supplement his income, he worked for a while as a ghostwriter for Billy Rose's syndicated column. Here his words issued from a body of print signed with someone else's name. He also wrote for popular science magazines including *Mechanix Illustrated,* frequently contributing to articles published under someone else's byline. In addition, he collaborated on low-level popular science books. One of these, *Plastics, What Everyone Should Know,* appeared under Wolfe's name, although it was written by someone else.[21] The synthetic chemical product that came of age in World War II and that Wolfe envisions as the substance of choice for prostheses thus functions as a kind of prosthesis for his corpus, extending his name through a body of print ventriloquized by someone else.

To explore the complex play between Martine's voiced narrative and the drawings, nonverbal lines, and punning neologisms that serve as prostheses to the textual trunk, I want to consider one of the drawings in more detail. It shows a nude woman with three prosthetic legs—the Immob logo—extruding from each of her nipples.[22] She wears glasses, carries a huge hypodermic needle, and has around her neck a series of tiny contiguous circles, which can be taken to represent the popular 1950s necklace known as a choker. To the right of her figure is a grotesque and diapered male torso, minus arms and legs, precariously perched on a flat carriage outfitted with Immob prosthetic legs instead of wheels. He has his mouth open in a silent scream, perhaps because the woman appears to be aiming the needle at him. In the text immediately preceding the drawing, Rosemary is mentioned. Although the truncated text does not acknowledge the drawing and indeed seems unaware of its existence, the proximity of Rosemary's name indicates that the drawing is of her, with the needle presumably explained by her profession as a nurse.

In a larger sense the drawing depicts the Immob woman. The voiced narrative ventriloquizes her body to speak of the injustices she has inflicted on men, constructing her retrospectively as a cyborg who nourishes and emasculates cyborg sons. It makes her excess, signified by the needle that she brandishes and the legs that sprout from her nipples, responsible for her lover/son's lack. In this deeply misogynistic writing, it is no surprise to read that woman are raped because they want to be. Female excess is represented as stimulating and encouraging male violence, and rape is poetic revenge for the violence women do to men when they are too young and helpless to protect themselves. The voiced narrative strives to locate the origin of the relentless dynamic of splitting and truncation within the female body. According to this textual trunk, the refusal of the woman's body to respect decent boundaries between itself and another initiates the downward spiral into amputation and eventual holocaust.

Countering these narrative constructions are other interpretations authorized by the drawings, nonverbal lines, puns, and lapses in narrative continuity. From these semiotic spaces, which Julia Kristeva has associated with the feminine, come inversions and disruptions of the hierarchical categories that the narrative uses to construct maleness and femaleness.[23] Rosemary, written into nonexistence by her suicide within the text's represented world, returns in the prosthetic space of the drawing and demands to be acknowledged. On multiple levels, the drawing deconstructs the narrative's gender categories. In the represented world, women are not allowed to be cyborgs, yet this female figure has more pros attached to her body than does any man. Women rank after men in the represented world, but here the woman's body is on the left and is thus "read" before the man's. Above all, women and men are separate and distinct in the represented world, but in this space, parts of the man's body have attached themselves to her. Faced with these disruptions, the voiced narrative is forced to recognize that it does not unequivocally control the textual space. The semiotic intrusions contest its totalizing claims to write the world.

The challenge is reflected within the narrative by internal contradictions that translate into pros/e the intimations of the semiotic disruptions. As the voiced narrative tries to come to grips with these contradictions, it cycles closer to the realization that the hierarchical categories of male and female have imploded into the same space. The lobotomies that Martine performs suggest the depth of this collapse. To rid the psyche (coded male in *Limbo*) of subversive (female) elements, it is necessary to amputate. For a time the amputations work, allowing male performance to be enhanced by prostheses that bestow new potency. But eventually these must be shed and the woman encountered again. Then the subvocal feminine within merges with the prostheses without, initiating a new cycle of violence and amputation. No matter how deeply the cuts are made, they can never excise the ambiguities that haunt and constitute these posthuman and post-typographic bodies. *Limbo* envisions cybernetics as a writing technology that inscribes over the hierarchical categories of traditional sexuality the indeterminate circuitry of cyborg gender.

As a white male writing in the early 1950s, Wolfe was aware that the politics of gender relations were beginning to shift. Several times, the narrator mentions "women's liberation," quarantined by quotation marks and authorial scorn. Nevertheless, even he cannot escape the feminine within. After ending his first notebook with a huge "NO" inscribed across the page, the narrator ends the second notebook with an equally vehement "YES," which he intends as an affirmation of humankind's hyphenated nature. His mother's birth name was Noyes ("No-yes"), and he dimly senses the connection between matrilineal heritage and the affirmation he seeks. But the hyphen is not the same as the splice. By inscribing Noyes as No-yes, he seeks to draw a line that will preserve each half of the hyphenation as a distinct entity. His voiced concessions to sexual politics are similarly limited to realizing that women are not entirely monsters. The real power relations at stake in sexist relations remain opaque to him, just as do the deeper implications of being wired into a cybernetic circuit.

But *Limbo* knows more than it can say, a paradox inscribed within the text by the narrator's image of a "screen of words" that hides something from him. Throughout, there are flashes of insight that exceed his formulations and that are never adequately accounted for by his theorizing. The effect is finally of another voice trying to emerge, authored not so much by Wolfe as by the cybernetic circuit he can imagine but not fully articulate. Just as Martine's first notebook has been ventriloquized by Helder, so the narrative as a whole is ventriloquized by a constellation of forces that make it speak of a future in which hyphenation gives way to the spliced pros/e that both signifies and is the cyborg. If the ownership of the writing with which the prosthesis signifies is unclear, the obscurity is appropriate, for it indicates that control in a cybernetic circuit is not a localized function but is an emergent property. Neither entirely in control nor out of control, *Limbo* teeters on the edge of an important recognition.

In one sense, the bodies of *Limbo* are the cyborgs who populate the imaginative world of Immob. In another, more literal sense, the body of *Limbo* is constituted through the typefaces that march across the page. Normally readers attend to the represented world and only peripherally notice the physical body of the text. When the pros/e of *Limbo* itself becomes a cyborg, however, the splice operates to join the imaginative world of the signifier with the physical body of print. Parallels between a text's physical form and its represented world have a long history in literature, from the seventeenth-century iconographic poems of George Herbert to the maps, tattoos, and body writing that litter the surfaces of Kathy Acker's contemporary novels. What is distinctive about Wolfe's use of the correlation is the suggestion that the bodies in the text and the body of the text not only represent cyborgs but also together compose a cyborg in which the neologistic splice operates to join imaginative signification with literal physicality. In this integrated circuit, the physical body of the text and the bodies represented within the text evolve together toward a posthuman, posttypographic future in which human and intelligent machine are spliced together in an integrated circuit, subjectivity is dispersed, vocalization is nonlocalized, bodies of print are punctuated with prostheses, and boundaries of many kinds are destabilized. More than a conduit through which ideas from cybernetics boiled into the wider U.S. culture in the 1950s, *Limbo* is a staging of the complex dynamics between cyborg and literary bodies. As such, it demonstrates that neither body will remain unchanged by the encounter.

THE SECOND WAVE OF CYBERNETICS:
FROM REFLEXIVITY TO SELF-ORGANIZATION

It all started with a frog. In a classic article entitled "What the Frog's Eye Tells the Frog's Brain," central players in the Macy group—including Warren McCulloch, Walter Pitts, and Jerry Lettvin—did pioneering work on a frog's visual system. They demonstrated, with great elegance, that the frog's visual system does not so much *represent* reality as *construct* it.[1] What's true for frogs must also hold for humans, for there's no reason to believe that the human neural system is uniquely constructed to show the world as it "really" is. Not everyone in the research group was interested in pursuing the potentially radical epistemological implications of this work. McCulloch, for example, remained wedded to realist epistemology. But a young neurophysiologist from Chile, Humberto Maturana, was also on the research team, and he used it as a springboard into the unknown. Pushing the envelope of traditional scientific objectivity, he developed a new way of talking about life and about the observer's role in describing living systems. Entwined with the epistemological revolution he started are the three stories we have been following: the reification of information, the cultural and technological construction of the cyborg, and the transformation of the human into the posthuman. As a result of work by Maturana and his collaborator, Francisco Varela, all three stories took decisive turns during the second wave of cybernetics, from 1960 to 1985. This chapter follows the paths that Maturana and Varela took as they probed deeply into what it means to acknowledge that the observer, like the frog, does not so much discern preexisting systems as create them through the very act of observation.

Central to the seriated changes connecting these second-wave developments to the first wave is the difficult and protean concept of reflexivity. As we saw in chapter 3, participants in the Macy Conferences wrestled with reflexivity, without much success. The particularities of the situation—the embedding of reflexivity within psychoanalytic discourse, Kubie's halitosis of the personality, the unquantifiability of reflexive concepts—put a spin on reflexivity that affected its subsequent development.[2] Gregory Bateson's 1968 conference had made clear that the problems posed by including the observer could be addressed only if a substantial reworking of realist epistemology was undertaken. The intuitive leap made by Bateson in concluding that the internal world of subjective experience is a metaphor for the external world remained a flash of insight rather than a quantitatively reliable inference that experimentalists like Warren McCulloch could endorse. The problem was how to make the new epistemology operational by integrating it with an experimental program that would replace intuition with empirical data.

At issue in this evolving series of events are questions crucially important to the technoscientific concepts of information, the cyborg, and the posthuman. Like Norbert Wiener, Maturana has strong ties with liberal humanism. At stake for him was how to preserve the central features of autonomy and individuality while still wrenching them out of the Cartesian and Enlightenment frameworks in which they are embedded. Even as he struggled mightily to "say something new," his work replicates some assumptions of the first wave at the same time that it radically revises others.[3] We can see an early form of the struggle in the essays of Heinz von Foerster, the genial and well-connected Austrian émigré who functions as a transitional figure linking first- and second-wave cybernetics. From this beginning, we will trace the epistemological revolution that Maturana fomented, delineate its connections with the three stories we have been following, and finally explore the differing assumptions that led Varela, Maturana's collaborator, to set off in a new direction.

Reflexivity Revisited

Von Foerster left Austria in 1948, after working on microwave electronics for Germany during World War II, work that had important applications in radar (his 1949 vita lists much of this research as "secret").[4] In the spring of 1949 von Foerster wrote McCulloch, renowned for his generosity in helping younger men, to seek his help in finding a job in North America.[5] McCulloch found the Austrian a position at the University of Illinois; he also introduced von Foerster into the Macy group. Soon afterward, McCulloch and Mead asked von Foerster if he would serve as principal editor of the published transcripts. Although he had some misgivings because

English was not his native language, he agreed. With his name emblazoned on the title pages, the published transcripts are associated with him as much as with anyone.

It was not until the Macy Conferences had run their course, however, that von Foerster tried to develop more fully the epistemological implications of including the observer as part of the system. The punning title of his essay collection, *Observing Systems,* announces reflexivity as a central theme. "Observing" is what (human) systems do; in another sense, (human) systems themselves can be observed. The earliest essay ("On Self-Organizing Systems and Their Environments"), taken from a presentation given in 1960, shows von Foerster thinking about reflexivity as a circular dynamic that can be used to solve the problem of solipsism. How does he know other people exist, he asks. Because he experiences them in his imagination. His experience leads him to believe that other people similarly experience him in their imaginations. "If I assume that I am the sole reality, it turns out that I am the imagination of somebody else, who in turn assumes that *he* is the sole reality."[6] In a circle of intersecting solipsisms, I use my imagination to conceive of someone else and then of the imagination of that person, in which I find myself reflected.[7] Thus I am reassured not only of the other person's existence but of my own as well. Although charmingly posed, the argument is logically nonsensical, for there is no assurance that other imaginations are conceiving of me any more than I am conceiving of them. Maybe I am thinking not about von Foerster but about a Big Mac. That even a fledgling philosopher could reduce the argument to shreds is perhaps beside the point. Von Foerster himself seemed to recognize that the argument was the philosophical equivalent to pulling a rabbit from a hat, for he finally "solves" the paradoxical circling between solipsistic imaginations by asserting what he was to prove, namely the existence of reality.

Although the argument is far from rigorous, it is interesting for the line of thought it suggests. Its implications are illustrated by a cartoon (drawn by Gordon Pask at von Foerster's request) of a man in a bowler hat, in whose head is pictured another man in a bowler hat, in whose head is yet another man in a bowler hat.[8] The potentially infinite regress of men in bowler hats does more than create an image of the observer who observes himself by observing another. It also visually distinguishes the observer as a discrete system inside the larger system of the organism. In the aftermath of the Macy Conferences, one of the central problems with reflexivity was how to talk about it without falling into solipsism or resorting to psychoanalysis. The message from the Macy Conferences was clear: if reflexivity was to be credible, it had to be insulated against subjectivity and presented in a context in which it had at least the potential for rigorous (preferably mathematical) formulation. As Norbert Wiener was later to proclaim, "Cybernetics is nothing if it is not mathematical."[9] Distinguishing the observer as a system separate from the organism was one way to make reflexivity more manageable, for it reduced the problem of the observer to a problem of communication among systems.

Throughout the 1960s, von Foerster remained convinced of the importance of reflexivity, and he experimented with various ways to formulate it. A breakthrough occurred in 1969, when he invited Maturana to speak at a conference at the University of Illinois. There Maturana unveiled his ideas about treating "cognition as a biological phenomenon."[10] The power of Maturana's theory must have deeply affected von Foerster, for his thinking about reflexivity takes a quantum leap in complexity after this date. The increased sophistication can be seen in his 1970 essay "Molecular Ethology: An Immodest Proposal for Semantic Clarification," in which he criticizes behaviorism by making the reflexive move of turning the focus from the observation back onto the observer. Behaviorism does not demonstrate that animals are black boxes that give predictable outputs for given inputs, he argues. Rather, behaviorism shows the cleverness and power of the experimenter in getting animals to behave as such. "Instead of searching for mechanisms in the environment that turn organisms into trivial machines, we have to find the mechanisms within the organisms that enable them to turn their environment into a trivial machine."[11] Here reflexivity moves from men in bowler hats to the beginning of a powerful critique of objectivist epistemology. By 1972, von Foerster had been so thoroughly convinced by Maturana's theory that one of the latest essays in *Observing Systems*, "Notes on an Epistemology of Living Things" (pp. 258–71), recasts the theory in the form of a circular set of numbered quasi-mathematical propositions, in which the last repeats the first.

To trace the evolution of Maturana's epistemology, let us turn now to the seminal paper "What the Frog's Eye Tells the Frog's Brain." In it, Maturana and his coauthors demonstrate that the frog's sensory receptors speak to the brain in a language highly processed and species-specific. To arrive at this conclusion, the authors implanted microelectrodes in a frog's visual cortex to measure the strength of neural responses to various stimuli. At this point the frog's brain became part of a cybernetic circuit, a bioapparatus reconfigured to produce scientific knowledge. Strictly speaking, the frog's brain had ceased to belong to the frog alone. I will therefore drop the possessive and follow the authors by referring to the frog's brain simply as "the brain" (a phrase that eerily echoes the title of Norbert Wiener's short story discussed in chapter 5). From the wired-up brain, the researchers discovered that small objects in fast, erratic motion elicited maximum response, whereas large, slow-moving objects evoked little or no response. It is easy to see how such perceptual equipment is adaptive from the frog's point of view, because it allows the frog to perceive flies while ignoring other phenomena irrelevant to its interests. The results implied that the frog's perceptual system does not so much register reality as *construct* it. As the authors noted, their work "shows that the [frog's] eye speaks to the brain in a language already highly organized and interpreted instead of transmitting some more or less accurate copy of the distribution of light upon the receptors."[12] The work led Maturana to the maxim fundamental to his epistemology: "Everything said is said by an observer" (*AC*, p. xxii). No wonder the article was quickly recognized as a classic, for it blew a frog-sized hole in realist epistemology.

Despite the potentially radical implications of the article's *content*, however, its *form* reinscribed conventional realist assumptions of scientific discourse. The results are reported in an objectivist rhetoric that masks the fact they are interpreted through the sensory and cognitive interfaces of embodied researchers, whose perceptions were at least as transformative as the frog's. Years later, Maturana would recall that he and Lettvin continued to work in an objectivist framework even as that framework was being called into question by their research. In the preface to *Autopoiesis and Cognition: The Realization of the Living*, Maturana recalled: "When Jerry Y. Lettvin and I wrote our several articles on frog vision . . . we did it with the implicit assumption that we were handling a clearly defined cognitive situation: there was an objective (absolute) reality, external to the animal, and independent of it (not determined by it), which it could perceive (cognize). . . . But even there the epistemology that guided our thinking and writing was that of an objective reality independent of the observer" (*AC*, p. xiv). Faced with this inconsistency, Maturana had a choice. He could continue to work within the prevailing assumptions of scientific objectivity, or he could devise a new epistemology that would construct a worldview consistent with what he thought the experimental work showed.

The break came with his work on color vision in other animals, including birds and primates. He and his coauthors (not the Macy group this time) found they could not map the visible world of color onto the activity of the nervous system.[13] There was no one-to-one correlation between perception and the world. They could, however, correlate activity in an animal's retina with its *experience* of color. If we think of sense receptors as constituting a boundary between outside and inside, this implies that organizationally, the retina matches up with the inside, not the outside. From this and other studies, Maturana concluded that perception is not fundamentally representational. He argued that to speak of an objectively existing world is misleading, for the very idea of a world implies a realm that preexists its construction by an observer. Certainly there is something "out there," which for lack of a better term we can call "reality." But it comes into existence for us, and for all living creatures, *only through interactive processes determined solely by the organism's own organization.* "No description of an absolute reality is possible," he and Varela wrote in *Autopoiesis and Cognition,* for such a description "would require an interaction with the absolute to be described, but the representation that would arise from such an interaction would necessarily be determined by the autopoietic organization of the observer . . . hence, the cognitive reality that it would generate would unavoidably be relative to the observer" (*AC*, p. 121). Thus he was led to a premise fundamental to his theory: living systems operate within the boundaries of an organization that closes in on itself and leaves the world on the outside.

With Varela, Maturana developed the implications of this insight in *Autopoiesis and Cognition.* He arrived at his theory, he explains in the introduction, by deciding to treat "the activity of the nervous system as determined by the nervous system itself, and not by the external world; thus the external world would have only a triggering role in the release of the internally-determined activity of the nervous system" (*AC*, p. xv). His key insight was to realize that if the action of the nervous system is determined by its organization, the result is a circular, self-reflexive dynamic. A living system's organization causes certain products to be produced, for example, nucleic acids. These products in turn produce the organization characteristic of that living system. To describe this circularity, he coined the term *autopoiesis* or self-making. "It is the circularity of its organization that makes a living system a unit of interactions," he and Varela wrote in *Autopoiesis and Cognition,* "and it is this circularity that it must maintain in order to remain a living system and to retain its identity through different interactions" (*AC*, p. 9). Building on this premise of autopoietic closure, Maturana developed a new and startlingly different account of how we know the world.[14]

What is this account? One path into it is to regard the account as an attempt to counteract anthropomorphic projection by clearly distinguishing between two domains of description. On the one hand, there is what one can say about the circularity of autopoietic processes in themselves, taking care not to attribute to them anything other than what they exhibit. On the other hand, there are the inferences that observers draw when they place an autopoietic system in the context of an environment. Seeing system and medium together over a period of time, observers draw connections between cause and effect, past and future. But these are the observers' inferences; they are not intrinsic to the autopoietic processes in themselves. Let's say I see a blue jay flash through the trees and settle on the birdbath. I may think, "Oh, it's getting a drink." Other species, for example those lacking color vision, would react to this triggering event with different constructions. A frog might notice the quick, erratic flight but be oblivious to the blue jay at rest. Each living system thus constructs its environment through the "domain of interactions" made possible by its autopoietic organization. What lies outside that domain does not exist for that system. Maturana, realizing that he was fighting a long tradition of realist assumptions deeply embedded in everyday language, developed an elaborate vocabulary as a prophylactic against having anthropomorphism creep back in. The necessity of finding a new language in which to express his theory was borne home to him during the student revolution in Chile in May 1968. It was then, he wrote in *Autopoiesis and Cognition,* that he discovered that "language was a trap, but the whole experience was a wonderful school in which one could discover how mute, deaf and blind one was . . . one began to listen and one's language began to change; and then, but only then, new things could be said" (*AC,* p. xvi).

Shortly we will analyze places where Maturana, like the participants in the Macy Conferences, seems unable to escape from the tar baby of self-reflexive language. For the moment, however, let us explore the "new things" he tried to say. No doubt the cumbersome—many would not hesitate to call it tortured—quality of his language will be immediately apparent to the reader.[15] Before we judge it harshly, however, we should remember that Maturana was attempting nothing less than to give a different account of how we know the world. Since it is partly through language that humans bring worlds into being for themselves, he was in the impossible position of pulling himself up by his own bootstraps, trying to articulate the new by using the only language available, the *lingua franca* whose meanings had long ago settled along lines very different from those he was trying to envision.

We can start with that most problematic of constructions, the observer. From Maturana's point of view, the "fundamental cognitive operation that an observer performs is the operation of distinction" (*AC,* p. xxii). Influenced by G. Spencer-Brown, Maturana (and even more so Varela in his work *Principles of Biological Autonomy*)[16] sees the operation of distinction as marking space so that an undifferentiated mass is separated into an inside and an outside or, in Maturana's terminology, into a unity and a medium in which the unity is embedded. Unities distinguished by the observer can be of two types, simple and composite. A simple unity "only has the properties with which it is endowed by the operations of distinction through which it becomes separated from a background." Composite unities, by contrast, have "structure and organization," (*AC*, p. xx), terms that Maturana uses in special senses and that require further explanation.

A composite unity's organization is the complex web of all possible relationships that can be realized by the autopoietic processes as they interact with one another. When Maturana speaks of a system's organization, he does not mean how this web of relationships might be described in abstract form. Rather, he intends *organization* to denote the relations actually instantiated by the autopoietic unity's circular processes. Structure, by contrast, is the particular instantiation that a composite unity enacts at a particular moment. For example, when a female human is born, she has one kind of structure; when she enters puberty, she has another; if she contracts a disease, she has still another. But throughout her lifetime, her organization remains the same: that which is characteristic of a living human. Only when death occurs does her organization change. According to Maturana, this ability of living organisms to conserve their autopoietic organization is the necessary and sufficient condition for them to count as living systems. All living systems are autopoietic, and all physical systems, if autopoietic, can be said to be living (*AC*, p. 82). Thus life and autopoiesis are coextensive with one another. Here's how that proposition sounds in Maturana's terminology. "The living organization is a circular organization which secures the production or maintenance of the components that specify it in such a manner that the product of their functioning is the very same organization that produces them" (*AC*, p. 48).

To account for a system's embeddedness in an environment, Maturana uses the concept of structural coupling. All living organisms must be structurally coupled to their environments to continue living; humans, for example, have to breathe air, drink water, eat food (*AC*, pp. x–xi). In addition, systems may be structurally coupled to each other. For example, a cell within my body may be considered as a system in itself, but it relies for its continued existence on its structural coupling to my body as a whole. Here again the role of the observer becomes important, for Maturana is careful to distinguish between the triggering effect that an event in the medium has on a system structurally coupled with it and the causal relationship that observers construct in their mind when they perceive the system interacting with the environment. When my bird dog sees a pigeon, I may think, "Oh, he's pointing because he sees the bird." But in Maturana's terms, this is an inference I draw from my position in the "descriptive domain" of a human observer (*AC,* p. 121). From the viewpoint of the autopoietic processes, there is only the circular interplay of the processes as they continue to realize their autopoiesis, always operating in the present moment and always producing the organization that also produces them. Thus, time and causality are not intrinsic to the processes themselves but are concepts inferred by an observer. "The present is the time interval necessary for an interaction to take place," Maturana and Varela wrote. "Past, future and time exist only for the observer" (*AC,* p. 18).

Information, coding, and teleology are likewise inferences drawn by an observer rather than qualities intrinsic to autopoietic processes. In the autopoietic account, there are no messages circulating in feedback loops, nor are there even any genetic codes. These are abstractions invented by the observer to explain what is seen; they exist in the observer's "domain of interactions" rather than in autopoiesis itself. "The genetic and nervous system are said to code information about the environment and to represent it in their functional organization. This is untenable," Maturana and Varela noted. "The genetic and nervous systems code processes that specify series of transformations from initial states, which can be decoded only through their actual implementation, not *descriptions* that the observer makes of an environment which lies exclusively in *his* cognitive domain" (*AC,* p. 53). Similarly, "the notion of information refers to the observer's degree of uncertainty in his behavior within a domain of alternatives defined by him, hence the notion of information only applies within his cognitive domain" (*AC,* p. 54). The same applies to teleology. "A living system is not a goal-directed system; it is, like the nervous system, a stable state-determined and strictly deterministic system closed on itself and modulated by interactions not specified by its conduct. These modulations, however, are apparent as modulations only for the observer who beholds the organism or the nervous system externally, from his own conceptual (*descriptive*) perspective, as lying in an environment and as elements in his domain of interactions" (*AC,* p. 50).

One implication of letting go of causality is that systems always behave as they should, which is to say, they always operate in accord with their structures, whatever those may be. In Maturana's world, my car always works, whether it starts or not, because it operates only and always in accord with its structure at the moment. It is I, as an observer, who decides that my car is not working because it will not start. Such "punctuations," as Maturana and Varela call them, belong to the "domain of the observer" (AC, pp. 55–56). Because they are extrinsic to the autopoietic processes, they are also extrinsic to the biological description that Maturana aims to give of life and cognition. In an important essay entitled "Biology of Language," Maturana remarks that the "operation of a structure-determined system is necessarily perfect: that is, it follows a course determined only by neighborhood relations in its structure and by nothing else. It is only in a referential domain, such as the domain of behavior, that an observer can claim that an error has occurred when his or her expectations are not fulfilled."[17]

To assess the changes that the autopoietic view entails, let us turn now to compare its account of living systems with that given by first-wave cybernetics. A convenient focal point for the comparison of the two theories is liberal humanism, where their implications for the construction of subjectivity will become apparent. Having traced these implications, we will then consider the impact of second-wave cybernetics on the entwined stories we have been following: the reification of information, the construction of the cyborg, and the transformation of the human into the posthuman.

Reconfiguring the Liberal Humanist Subject

As we saw in chapter 4, Norbert Wiener had a complex relation to the liberal humanist subject. Father of a theory that put humans and machines into the same category, he was nevertheless committed to creating a cybernetics that would preserve autonomy and individuality. His nightmare was the human reduced to a cog in a rigid machine, losing the flexibility and autonomous functioning that Wiener regarded as the birthright of a cybernetic organism. Echoes of this cybernetic tradition linger in Maturana's description of composite unities as "autopoietic machines" (AC, p. 82). Fully aware of the implications of calling autopoietic systems "machines," Maturana makes clear that there is nothing in his theory to prohibit artificial systems from becoming autopoietic unities. "If living systems were machines, they could be made by man," he and Varela point out (AC, p. 83). They pooh-pooh the idea that life cannot or should not be created by humans. "There seems to be an intimate fear that the awe with respect to life and the living would disappear if a living system could be not only reproduced, but designed by man. This is nonsense. The beauty of life is not a gift of its inaccessibility to our understanding" (AC, p. 83). When Maturana objects to first-wave projects that attributed biological properties to machines, his criticism addresses how life is defined, not the idea that machines can be alive. For example, he criticizes John von Neumann's proposal to create a self-reproducing machine by arguing that von Neumann modeled descriptions that biologists had made rather than autopoietic processes in themselves. Von Neumann modeled inferences about "what appeared to take place in the cell in terms of information content, program and coding. By modeling the processes expressed in these descriptions he produced a machine that could make another machine but he did not model the phenomena of cellular reproduction, heredity and genetics as they take place in living systems."[18]

This critique points to an important change between Maturana's position and that announced by Wiener and his coauthors in their cybernetic manifesto. Whereas the latter argued that it is the system's *behavior* that counts, Maturana argues that it is the *autopoietic processes generating behavior* that count. As we have seen, first-wave researchers concentrated on building artifacts that would behave as cybernetic mechanisms: John von Neumann's self-reproducing machines; Claude Shannon's electronic rat; Ross Ashby's homeostat. By contrast, Maturana and others in the second wave look to systems instantiating processes that count as autopoietic. The homeostat might behave cybernetically, for example, but it does not count as an autopoietic machine because it does not produce the components that produce its organization. Perhaps because of this emphasis on process, autopoietic theory has proven readily adaptable to the analysis of social systems. In autopoietic theory, the machine of interest is much more likely to be the state than Robocop or Terminator.[19]

In first-wave cybernetics, questions of boundary formation were crucial to its constructions of subjectivity. Boundary questions are also important in autopoietic theory. Wiener's anxieties recirculate in discussions about what happens when one autopoietic unity is encapsulated within the boundaries of a larger autopoietic unity, for example when a cell functions as part of a larger machine. Can the cell continue to function as an autonomous entity, or must its functioning be subordinated to the larger unity? To distinguish these two cases, Maturana introduces the term *allopoietic*. Whereas autopoietic unities have as their only goal the continuing production of their autopoiesis, allopoietic unities have as their goal something other than producing their organization. When I drive my car, its functioning is subordinated to the goals I set for it. Instead of the pistons using their energy to repair themselves, for example, they use their energy to turn the drive shaft so that I can get to the store. I function autopoietically, but the car functions allopoietically.

We saw in chapter 4 that cybernetic boundary questions often involve deep ethical and psychological issues, such as those that troubled Wiener when he envisioned the dissolution of the autonomous liberal subject. In autopoietic theory, one of the principal effects of autopoiesis is to secure for a living system the crucial qualities of autonomy and individuality. Consequently, boundary issues are often played out in discussions of how much autonomy autopoietic systems will retain for themselves and how much autonomy they will demand from the systems with which they are structurally coupled. The distinction between allopoietic and autopoietic gives Maturana a way to talk about power struggles within society. In autopoietic theory, the idea corresponding to Wiener's horror at a man being forced to act as a cog in a machine is a system that is capable of autopoiesis being forced instead to function allopoietically, especially for humans. Maturana's ideal is a human society in which one would "see all human beings as equivalent to oneself, and to love them . . . without demanding from them a larger surrender of individuality and autonomy than the measure that one is willing to accept for oneself while integrating it as an observer" (AC, p. xxix). Such a society, he adds, "is in its essence an anarchist society, a society made for and by observers that would not surrender their condition of observers as their only claim to social freedom and mutual respect" (AC, p. xxx). In such rhetoric, we can easily hear a reinscription of liberal humanist values, even though the epistemology that Maturana advocates is very different from that which gave rise to the Enlightenment subject.

Yet it would be a mistake to think that Maturana's radicalism can be so easily recuperated back into liberal subjectivity. The split between his position and liberal philosophy becomes obvious when questions of objectivity arise. Consider, for example, his insistence that ethics cannot be separated from scientific inquiry. Instead of accepting the proposition that the scientist simply reports what he or she sees and in this sense remains aloof from ethical considerations, Maturana envisions autopoietic theory as a way to reconnect ethics and science. Emphasizing that autonomy always takes place in the context of structural coupling, autopoiesis rejects the objectivism that drives a wedge between the scientist-observer and the world being observed. For Maturana, observation does not mean that the observer remains separate from what is being observed; on the contrary, the observer can observe only because the observer is structurally coupled to the phenomenon she sees. Expanded to social ethics, this implies "in man as a social being . . . all actions, however individual as expressions of preferences or rejections, constitutively affect the lives of other human beings and, hence, have ethical significance." Structural coupling requires that human beings "as components of a society, necessarily realize their individual worlds and contribute to the determination of the individual worlds of others" (*AC*, p. xxvi).

Although Maturana thus follows in the liberal tradition of cyberneticians like Wiener in placing a high value on the autonomous individual, the meaning of autonomy has undergone significant change. Autonomy as Maturana envisions it is not consistent with laissez-faire capitalism; it is not consistent with the idea that each person is out for himself and devil take the hindmost; and it is not consistent with the ethical position that a scientist could undertake a research program without being concerned about how the results of the research would be used. In these respects, the individualism and autonomy that Maturana champions challenge the premises embodied in liberal subjectivity at least as much as they reinscribe those premises.

To explore further how liberal subjectivity is both contested and reinscribed in autopoietic theory, let us turn now to Maturana's account of the observer. Nowhere does Maturana depart more clearly from first-wave philosophies than in his insistence that the observer must be taken into account. *"The observer is a living system and any understanding of cognition as a biological phenomenon must account for the observer and his role in it"* (*AC*, p. 48). The act of observation necessarily entails reflexivity, for one of the systems that an observer can describe is the observer as an autopoietic system. But reflexivity as Maturana envisions it is very different from the psychoanalytic reflexivity that Lawrence Kubie introduced into the Macy Conferences (see chapter 3). In contrast to Kubie's emphasis on unconscious symbolism, Maturana's observer does not have psychological depth or specificity. Rather, Maturana's observer is more like the observer that Albert Einstein posits in the special theory of relativity. The one who sees is always called simply "the observer," without further specification, implying that any individual of that species occupying that position would see more or less the same thing. Although the observer's perceptions construct reality rather than passively perceive it, for Maturana this construction depends on *positionality* rather than *personality*. In autopoietic theory, the opposite of objectivism is not subjectivism but relativism.

If the interplay between conscious and unconscious processes is not important for Maturana, how is the observer produced? The observer begins as an autopoietic unity, as all living systems are said to be. As a particular kind of autopoietic unity capable of becoming an observer, the observer-system can generate representations of its own interactions. When the system recursively interacts with these representations, it becomes an observer. The system can then recursively generate representations of these representations and interact with them, as when an observer thinks, "I am an observing system observing itself observing." Each twist of this reflexive spiral adds additional complexity, enlarging the domain of interactions that specify the world for that autopoietic unity. Maturana and Varela summarize the situation thus in *Autopoiesis and Cognition:* "We become *observers* through recursively generating representations of our interactions, and by interacting with several representations simultaneously we generate relations with the representations of which we can then interact and repeat this process recursively, thus remaining in a domain of interactions always larger than that of the representation" (*AC*, p. 14). Reflexivity is thus fundamental to Maturana's account not only because the autopoietic operations of a unity specify for it a world but also because the system's reflexive doubling back on its own representations generates the human subject as an observer.

What about consciousness? Maturana seldom uses this word, preferring to talk instead about "thinking" and "self-consciousness." Thinking occurs in a state-determined nervous system when neurophysiological processes can interact "with some of its own internal states as if these were independent entities." This recursive circling "corresponds to what we call thinking" (*AC*, p. 29). To get from "thinking" to "self-consciousness" requires language, according to Maturana. In the same way that perception does not consist of information from the environment passing into the organism, so language does not consist of someone giving information to someone else. Rather, when an observer uses language, this acts as a trigger for the observer's interlocutor, allowing the interlocutor to establish an orientation within his or her domain of interactions similar to the orientation of the speaker. Only when two entities have largely overlapping domains—for example, when they are both humans sharing similar cultures and beliefs—is it possible for them to achieve the illusion that communication between them has occurred.

From this description, it is apparent that Maturana explains language by simply extending to the linguistic realm the same ideas and terminology he uses to explain perception—an explanation that, in my view, fails to account for some of the distinctive features of language. Shortly we will have an opportunity to look critically at this view of language. For the moment, it permits us to understand Maturana's view of self-consciousness. Self-consciousness arises when the observer "through orienting [linguistic] behavior can orient himself towards himself, and then generate communicative *descriptions* that orient him toward his *description* of this self-orientation." The observer generates self-consciousness, then, when he endlessly describes himself describing himself. "Thus discourse through communicative *description* originates the apparent paradox of self-description: *self-consciousness,* a new domain of interactions" (*AC,* p. 29). Because Maturana understands self-consciousness solely in linguistic terms, seeing it as an emergent phenomenon that arises from autopoietic processes when they recursively interact with themselves, consciousness for him becomes a epiphenomenon rather than a defining characteristic of the human as an autopoietic entity. The activity of cerebration represents only a fraction of the total autopoietic processes, and self-consciousness represents only a fraction of cerebration. Thus the theory implicitly assigns to consciousness a much more peripheral role than it does to autonomy and individualism. In this respect, autopoietic theory points toward the posthuman even as it reinscribes the autonomy and individuality of the liberal subject.

The complex relation of autopoietic theory to liberal humanism becomes even more apparent when we ask how the theory attempts to establish a foundational ground for itself. As we saw in chapter 1, liberal humanism (in C. B. Macpherson's reading of it) grounds itself on the notion of possessive individualism, the idea that subjects are individuals first and foremost because they own themselves. The equivalent foundational premise in autopoietic theory is the idea that living systems are living because they instantiate organizational closure. It is precisely this closure that guarantees the subject will operate as an autonomous individual. But how is it that Maturana (or anyone else) knows that organizational closure exists? Is the claim that autopoietic closure is *intrinsically* a feature of living systems, or is it how a human observer *perceives* living systems, including itself? This question lies coiled around the brainstem of autopoietic theory, layered into its evolutionary history through its founding distinctions between qualities intrinsic to autopoietic processes and qualities attributed to them by an observer. If the theory says that the observer creates the system by drawing distinctions, it risks undercutting the ontological primacy of organizational closure. If it says that autopoietic processes are an essential feature of reality, it risks undercutting its epistemological radicalism. Faced with this Scylla and Charybdis, Maturana at first steered toward relativism and then, as its dangers loomed closer, changed course and steered toward the absolutism of autopoietic processes existing in reality as such.

So in "Biology of Cognition," the earlier essay in *Autopoiesis and Cognition,* Maturana often wrote as if it is the observer's action that distinguishes an autopoietic unity from its background or medium. "Although a distinction performed by an observer is a cognitive distinction and, strictly, the unity thus specified exists in his cognitive domain as a description, *the ob-*

server in his discourse specifies a metadomain of descriptions from the perspective of which he established a reference that allows him to speak as if a unity... existed as a separate entity" (*AC,* p. xxii, emphasis added). This implies the autopoietic unity exists as a distinction that is performed by the observer rather than as an entity that could exist in the absence of an observer. However, in "Autopoiesis: The Organization of the Living," the second and later essay in *Autopoiesis and Cognition,* Maturana and Varela wrote as if an autopoietic unity has the ability to constitute itself independent of an observer. Autopoietic machines, through *"their interactions and transformations ... continuously regenerate and realize the network of processes (relations) that produced them,"* in the process constituting themselves *"as a concrete unity in the space in which they (the components) exist by specifying the topological domain of* [the autopoietic machine's] *realization of such a network"* (*AC,* p. 79). Here the operation of the autopoietic entity itself—rather than a distinction drawn by an observer—creates the space in which the entity exists. Even more explicit is the claim that individuality comes from the processes themselves rather than from the actions of an observer. "Autopoietic machines have individuality; that is, by keeping their organization as an invariant through its continuous production they actively maintain an identity which is independent of their interactions with an observer" (*AC,* p. 80).

It is not surprising that the issue continues to be debated in autopoietic theory, for it admits of no easy solution. In Maturana's desire to found autopoiesis on something more than an observer's distinction, we can see him trying to pull away from the tar baby of his own reflexive language. Relevant for our purposes is not so much the resolution to this dilemma (as if there could be a definitive resolution!) or even the demonstration that the theory's founding moves make it vulnerable to deconstructive critique. Rather, the important point here is that the foundational ground for establishing the subject's autonomy and individuality has shifted from self-possession, with all of its implications for the imbrication of the liberal subject with industrial capitalism. Instead, these privileged attributes are based on organizational closure (the system closes on itself, by itself) or on the reflexivity of a system recursively operating on its own representations (the observer's distinctions close the system). Closure and recursivity, then, play the foundational role in autopoietic theory that self-possession played in classic liberal theory. The emphasis on closure is visually apparent in the computer simulations, called tessellation automata, that Varela created to illustrate autopoietic dynamics. In contrast to the artificial-life programs that will be discussed in chapter 9, the point of tessellation simulations is to find out how boundaries close on themselves, how they are maintained when interacting with other tessellation automata, and how and when boundaries break down, which in autopoietic theory is equivalent to death. In this description we see the affinity of autopoiesis not for industrial capitalism (which Maturana frequently excoriates) but for utopian anarchy. Autonomy is important not because it serves as the (paradoxical) foundation for market relations but because it establishes a sphere of existence for the individual, a location from which the subject can ideally learn to respect the boundaries that define other autopoietic entities like itself. This emphasis on closure, autonomy, and individuality also changes what count as primary concerns. When the existence of the world is tied to an observer, the urgent questions revolve around how to maintain boundaries intact and still keep connection with a world that robustly continues to exist regardless of what we think about it.

These changes from liberal humanism also bring with them limitations that are distinctively different from those of first-wave cybernetics. Whereas first-wave philosophies tended to obscure the importance of embodiment and the observer, autopoietic theory draws its strength precisely from its emphasis on these attributes. Its Achilles' heel, by contrast, is accounting for living systems' explosive potential for transformation. The very closure that gives autopoietic theory its epistemological muscle also limits the theory, so that it has a difficult time accounting for dynamic interactions that are not circular in their effects. A prime example, in my view, is the convoluted and problematic way that Maturana treats language. Consistent with his emphasis on circularity, he prefers to talk not about language but "languaging," a process whereby observers, acting solely within their own domain of interactions, provide the triggers that help other observers similarly orient themselves within their domains. Autopoietic theory sees this exchange as a coupling between two independent entities, each of which is formed only by its own ongoing autopoietic processes. As this description shows, the theory is constantly in danger of solipsism, a danger it both acknowledges and attempts to avoid by protesting that it is *not* solipsistic. The main reason the theory adduces for not being solipsistic is its acknowledgment of "structural coupling," the phrase used to denote an organism's interaction with the environment. Even if we grant that this move rescues the theory from solipsism, the theory still seriously understates the transformative effects that language has on human subjects. We have only to recall the term that Maturana employs for a language-using subject—"the observer"—to see how curiously inert and self-enclosing is his view of language.

What drops from view in Maturana's account is the *active* nature of linguistic interactions. Researchers from Jean Piaget on have shown that a child's neural hardware continues to develop after birth in conjunction with the linguistic and social environment in which the child is embedded. In light of this work, it is misleading to talk about the process of active shaping through language simply as an entity "orienting" itself with the aid of an environmental trigger. To appreciate just how active this process is, we can look at instances where it has been short-circuited and the child thus consequently fails to develop normally. In *Mindblindness: An Essay on Autism and Theory of Mind*, Simon Baron-Cohen argues this is what happens with autistic children.[20] Somehow the shaping mechanisms fail to direct neural development, and as a result the child is unable to create an internal scenario that would explain why others act as they do. For such children, Baron-Cohen argues, the world of social interactions is chaotic and unpredictable because they suffer from "mindblindness," an inability to imagine for others the emotions and feelings they themselves have. Autopoietic theory, in its zeal to construct an autonomous sphere of action for self-enclosing entities, formulates a description that ironically describes autistic individuals more accurately than it does normally responsive people. For the autistic person, the environment is indeed merely a trigger for processes that close on themselves and leave the world outside.

In the next section, we will turn to a discussion of how autopoietic theory treats evolution. Like language, evolution represents another area where Maturana's version of autopoietic theory fails to come to terms with the dynamic, transformative nature of the interactions between living systems and their environments. From there we will explore the split that develops between Maturana and Varela. While Maturana continued to replicate his original formulation of the theory, Varela and others became increasingly interested in changing the theory so that it could better account for dynamic interactions. Keeping many of the central insights of autopoietic theory, Varela added new material and reworked some assumptions in the seriated pattern of innovation and replication we have seen at work in other sites. One effect of these changes was to allow elements of autopoietic theory to be integrated into contemporary cognitive science and especially artificial life, which will be the focus of my discussion of third-wave cybernetics in chapter 9.

At this point, a summary may be useful of how autopoietic theory contributes to our evolving stories of (1) the reification of information, (2) the construction of the cyborg, and (3) the transformation of the human into the posthuman. First, whereas first-wave cybernetics played a large role in divesting information of its body, autopoietic theory draws attention to the fact that "information," so defined, is an abstraction that has no basis in the physically embodied processes constituting all living entities. Autopoiesis thus swerves from the trajectory traced in chapters 3 and 4 with regard to information, insisting that information without a body does not exist other than as an inference drawn by an observer. Second, whereas first-wave cybernetics envisioned the cyborg mostly as an amalgam between the organic and the mechanical, autopoietic theory uses its expanded definition of life to speculate on whether social systems are alive. The paradigmatic cyborg for autopoiesis is the state, not the kind of mechanical human imagined by Bernard Wolfe or Philip K. Dick. Third, autopoietic theory preserves the autonomy and individuality characteristic of liberal humanism, but it sees thinking as a secondary effect that arises when an autopoietic entity interacts with its own representations. Self-consciousness, a subset of thinking, is relegated to a purely linguistic effect. The grounding assumptions for individuality shift from self-possession to organizational closure and the reflexivity of a system recursively operating on its own representations.

A status report, then: information's body is still contested, the empire of the cyborg is still expanding, and the liberal subject, although more than ever an autonomous individual, is literally losing its mind as the seat of identity.

Autopoiesis and Evolution

It is no accident that evolution is a sore spot for autopoietic theory, for the theory was designed to correct what Maturana and Varela saw as an overemphasis on evolution and reproduction as the defining characteristics of life. Over and over, they argue that evolution and reproduction are logically and practically subordinate to autopoiesis. "Reproduction and evolution are not essential for living organisms," they assert in *Autopoiesis and Cognition* (*AC*, p. 11). They are even more opposed to defining living organisms in terms of genetic code. As Varela made clear in a retrospective assessment, he and Maturana were consciously aware of wanting to provide an alternative account of life, an account that would *not* depend in any important way on the idea of a genetic code. "The notion of autopoiesis was proposed . . . with the intention of redressing what seemed to us to be a fundamental imbalance in the understanding of living organization." In correcting this imbalance, they had two interrelated goals. Along with creating a theory of the living that would debunk the current emphasis on DNA as the "master molecule" of life, they also wanted to insist on the holistic nature of living systems.[21]

Varela is willing to admit that perhaps they erred on the side of overemphasizing autopoiesis at the expense of genetics. By contrast, Maturana became if anything more confirmed in his opposition as time went on. Many critics, including Richard Lewontin, Evelyn Fox Keller, Lily Kay, Richard Doyle, and others, have commented on the distortions created in modern biology by the present overemphasis on DNA.[22] But few go as far as Maturana. In the 1980 article "Autopoiesis: Reproduction, Heredity, and Evolution," a recapitulation of autopoietic theory, he wrote, "I claim that nucleic acids do not determine hereditary and genetic phenomena in living systems, and that they are involved in them, like all other cellular components, according to the particular manner in which they integrate the structure of the living cell and participate in the realization of its autopoiesis."[23] Let us grant that modern biology overemphasizes the role of DNA and that DNA is, as Maturana points out in this passage, only one of many cellular components involved in reproduction. Does he nevertheless go too far in the other direction by insisting that everything be subordinated to autopoiesis?

The problems created by subordinating everything to autopoiesis can be seen in *The Tree of Knowledge,* an account of autopoiesis written for a general audience.[24] As the opening diagram indicates, Maturana and Varela envision each chapter leading into the next, with the final one coming back to the beginning, so that the form of the book recapitulates the circularity of autopoietic theory. "We shall follow a rigorous conceptual itinerary," they announce in the introduction, "wherein every concept builds on preceding ones, until the whole is an indissociable network" (p. 9). In *Autopoiesis and Cognition,* Maturana commented that he and Varela were unable to agree on how to contextualize the theory, so he wrote the introduction by himself. Now, seven years later, Varela is less his student and more an accomplished figure in his own right. This is the last work the two men will coauthor together; Varela has already begun to head in a different direction. The divergences in their viewpoints are accommodated through a clever visual device. Certain key ideas are separated from the text and put into boxes. Each box has a cartoon figure representing the speaker. Maturana's figure wears heavy glasses and is noticeably older than Varela's, so it is easy to identify which is which. Sometimes Maturana's figure authorizes the boxed comments, sometimes Varela's, and sometimes both together. Even without the boxes, it is not difficult to discern that Varela's voice is stronger in *The Tree of Knowledge* than in *Autopoiesis and Cognition.*

I take Varela's Buddhist orientation to be the inspiration behind what the authors announce as a central idea, "all doing is knowing and all knowing is doing" (p. 27). They illustrate the concept by constructing the book as a circle, starting their discussion with unicellular organisms (first-order systems), progressing to multicellular organisms with nervous systems (second-order systems), and finally coming to cognitively aware humans who interact through language (third-order systems). Pointing out that humans in turn are composed of cells, they close the circle by nesting first- and second-order systems within third-order systems, thus joining the doing of autopoiesis with the knowing of cognitively aware creatures. Autopoiesis is the governing idea connecting systems at all levels, from the single cell to the most complex thinking being. "What defines [living systems] is their autopoietic organization, and it is in this autopoietic organization that they become real and specify themselves at the same time" (p. 48). Traversing this path, the "doing" of the reader—the linear turning of pages during the reading—is to become a kind of "knowing" as the reader experiences the organization characteristic of autopoiesis through a textual structure that circles back on itself.

The problem comes when the authors try to articulate this circular structure together with evolutionary lineages. In evolution, lineage carries the sense both of continuity (traced far enough back, all life originates in single-cell organisms) and of qualitative change (different lines branch off from one another and follow separate evolutionary pathways). Whereas in autopoiesis the emphasis falls on circular interactions, in evolution lines proliferate into more lines as speciation takes place through such mechanisms as genetic diversity and differential rates of reproductive success. The tension between evolutionary lines of descent and autopoietic circularity becomes apparent in the authors' claim that autopoiesis is conserved at every point as organisms evolve. To describe the changes taking place, the authors use the term "natural drift." There seems to be a natural drift in "natural drift," however, and in later passages "natural drift" becomes "structural drift." If structure changes, what does it mean to say that autopoiesis is conserved? Here they fall back on the structure/organization distinction that they had previously used in *Autopoiesis and Cognition*. "*Organization* denotes those relations that must exist among the components of a system for it to be a member of a specific class. *Structure* denotes the components and relations that actually constitute a particular unity and make its organization real" (p. 47). Interestingly, they use a mechanical rather than a biological analogy to illustrate the distinction. A toilet's parts can be made of wood or plastic; these different materials correspond to differences in structure. Regardless of the material used, however, the toilet will still be a toilet if it has a toilet's organization. The analogy is strangely inappropriate for biology. All life is based on the same four nucleotides; hence for living organisms, it is not the material that changes but the way the material is organized.

What does it mean, then, to say that autopoiesis is conserved? According to the authors, it means that organization is conserved. And what is organization? Organization is "those relations that must exist among the components of a system for it to be a member of a specific class" (p. 47). These definitions force one to choose between two horns of a dilemma. Consider the case of an amoeba and a human. Either an amoeba and a human have the same organization, which would make them members of the same class, in which case evolutionary lineages disappear because all living systems have the same organization; or else an amoeba and a human have different organizations, in which case organization—and hence autopoiesis—must not have been conserved somewhere (or in many places) along the line. The dilemma reveals the tension between the conservative circularity of autopoiesis and the linear thrust of evolution. Either organization is conserved and evolutionary change is effaced, or organization changes and autopoiesis is effaced.

The strain of trying to articulate autopoiesis with evolution is perhaps most apparent in what is not said. Molecular biology is scarcely mentioned and then only in contexts that underplay its importance—a choice consistent with Maturana's claim that heredity does not depend on nucleic acids. There is an additional problem in bringing up molecular biology, for any discussion of DNA coding would immediately reveal that the distinction between structure and organization cannot be absolute—and if this distinction goes, autopoiesis is no longer conserved in evolutionary processes. For if organization is construed to mean the biological classes characterized as species, then it is apparent that organization changes as speciation takes place. If organization means something other than species, then organization ceases to distinguish between different kinds of species and simply becomes the property of any living system. Conserving organization means conserving life, a fact that may be adequate for autopoiesis to qualify as a property of living systems but does nothing to articulate autopoiesis with evolutionary change.

The essential problem here is not primarily one of definitions, although the problem becomes manifest at these sites in the text because definitions are used to anchor the argument, which otherwise drifts off into such nebulous terms as "natural drift." Rather, the difficulties arise because of Maturana's passionate desire to have something conserved in the midst of continuous change. Leaving aside the problems with his explanation of structure and organization, that something is basically the integrity of a self-contained, self-perpetuating system that is operationally closed to its environment. In Maturana's metaphysics, the system closes on itself and leaves historical contingency on the outside. Even when he is concerned with the linear branching structures of evolution, he turns this linearity into a circle and tries to invest it with a sense of inevitability. Seen as a textual technology, *The Tree of Knowledge* is an engine of knowledge production that vaporizes contingency by continuously circulating it within the space of its interlocking assumptions.[25]

Nowhere is the divergence of Varela and Maturana since 1980 clearer than on this point. While Varela moved on to other issues and ways of thinking about them, Maturana continued to occupy essentially the same position and to use the same language as in *Autopoiesis and Cognition*. Clearly Maturana has a more intense and long-lasting commitment to the original formulation of autopoiesis than does Varela. Not coincidentally, Maturana regards himself as the father of the theory, whereas he sees Varela's role as more tangential. In a 1991 article titled "The Origin of the Theory of Autopoietic Systems," he claims credit as the creator of the theory and says that Varela was very much a collaborator who appeared on the scene after the basic ideas had been formulated. "Strictly, Francisco Varela did not contribute to the development of the notion of autopoiesis," Maturana wrote. "This notion was developed between 1960 and 1968. Francisco was my student as an undergraduate during the years 1966 and 1967 in Santiago, then he went to Harvard where he was from 1968 to 1970, when he returned to Chile to work with me in my laboratory in the Faculty of Sciences in Santiago." Although Varela's *Principles of Biological Autonomy* clearly shows that Varela did most of the actual computer work in creating tessellation automata, Maturana claims credit for this idea too. He wrote, "During the year 1972, I proposed one day to make a computer program that would generate an autopoietic system in a graphic space as the result of generating in that space certain elements like molecules."[26] In *Principles,* Varela acknowledged that Maturana was among those "who have influenced this book so pervasively" that their thought was woven into it throughout, but he also wrote in "Describing the Logic of the Living," his 1981 retrospective assessment of autopoiesis, that "the notion of autopoiesis was proposed by Humberto Maturana *and myself.*"[27] This jostling for position, especially when a theory has proven to be historically important, is of course common in almost every field, and particularly in scientific communities, where great emphasis is placed on being the first to discover something. I mention it here not in any way to diminish the contributions of either Maturana or Varela but to contextualize the fact that Varela moved on to other ways of thinking about autopoiesis while Maturana continued to write in much the same vein as when he had started.

The Voice of the Other: Varela and Embodiment

After *The Tree of Knowledge,* Varela increasingly moved away from the closure that remains a distinctive feature of autopoiesis. The change can be seen in "Describing the Logic of the Living: The Adequacy and Limitations of the Idea of Autopoiesis," his contribution to the important 1981 collection edited by Milan Zeleny: *Autopoiesis: A Theory of Living Organization.* While stressing that he continues to see autopoiesis as very valuable because it "pointed to a neglecting of *autonomy* as basic to the living individual," Varela also criticizes autopoiesis for going both too far and not far enough (p. 37). It went too far, in his view, in becoming a paradigm not just for biological organisms but for social systems as well. Insisting that autopoiesis should not be confused with organizational closure in general, he points out that "the definition of autopoiesis has some precision because it is based on the idea of *production* of components, and this notion of production cannot be stretched indefinitely without losing all of its power" (pp. 37–38). Although cells and animals clearly do physically produce the components that instantiate their organization, social systems do not. Departing from Maturana on this point, Varela would restrict autopoiesis to where, in his view, it is most applicable, to the "domain of cells and animals" (p. 38).

Autopoiesis did not go far enough in building a bridge between its approach and the first-wave emphasis on information flow, teleology, and behavior. "We did not take our criticism far enough to *recover* a non-naive and useful role of information notions in the descriptions of living phenomena," he wrote. Conceding that information, coding, and messages can be "valid explanatory terms," he suggests that they might serve as complementary modes of description for autopoiesis (p. 39). Although he continues to maintain that autopoiesis is logically *necessary* to a complete explanation, it may not be "sufficient to give a satisfactory explanation of living *phenomena* on both logical *and* cognitive grounds" (p. 44). "There was, evidently, a need in [*Autopoiesis and Cognition*] to overemphasize a neglected side of a polarity" (p. 39). To posit an analogous situation in literature, imagine trying to explain how to read a Shakespearean sonnet by starting out with a description of cellular processes. Logically, it is true that the behavior resulting in reading the sonnet has to originate in cellular processes, but one does not need to be a literature teacher to see that a "chunked," higher-level description would be much more useful.

What Varela argues for, finally, is a dual system of explanation. The operational explanation would emphasize the physical concreteness of actual processes; the symbolic or systems-theoretic explanation would emphasize more abstract ideas that help to construct the system at a higher level of generality. Even so, this "duality of explanation" should "remain in full view" as an antidote to those in computer science and systems engineering who mistake a symbolic description for an operational one, for example by considering that "information and information processing are in the same category as matter and energy." In this respect Varela remains fiercely loyal to autopoiesis. "To the extent that the engineering field is prescriptive by design, this kind of epistemological blunder is still workable. However, it becomes unbearable and useless when exported from the domain of prescription to that of description of natural systems. . . . To assume in these fields that information is some *thing* that is transmitted, that symbols are *things* that can be taken at face value, or that purposes and goals are made clear by the systems themselves is all, it seems to me, nonsense. . . . Information, *sensu strictu,* does not exist. Nor do, by the way, the laws of nature" (p. 45).

In more recent work, Varela and his coauthors provide a positive dimension to this critique of disembodied information. They explore the *constructive* role of embodiment in ways that go importantly beyond autopoiesis. Although autopoietic theory implicitly privileges embodiment in its emphasis on actual biological processes, it has little to say about embodied action as a dynamic force in an organism's development. It is precisely this point that is richly elaborated by Varela and his coauthors in their concept of "enaction."[28] Enaction sees the active engagement of an organism with the environment as the cornerstone of the organism's development. The difference in emphasis between enaction and autopoiesis can be seen in how the two theories understand perception. Autopoietic theory sees perception as the system's response to a triggering event in the surrounding medium. Enaction, by contrast, emphasizes that perception is constituted through perceptually guided actions, so that movement within an environment is crucial to an organism's development. As Varela further explained in "Making It Concrete: Before, During, and After Breakdowns," enaction concurs with autopoiesis in insisting that perception must not be understood through the viewpoint of a "pre-given, perceiver-independent world." Whereas autopoietic theory emphasizes the closure of circular processes, however, enaction sees the organism's active engagement with its surroundings as more open-ended and transformative. A similar difference informs the views of cognition in the two theories. For autopoiesis, cognition emerges from the recursive operation of a system representing to itself its own representations. Enaction, by contrast, sees cognitive structures emerging from "recurrent sensory-motor patterns."[29] Hence, instead of emphasizing the circularity of autopoietic processes, enaction emphasizes the links of the nervous system with the sensory surfaces and motor abilities that connect the organism to the environment.

Embedded in the idea of enaction is also another story about what consciousness means, a story different from that articulated by autopoietic theory. In *The Embodied Mind: Cognitive Science and Human Experience*, Varela and his coauthors take the Buddhist-inspired point of view that the "self" is a story consciousness tells itself to block out the fear and panic that would ensue if human beings realized there is no essential self. Opposed to the false unity and self-presence of grasping consciousness is true awareness, which is based on actualizing within the mind an embodied realization of the person's ongoing processes. We saw that autopoietic theory invokes the "domain of the observer" as a way to integrate common-sense perceptions with the theory's epistemological radicalism, a move that ended up deconstructing the liberal humanist subject in some respects but recuperating it in others. By contrast, in enaction, consciousness is seen as a cognitive balloon that must be burst if humans are to recognize the true nature of their being. The thrust of *The Embodied Mind* is to show that cognitive science has already been headed in this direction and to interpret the significance of this trajectory in the framework of Buddhist philosophies of emptiness and the not-self. Here the boundaries of the liberal subject are not so much penetrated, stretched, or dissolved as they are revealed to have been an illusion all along. In contrast to the anxiety and nostalgia that Wiener and Maturana expressed when confronted with the loss of the liberal subject, Varela, speaking in a voice now not conjoined with his teacher and mentor, celebrates the moment when the self drops away and awareness expands into a realization of its true nature. No longer Wiener's island of life in a sea of entropy or Maturana's autonomous circularity, awareness realizes itself as a part of a larger whole—unbounded, empty, and serene.

What marks this realization as something other than a mystical vision is Varela's insistence that the most advanced research in Western cognitive science points toward the same conclusion. Referencing such works as R. Jackendoff's *Consciousness and the Computational Mind* and Marvin

Minsky's *Society of Mind* (about which we will hear more in chapter 9), he and his coauthors show that contemporary models of cognition implicitly deconstruct the notion of a unified self by demonstrating that cognition can be modeled through discrete and semiautonomous agents. Each agent runs a modular program designed to accomplish a specific activity, operating relatively independent of the others. Only when conflicts occur between agents does an adjudicating program kick in to resolve the problem. In this model, consciousness emerges as an epiphenomenon whose role it is to tell a coherent story about what is happening, even though this story may have little to do with what is happening processually. These models posit the mind, Varela wrote, "not as a unified, homogenous unity, nor even as a collection of entities, but rather as *a disunified, heterogeneous, collection of processes*" (p. 100).

In "Making It Concrete," Varela expands this line of thought by showing how Minsky's "society of mind" model can be combined with nonlinear dynamics to give an account of living systems in action. He continues to insist on the importance of the concrete and embodied. "The concrete is not a step towards anything: it is how we arrive and where we stay." Reminiscent of the autopoietic theory's claim that processes happen always and only in the present, he remarks that "it is in the immediate present that the concrete actually lives" (p. 98). To show how Minsky's model is incomplete, he points out that "it is not a model of neural networks or societies; it is a model of the cognitive architecture that abstracts (again!) from neurological detail and hence from the web of the living and of lived experience." "What is missing here," he continues, "is the detailed link between such agents and the incarnated coupling, by sensing and acting, which is essential to living cognition" (p. 99).

The question he poses is how the mind can move smoothly from one agent processing its program to another agent running quite another program. To answer this question satisfactorily, he suggests, we need to link these abstractions with embodied processes. He proposes a "readiness to action" that in effect constitutes a microidentity. As an example, he imagines a man walking down the street, and Varela sketches the kind of behavior associated with this microidentity. Suddenly the man realizes he has left his billfold behind in the last store he visited. Instantly a different microidentity kicks in, geared toward a search operation rather than a leisurely stroll down the street. How does one get from the microidentity of "stroll" to the microidentity of "intense search"? The answer, Varela speculates, involves chaotic, fast dynamics that allows emergent self-organizing structures to arise. In linking the dynamics of self-organizing structures with microidentities, Varela is following a line of thought vigorously pursued by Zeleny and others, who want to join autopoietic theory with the dynamics of self-organizing systems.[30] The idea is to supplement autopoietic theory so that it can also more adequately account for change and transformation and also to specify the mechanisms and dynamics through which an autopoietic system progresses from one instant in the present to another. These revisions aim to jog autopoietic theory out of its relentless repetitive circularity by envisioning a living organism as a fast, responsive, flexible, and self-organizing system capable of constantly reinventing itself, sometimes in new and surprising ways. In this turn toward the new and unexpected, autopoietic theory begins to look less like the homeostasis of the first wave and more like the self-evolving programs that will be discussed in chapter 9 as exemplars of third-wave cybernetics.

As autopoietic theory continues to evolve, what are likely to be the enduring contributions of autopoiesis as Maturana originally formulated it? In my view, these will certainly include the following: his emphasis on the concreteness and specificity of embodied processes; his insistence that the observer must be taken into account, with all the implications this has for scientific objectivism; his distinction between allopoietic and autopoietic systems, and the ethical implications bound up with making this distinction; and his insight that, in a literal sense, we make a world for ourselves by living it.

In one of his more radical moments, Maturana used the insights of autopoiesis to push toward a formulation that, taken out of context, sounds solipsistic indeed: "We do not see what we do not see, and what we do not see does not exist."[31] In context, he is always careful to qualify this apparent solipsism by pointing out that a world outside the domain of one observer may exist for others, as when I see a large stationary object that a frog cannot perceive. In this way, the world's existence is recuperated in a modified sense—not as an objectively existing reality but as a domain that is constantly enlarging as self-conscious (scientific) observers operate recursively on their representations to generate new representations and realizations. If this isn't exactly the "scientific quest for new knowledge," it nevertheless allows for a qualified vision of scientific progress.

But what if "the observer" ceases to be constructed as a generic marker and becomes invested with a specific psychology, including highly idiosyncratic and possibly psychotic tendencies? Will the domains of self-conscious observers fail to stabilize external reality? Will the uncertainties then go beyond questions of epistemology and become questions of ontology? Will the observation that "what we do not see does not exist" sink deep into the structure of reality, undermining not only our ability to know but the ability of the world to be? To entertain these suppositions is to enter into the world as it is constructed in the literary imagination of Philip K. Dick. Writing contemporaneously with Maturana but apparently with no knowledge of autopoietic theory, Dick is obsessed with many of the same issues. In turning from Maturana's radical epistemology to Dick's radical ontology, we will follow our evolving stories of the reification of information, the construction of the cyborg, and the emergence of the posthuman into a phantasmagoric territory that continues to exist only as long as an observer thinks it does. And what observers Dick's characters turn out to be!

TURNING REALITY INSIDE OUT AND
RIGHT SIDE OUT: BOUNDARY WORK
IN THE MID-SIXTIES NOVELS
OF PHILIP K. DICK

As much as any literary work in his generation, Philip K. Dick's fictions enact the progressively deeper penetration of cybernetic technologies into
the fabric of the world. For a public fascinated by the artificial life-forms
made famous in *Blade Runner* (adapted from his most famous novel, *Do
Androids Dream of Electric Sheep?*), his work demonstrates how potent
the android is as an object for cultural appropriation in the late twentieth
century. Consistently in his fictions, androids are associated with unstable
boundaries between self and world. The form that these associations take
may be idiosyncratic, but the anxieties that his texts express are not. Subterranean fears about the integrity of the subject under the cybernetic paradigm were already present in the subtext of Norbert Wiener's 1948
Cybernetics, as we saw in chapter 4. When system boundaries are defined
by information flows and feedback loops rather than epidermal surfaces,
the subject becomes a system to be assembled and disassembled rather
than an entity whose organic wholeness can be assumed.

For Humberto Maturana, the problem of system definition was solved
by positing a circular dynamic whereby the living continually produces and
reproduces the relations constituting its organization. In effect, he turned
first-wave cybernetics inside out. Instead of treating the system as a black
box and focusing on feedback loops between the system and the environment, he treated the environment as a black box and focused on the
reflexive processes that enable the system to produce itself as such. He developed the political implications of autopoietic theory by suggesting that
power struggles often take the form of an autopoietic system forcing another system to become allopoietic, so that the weaker system is made to
serve the goals of the stronger rather than pursuing its own systemic unity.

Dick's relation to the work of Maturana and Francisco Varela is almost certainly not a case of direct influence. Rather, both Dick and the creators of autopoiesis were responding to the problem of incorporating the observer into the system and, as a result, were experimenting with more or less radical espistemologies. Without using autopoietic terminology (indeed, there is no evidence that he knew of it), Dick explored the political dimension of android-human interactions in terms consistent with Maturana's analysis. In *Do Androids Dream,* Roy Baty understands full well that androids have been denied the status of the living and consequently forced to serve as slaves rather than function as the autopoietic systems they are capable of becoming. The struggle to achieve autopoietic status can be understood as a boundary dispute in which one tries to claim the privileged "outside" position of an entity that defines its own goals while forcing one's opponent to take the "inside" position of an allopoietic component incorporated into a larger system. Working along apparently independent lines of thought, Dick understood that how boundaries are constituted would be a central issue in deciding what counts as living in the late twentieth century.

Especially revealing are the novels he wrote from 1962 to 1966, when he popped amphetamines like crazy and channeled the released energy into an astonishingly large creative output (eleven novels in one year alone!), including a series of major works that sought to define the human by juxtaposing it with artificial life-forms.[1] Drawing on the scientific literature on cybernetics, Dick's narratives extend the scope of inquiry by staging connections between cybernetics and a wide range of concerns, including a devastating critique of capitalism, a view of gender relations that ties together females and androids, an idiosyncratic connection between entropy and schizophrenic delusion, and a persistent suspicion that the objects surrounding us—and indeed reality itself—are fakes.

At the center of this extraordinarily complex traffic between cultural, scientific, and psychological implications of cybernetics stands what I will call the "schizoid android," a multiple pun that hints at the splittings, combinations, and recombinations through which Dick's writing performs these complexities. In Dick's fiction, the schizoid functions as if autistic. Typically gendered female, she is often represented as a bright, cold, emotionally distant woman. She is characterized by a flattening of affect and an inability to feel empathy, incapable of understanding others as people like herself. Whether such creatures deserve to be called human or are "things" most appropriately classified as androids is a question that resonates throughout Dick's fictions and essays. In one of its guises, then, the schizoid android represents the coming together of a person who acts like a machine with a literal interpretation of that person as a machine. In other instances, however, the android is placed in opposition to the schizoid. If some humans can be as unfeeling as androids, some androids turn out to be more feeling than humans, a confusion that gives *Do Androids Dream* its extraordinary depth and complexity. The capacity of an android for empathy, warmth, and humane judgment throws into ironic relief the schizoid woman's incapacity for feeling. If *even an android* can weep for others and mourn the loss of comrades, how much more unsympathetic are unfeeling humans? The android is not so much a fixed symbol, then, as a signifier that enacts as well as connotes the schizoid, splitting into the two opposed and mutually exclusive subject positions of the human and the not-human.

Implicated in these boundary disputes between human and android are the landscapes of Dick's mid-sixties novels. Typically these are highly commercialized spaces in which the boundaries between autonomous individual and technological artifact have become increasingly permeable. Circulating through them are not only high-end products such an intelligent androids but also a more general techno-animation of the landscape: artificial insects that buzz around spouting commercials; coffeepots that demand coins before they will begin to perk; and homeostatic apartment doors that refuse to open for the tenant until fed the appropriate credit. The interpellation of the individual into market relations so thoroughly defines the characters of these novels that it is impossible to think of the characters apart from the economic institutions into which they are incorporated, from small family firms to transnational operations. Moreover, the corporation is incorporated in multiple senses, employing people who frequently owe to the corporation not only their economic and social identities but also the very corporeal forms that define them as physical entities, from organ implants and hypertrophied brains to completely artificial bodies. Given this dynamic, it is no surprise that the struggle for freedom often expresses itself as an attempt to get "outside" this corporate encapsulation. The ultimate horror for the individual is to remain trapped "inside" a world constructed by another being for the other's own profit.

The figure of the android thus allows Dick to combine a scathing critique of the politics of incorporation with the psychological complexities of trying to decide who qualifies as an "authentic" human. Gender dynamics is central to these complexities, for when the schizoid woman is brought into close proximity with a male character, he reacts to the androidism in her personality by experiencing a radical instability in the boundaries that define him and his world. With the issue of what is "outside" someone else's "inside" already supercharged with psychological and political tensions, these enfoldings further implicate capitalism with androidism, the human with the not-human, and the technological with the ontological—that is, cybernetics with the social, political, economic, and psychological formations that define the liberal subject. To unpack these complexities and relate them to the mid-sixties novels, let us turn first to Dick's essays and biography, where the genetic elements that make up these recombinant fictions can be found.

The Schizoid Woman and the Dark-Haired Girl

In "How To Build a Universe That Doesn't Fall Apart Two Days Later," a speech written in 1978 and first published in 1985, Dick linked the "authentic human" with the "real," a construction that also implies its inverse. "Fake realities will create fake humans. Or, fake humans will generate fake realities and then sell them to other humans, turning them, eventually, into forgeries of themselves."[2] The ontology of the human and the ontology of the world mutually construct each other. When one is fake, the other is contaminated by fakery as well; when one is authentic, the authenticity of the other is, if not guaranteed, at least held out as a strong possibility.

With so much riding on the "authentic" human, the qualities defining it take on special significance. Authenticity does not depend on whether the being in question has been manufactured or born, made of flesh and blood or of electronic circuits. The issue cannot be decided by physiological criteria. Here Dick would agree with Maturana and Varela, who argued that artificially created systems can certainly qualify as living. Unlike Maturana and especially Varela, however, he leaps over the importance of embodiment. His fiction displays the same orientation, for it shows almost no interest in how intelligent machines are constructed, dismissing the issue with a few words of hand-waving "explanations" about homeostatic mechanisms. The important point for Dick is not how intelligent machines are built but that they could be build. Descriptions of bodies (except those of women, where the bodies serve as sexual markers) also rarely appear in Dick's fiction. The emphasis falls almost entirely on perception and thinking. Without embodiment to stabilize his epistemological skepticism, the cracks he opens in the perceiver's knowledge widen into rifts in the fabric of the world.

The differences between his ontological skepticism and autopoietic stability can be seen in how his fictions *enact* the human, as distinct from how he *defines* the human in his essays. In a formulation strikingly reminiscent of Maturana and Varela, he suggests that the human is that which can create its own goal. He goes on to develop other characteristics that, for him, set the human apart from the android: being unique, acting unpredictably, experiencing emotions, feeling vital and alive. The list reads like a compendium of qualities that the liberal humanist subject is supposed to have. Yet every item on the list is brought into question by the humans and androids of Dick's fiction. Human characters frequently feel dead inside and see the world around them as dead. Many are incapable of love or empathy for other humans. From the android side, the confusion of boundaries is equally striking. The androids and simulacra of Dick's fiction include characters who are empathic, rebellious, determined to define their own goals, and as strongly individuated as the humans whose world they share. What does this confusion signify?

Here I want to draw connections between Dick's biography and the female characters in his fiction, a topic so obviously important to Dick's work that its absence from much of the criticism on Dick amounts to a scandal.[3] In "The Evolution of a Vital Love," Dick documents his fascination for a certain type of woman: slender, lithe, and young (younger and younger as he grew older), with long dark hair. Repeatedly he hooked up with such women in his life and wrote about them in his fiction. He calls this type "the dark-haired girl."[4] Although the physical type remained the same, in "The Evolution of a Vital Love" he wrote about what he sees as a continuing development in these women and his relationships with them. Whereas the first women (fictional and actual) are schizoid, cruel, unfeeling, and unempathic—in short, androids by his definition—later he meets and has relationships with women who, although they match the physical type, are much warmer and more supportive of him and his goals. For Dick, the progression of the dark-haired girl from schizoid to empathic is vitally important to defining the human and, by implication, the real. "To define what is real is to define what is human, if you care about humans. If you don't you are schizoid and like Pris [a character in his novel *We Can Build You*] and the way I see it, an android: that is, not human and hence not real."[5] In Dick's reading of his life, then, the dark-haired girl started out being allied with the android, but as time went on she became polarized against the android, a stay against the unreality with which the android is persistently linked.

It goes without saying, I think, that Dick regards himself as human. Why, then, does he repeatedly refer to his attraction for the dark-haired girl as a programmed "tropism," a word he picked up from Norbert Wiener's account of cybernetic creations such as the Moth, which had build into it a "tropism" for light. If programmed behavior marks the difference between the human and the android, Dick's tropism for dark-haired girls puts him in the paradoxical position of acting most machinelike when he repeatedly seeks out the woman who, he says, "evolved" until she represented the authentically human. These subterranean connections between the dark-haired girl, machine behavior, and the construction of masculine subjectivity are explored repeatedly in the fiction through configurations that link androidism in an attractive dark-haired woman with a radical confusion of boundaries between "inside" and "outside" for a male subject. The linkage also has implications for female subjectivity. Replication, the mark of the machine, is injected back into the dark-haired girl even after she has supposedly evolved beyond androidism, because the male subject's "tropism" converts her into one of a series, a succession of brunette women who are at once different and the same—hence the multiple ambiguities of the descriptor that Dick constructs for them: "the dark-haired girl." The phrase points both to their singularity (each takes the definite article) and to their identity with one another. Each is unique and uniquely remembered by Dick but remembered as one of a repeating series that stretches back to the early stages of his erotic life.

Many of the critics who have written on Dick's critique of capitalism have scorned psychological explanations, as if they were trivial or unrelated to Dick's satiric view of economic exploitation.[6] But in Dick's fictional worlds, psychology interpenetrates social structure. Contradictions in social structure manifest themselves as aberrant psychology, and aberrant psychology has social consequences. Understanding the relation of Dick's life to his fictional constructions need not reduce his social critique to private neurosis. On the contrary, it illuminates how he was able to fashion a synthesis that undermines precisely the distinctions that would keep the personal in a sanitized domain separate from the social, political, and economic structures constituting the individual.

If we look for a psychological explanation for Dick's tropism, its origins are not hard to find. His parents were divorced when he was six, and he was raised by his mother, Dorothy Kindred Dick (from whom he takes his middle name). Whatever she was in fact, Dick perceived her as an intellectually gifted but emotionally cold woman who denied him warmth and affection. Yet he was also extremely dependent on her and maintained an emotional closeness almost incestuous in its intensity. As his biographer Lawrence Sutin skillfully and sensitively shows, the combination of extreme need for affection and extreme fear of rejection also marked his adult relationships, especially his relationships with women.[7]

His anxiety toward his mother was brought into focus for him through his twin sister, Jane, who died at six weeks of age because Dorothy did not have enough milk for both infants and was too ignorant to realize that the twins, already underweight at birth, were becoming fatally malnourished. "Somehow I got all the milk," Dick told his friends.[8] The story of Jane's death became a family legend. Dick reported that Dorothy discussed it with him on several occasions, trying to explain that she had done the best she could under the circumstances. Her explanations ironically had the opposite effect, for they vividly burned Jane's existence into his consciousness and invested her with intense emotional significance. As a young adult, Dick developed a phobia about eating and could not consume food in public, as if eating were a deeply shameful act.[9]

Despite Dorothy's explanations, Dick blamed his mother for Jane's death, seeing it as evidence of her inability to care physically and emotionally for her children. The blame was all the more intense because he must have felt that on some level he shared it, having taken from Jane the milk she needed to survive. He fantasized that if Jane had lived, she would have become a lovely dark-haired girl. He came to believe that she was a lesbian, a sexual orientation reflected in the character of Alys in *Flow My Tears, the Policeman Said*.[10] And he intuited that in some sense he continued to carry this shadowy Other within his body, a figuration that reflected the fact that Jane no longer had an autonomous existence apart from his imagination of her. Through no fault of her own, she was fated to occupy the subordinate "inside" position while he, the surviving twin, had an "outside" subject position that made him a recognized person in the world.[11]

With Jane as the first dark-haired girl (compounded with Dorothy, from whom she could scarcely be separated in Dick's mind), it is not difficult to see why the figure should be invested with so many conflicting emotional attributes. Like Dorothy, the dark-haired girls that Dick depicts in his fiction are intellectually brilliant but emotionally cold, capable of cutting the men around them to emotional shreds while feeling almost nothing themselves. In his account of the women in his life, he suggests that he was able to break away from this type of woman and find other "dark-haired girls" who were sympathetic. These are the figures he intends to rally to his cause to help him defeat the android. But his worst nightmare remains that the android will turn out to be none other than the dark-haired girl. The enemy is the ally, and the ally is the enemy: enemy mine.[12] It is no wonder that in his essays, the accounts of the human and the android often seem self-contradictory. For the complexities this entangled, he needed—and used—fiction to articulate them fully.

Dick's distinctive gift as a writer was to combine the personal idiosyncrasies of the schizoid woman/dark-haired girl configuration, along with the inside/outside confusions with which it is entangled, with much broader social interrogations into the inside/outside confusions of the market capitalism that incorporates living beings by turning humans into objects at the same time that it engineers objects so that they behave like humans. To explore further this complex nexus among the personal, the political, and the economic, let us turn now to the fiction.

Capitalism and the Schizoid Android

Elucidating the connection between Dick's fiction and capitalism is Carl Freedman's fine article arguing that Dick's fictional techniques reinscribe a post-Marxist view of the subject.[13] Freedman points out that the schizohrenic subject, as theorized by Lacan, Deleuze, Guattari, and others, evolves as an interplay between an alienated "I" and an alienating "not I." Under capitalism, these theorists argue, schizophrenia is not a psychological aberration but the normal condition of the subject. Freedman further argues that paranoia and conspiracy, favorite Dickian themes, are inherent to a social structure in which hegemonic corporations act behind the scenes to affect outcomes that the populace is led to believe are the result of democratic procedures. Acting in secret while maintaining a democratic facade, the corporations tend toward conspiracy, and those who suspect this and resist are viewed as paranoiac.

Dick's novel *The Simulacra* seems tailor-made to illustrate Freedman's point about the synergistic relation between capitalism and paranoid schizophrenia. Set in the USEA (United States of Europe and America), *The Simulacra* depicts a capitalist society that includes Germany as one of its most powerful states. Although national elections still exist, the president has been reduced to a nominal figurehead, "der Alte." The country appears to be run by the first lady, Nicole Thibodeaux, who takes as her husband whatever man the electorate chooses for her every four years. Unbeknownst to the electorate, "der Alte" is a simulacrum. Nicole herself is revealed to be a look-alike playing Nicole, who died several years earlier and has since been played by a succession of actresses. Behind Nicole is the shadowy Council, whose orders she follows, but even Nicole has never seen the Council. Thus the entire government is a fake, its real machinery hidden behind Nicole's beautiful face. The presidential simulacrum, far from being an anomaly, serves as a metaphor for the entire political process. Social classes are divided between the Ge (high status) and the Be (low status). The signifier generating the classes is the Geheimnis, the secret. Those who know the secret—that the government has become, in Dickian terms, a giant android rather than a human institution—are the Geheim-nisträger, bearers of the secret, in contrast to the Befahalträger, those who merely carry out instructions. Thus economic distinctions merge seam-lessly with the kind of social structure that a paranoid schizophrenic might imagine when constructing a system that brings everything together into a monolithic system of explanation.

Paranoid schizophrenia is enacted most dramatically through the char-acter of Richard Kongrosian, a psionic pianist who plays his instrument without touching the keys. Already unstable, Kongrosian is thrown into schizophrenia when he learns that "der Alte" is a simulacrum. The news shakes his faith in Nicole, who has served as his anchor in reality (a function that "the dark-haired girl" frequently plays for Dick's male characters). He suspects that Nicole is contaminated by the android government she serves (precisely the fear that haunts Dick's essays), and this intrusion of an-droidism tips his already fragile psyche into psychosis. *"Something terri-ble's happening to me,"* he warns Nicole. "I no longer can keep myself and my environment separate." As she watches, he makes a vase on the desk sail through the air and enter his body, telling her: "'I absorbed it. *Now it's me.* And—' He gestured at the desk. 'I'm it!'" The process also works in reverse. Where the vase had been, Nicole sees "forming into density and mass and colour, a complicated tangle of interwoven organic matter, smooth red tubes and what appeared to be portions of an endocrine system. . . . The or-gan, whatever it was, regularly pulsed; it was alive and active. . . . *'I'm turn-ing inside out!'* Kongrosian wailed. 'Pretty soon if this keeps up I'm going to have to envelop the entire universe and everything in it, and the only thing that'll be outside me will be my internal organs and then most likely I'll die!'"[14]

The conjunction in this scene of androidism, schizophrenia, and a pro-found confusion of "inside" and "outside" is more than coincidence. Kon-grosian enacts a confusion of boundaries not unlike commodity fetishism. Freedman recalls for his readers Marx's view of how commodities become fetishized under capitalism. Once objects are imbued with exchange value, they seem to absorb into themselves the vitality of the human relations that created them as commodities. Freedman reminds us that one definition for reification is the projection of social relations onto the relations of objects. The incidents that precipitate Kongrosian's turning inside out suggest that the specter of the android has somehow *caused* this bizarre phenomenon. In fact, the android performs an extraordinarily complex staging of reifica-tion in Dick's fiction. On the one hand, it is a commodity, an object created by humans and sold for money. In this guise it is reified in much the same way that any object capable of being bought and sold is reified, like the animals that bestow high status on their human owners in *Do Androids Dream* or like "der Alte" in *The Simulacra,* whose sole function as an object is to serve as a signifier for the democratic processes that are as fake as he/it is. In another sense, however, "der Alte" is an anomaly among Dick's androids, for most of them—Rachael Rosen, Abraham Lincoln, Edwin M. Stanton—are shown in scenes that make them virtually indistinguishable from humans. They think, feel outrage, bond with their fellows. Given their abilities, they should be able to participate in the social realm of human relations, but in such texts as *Do Androids Dream,* they can do so (legally) only as objects. In this view they are not objects improperly treated as if they were social beings but are social beings improperly treated as if they were objects. For them the arrow of reification points painfully in both directions.

The scene in which Kongrosian turns inside out is revealing in another respect as well, for it demonstrates the megalomaniac expansion of self that the paranoid schizophrenic experiences in the frenzy of delirium. The paranoid feels compelled to interpret all the surrounding mysterious signs and order them into a single coherent system. From here it is a small step to feeling responsible for the signs. If everything that happens is the paranoid's responsibility, the belief easily follows that the paranoid actually *caused* all these events. When Kongrosian states "I'm going to have to envelop the entire universe and everything in it," his actions can be understood as a literalization of this view. Gifted or cursed with telekinesis, he causes things to happen in the world by thinking about them. From this he moves on to believing that he orders the universe; then he progresses to the fantasy that he *is* the universe. Part of the guerrilla warfare that Dick stages on the everyday is to valorize such perceptions by rendering them as events that other characters witness and that the narrator reports as "actually" happening. In this way, the reader's perceptions undergo a transformation similar to Kongrosian's, for our relationship to the character is turned inside out. Instead of his world existing inside our minds, the textual world is rendered so as to make our perceptions work as if they were part of his internal world.

The battle to occupy the "external" position relative to other characters is waged over and over in Dick's fiction in different guises. The stakes are high, for if the self is unable to expand to megalomaniac proportions, it is likely to shrink so that it becomes merely a dot on the horizon, an atom in a cold, pitiless, inanimate landscape shaped by the dead forces of cause and effect and completely unresponsive to human desire. This is the tomb world, the landscape in which entropy rules. Frequently the pendulum will swing between dangerous hyperinflation and excruciating shrinkage of the ego without ever stabilizing at the middle position of everyday reality.

Scott Durham has perceptively pointed out that this alternation between the expanding self and the shrinking self is intimately related to the constitution of subjectivity under capitalism.[15] Capitalism encourages the inflation of desire, marketing its products by seducing the consumer with power fantasies. But when the realization sinks in that this is merely a capitalist ploy, the subject shrinks in inverse proportion to how much it had earlier inflated. So in Dick's novel *The Three Stigmata of Palmer Eldritch*, Chew-Z is marketed by linking it with claims of omnipotence through the slogan "GOD PROMISES ETERNAL LIFE; WE CAN DELIVER IT."[16] When the subject consumes the product (figuratively and also literally by ingesting it), he finds that he is catapulted into a world where Palmer Eldritch makes all the rules. Rather than taking the product inside him, he has been imprisoned inside the product. For he soon discovers that no matter what escape hatch he builds or invents, Palmer Eldritch remains exterior to his reality, determining its workings and glinting through the other characters as they begin to manifest Eldritch's telltale stigmata. The eternity delivered here is precisely not the apotheosis of the liberal autonomous subject capable of free thought and action but is the subject as pawn in a capitalist's game, imprisoned for eons in a universe that a terrifying and menacing alien other has created to increase his profits.

The boundary instability that the Kongrosian scene so vividly illustrates recurs repeatedly in Dick's mid-sixties novels. In one version of the drama, the [male] subject expands and contracts in an agonized dance between megalomania and victimization. This dance is intimately bound up with a related oscillation in the attractive female character between the schizoid woman and the dark-haired girl. In the promiscuous couplings these various subject positions permit, the android serves as an ambiguous term that simultaneously incorporates the liberal subject into the machine and challenges its construction in the flesh. To develop further the complex connections and disjunctions signified by the schizoid android, I turn now to pick up again the thread of the schizoid woman/dark-haired girl.

The Schizoid Woman (De)Constructs Male Subjectivity

Patricia Warrick has insightfully argued that Dick's fiction is structured as a series of reversals designed to defeat the reader's expectation that it is possible to discover what the situation "really" is.[17] Building on her argument and also modifying it, I want to demonstrate that the reversals are not arbitrary but follow an inner logic of their own. I shall take as my tutor text *We Can Build You,* in part because Warrick remarks, in passing, that its two themes—the construction of androids and the male character's fascination with the schizoid woman—are not connected with each other, which she sees as an aesthetic failing. I will argue that although the themes are not well integrated, they are deeply connected through the figure of the schizoid android.

As Dick recognizes in "The Evolution of a Vital Love," the prototype for the schizoid woman is Pris in *We Can Build You.* Louis Rosen, the little guy who serves as protagonist for the novel, finds Pris fascinating but also terrifying. Her most notable attribute is what he calls "emptiness dead center," an inability to feel empathy or indeed almost any emotion.[18] Talented, creative, and fiercely intelligent, Pris had a nervous breakdown while still in high school (as did Dick himself). When she takes up with the rich entrepreneur Sam Barrows, deserting the family firm to move in the glittering world of the rich and famous, Louis's attraction to her becomes compulsive. The more this dark-haired girl/schizoid woman recedes out of his realm, the more he yearns after her, finally becoming so obsessed that getting together with her becomes more important to him than anything else, including the family firm. "What a woman, what a *thing* to fall in love with," he thinks to himself, in a conflation that makes clear that Pris, as a schizoid woman, has more than a touch of androidism in her. "It was as if Pris, to me, were both life itself—and anti-life, the dead, the cruel, the cutting and rending, and yet also the spirit of existence itself" (p. 155). As her new name "Pris Womankind" indicates, she is at once uniquely herself and symbolic of the role that Dick assigned to the bright, cold, cutting women in his life, particularly his mother, Dorothy, and (after they divorced) his fourth wife, Nancy. More than a tie to life, Pris is Louis's anchor to reality, much as Nicole was for Kongrosian in *The Simulacra.* At the same time, Louis experiences Pris as the "anti-life," which in later novels takes shape as the tomb world. The flips that Warrick notices in Dick's text have an inner logic that make it impossible for the male character to do without the dark-haired girl/schizoid woman, even when he sees her as a source of powerful contamination driving him into psychosis.

When Louis and Pris rendezvous at a motel, the encounter fizzles because neither can abandon their cat-and-mouse games long enough to experience physical intimacy. Still obsessed, he falls into a delusional state in which he hallucinates that he is making love to her, although he is alone with his father and his brother in a bedroom and his "lovemaking" takes place while they look on bemusingly. By the novel's end, he traces Pris to the mental institution where she has been readmitted, and he himself becomes a patient. After months of drug therapy in which he hallucinates that they court, marry, and have a child, he finally has a chance to talk with her. She tells him she will soon be leaving and suggests how he can present his case to be dismissed also. He wins dismissal, only to learn that Pris has deceived him. "I lied to you, Louis," she tells him. "I'm not up for release; I'm much too sick. I have to stay here a long time more, maybe forever" (p. 252). For Louis, sanity means losing her and, with her, the vitality of life, which can be seen as a kind of mental illness, but having her means mental illness of another kind.

The ambiguity of Pris's motives in this final scene—does she trick Louis because she doesn't want him around or because she wants him to get on with his life?—indicates that even in a female protagonist figured mostly as a schizoid woman, flashes of the empathic dark-haired girl still appear. In *Do Androids Dream,* these instabilities in the female subject position are exacerbated as the schizoid woman is broken into twin characters, Rachael Rosen and Pris Stratton. The two are the same model of android, a Nexus-6, so they are physically identical. But they play very different roles in the plot. Rachael becomes for Deckard a particularly ambivalent version of the dark-haired girl.[19] At his low point she comes to him, and they end up in bed in a sexual liaison that Pris and Louis couldn't manage to bring off. During this scene and the one following, her characterization oscillates wildly between a desirable, empathic partner and a cold, calculating manipulator. The scenes are worth looking at in detail for what they reveal about the dynamic of the male character who experiences the schizoid woman as a splitting between an android (literally so with Rachael) and the dark-haired girl.

Deckard calls Rachael because she has offered to help him "retire" the escaped Nexus-6 androids. When he shows her his hit list, she turns pale because one of the androids ("andys"), Pris, is the same model as she is. Earlier that day Rick, working with Phil Resch, had helped to kill Luba Loft, an escaped android who was also a superb opera singer. After Deckard expressed regret at Luba's death, Resch (depicted as a cold-blooded killer who, unlike Deckard, feels no empathy for Luba or any of the androids he kills) interprets his regret as sexual desire. Although human-android sex is strictly illegal, he confesses that he once fancied a female android and advises Deckard that instead of killing the android and then wanting her, he should go to bed with her first and kill her afterward. Now, in the hotel room with Rachael, it occurs to Deckard that he is inadvertently about to follow Resch's advice, for he intends to go to bed with Rachael and kill Pris afterward. Shaken by the realization, he refuses to have sex with Rachael. In a last-ditch attempt to cajole him, she tells him she loves him and then, when he still refuses, offers to kill Pris herself. She explains, "I can't stand getting this close and then—" (p. 170).

After the sex, Rachael apparently feels free to reveal her motives. She tells Deckard that his career as a bounty hunter is over because no man who has gone to bed with her has found it possible afterward to kill any androids. She has had sex with several bounty hunters, she explains placidly, and it has worked every time. The one exception is a "very cynical man, Phil Resch" (p. 174). The euphemism that Deckard uses to describe killing androids— "retiring" them—ironically returns in a reversal that has a female android "retiring" bounty hunters.[20] Rachael's strategy implies that she feels empathy for her fellow androids, giving the lie to the government's position that androids feel no loyalty to each other. If she can care for her fellow androids, we may wonder if she also cares for Deckard, as she claims when she tells him that she loves him.

Enraged by her revelation that she has "retired" him, Deckard tries to kill her and cannot, whereupon she reproaches him for loving the Nubian goat he has acquired with his bounty money more than he does her—a response that works on multiple levels. It hints at the ironic fact that humans revere animal life but feel free to kill intelligent android life, and it also suggests that Rachael, although she uses Deckard for her own political purposes, cares about whether he cares for her. After Deckard succeeds in killing the last three andys, he returns home to discover that Rachael has pushed the goat off the roof, an act that conflates her jealousy of the goat with revenge for Deckard's killing her friends. The mixture of human passion and cold calculation in Rachael's responses shows that she combines within herself attributes of the dark-haired girl and of the android. The closer the relationship gets to intimacy, the wilder the oscillations between these subject positions become, in turn inducing alternating moods in Deckard: between despair and empowerment, ego shrinkage and inflation. *It is as if Deckard's attraction to her were destabilizing reality itself.*

That possibility, latent in Deckard's relationship with Rachael, becomes overt in J. R. Isidore's perceptions. Deckard's desire for Rachael is formally mirrored in the plot by J. R.'s desire for her twin, Pris. Rachael's character, split within itself, splits again into Rachael and Pris, a division in which Rachael is closer to the dark-haired girl and Pris to the schizoid woman. Whereas Rachael's manipulation of Deckard is relatively subtle, Pris's manipulation of J. R. is bald-faced and cold. Although J. R. fantasizes that perhaps they might have a relationship, Pris never indicates any feeling for him, and it is clear that sex between them isn't going to happen. In *We Can Build You*, Louis likens Pris to a spider, seeing her as an alien being who goes about her business oblivious of her effects on others. The image returns in *Do Androids Dream* in the scene in which Pris and Irmgard Baty, holed up in J. R.'s apartment, cut off the legs of a spider, which J. R. has found, to see how many legs it can do without and still walk. As a chicken-head (this society's term for someone who is degenerating mentally), J. R. lacks Deckard's analytical skills, and he often expresses his insights visually and intuitively, such as when he briefly hallucinates that Roy Baty is made of gears and coils rather than flesh and blood. Faced with this desecration of Mercer's decree that all life is sacred, J. R. perceives the force of "kipple" (a neologism that Dick uses for the entropic decay that has been nibbling away at the apartment building) suddenly becoming an avalanche of destruction. Chairs crumble; the table melts askew; gaping holes appear in the walls. From Pris's exclamations, the reader knows that J. R. has gone berserk and is causing the destruction himself. Nowhere, perhaps, is Dick's conflation of cybernetic concerns with idiosyncratic psychology more apparent than in this scene. The entropic decay that Wiener imagined could be forestalled by cybernetics is preternaturally accelerated until it is visibly apparent in the landscape, and this visibility also functions as a sign that system boundaries have become radically unstable. The moment is a finely realized piece of writing that performs cybernetic boundary disputes in a context that makes clear their relation to a male character's attachment to the female android/schizoid woman. The result is a deep confusion of boundaries between inside and outside.

Confronted with Pris's torture of the spider and thus implicitly with her emotional coldness, J. R. perceives the heat energy rushing from the room, as if the room's physical decay sprang directly from her lack of empathy. As this conflation of inside/outside suggests, his perception of the boundaries between himself and the outside world has become badly distorted. The psychological dynamic is clear from Dick's repeated use of the situation. The (male) self yearns to expand outward in a moment of union, but when the female android/schizoid woman rejects him, the result is a devastating instability in which it is difficult or impossible for him to establish robust boundaries between himself and the world. Louis Rosen, rejected by Pris Womankind, projects his fantasies of her into a hallucinatory love partner. J. R. Isidore, shocked by Pris Stratton's cruelty to his spider, perceives his own rampage as the impersonal force of kipple at work. In Dick's novels, the sudden collapse of an inside/outside distinction is often a sign that the male subject is plunging into psychosis. One of the sites where this dynamic plays itself out is the tomb world. Let us turn now to a closer examination of this surrealistic landscape to explore its connection with the schizoid android.

Wasting Time in the Tomb World

In *Do Androids Dream,* a compelling "proof" of the official ideology that androids occupy a category ontologically distinct from that of humans is the fact that androids cannot experience fusion with Mercer, a quasi-religious figure who appears when a human grips the handles of the empathy box. Androids, incapable of experiencing this fusion, are judged to be lacking in empathy, the touchstone of the "authentic" person. Staging a moment in human history when androids rival or surpass humans intellectually, *Do Androids Dream* shows the essential quality of "the human" shifting from rationality to feeling. Animals, evoking feeling in their owners and capable of feeling themselves, occupy the privileged position of fellow creatures whose lives, like human lives, are sacred, whereas the rational androids are denied the status of the living. The change comes when nonhuman animals, rapidly fading into extinction, have ceased to pose any conceivable threat to human domination. Since the real threat now comes from the androids, the shift in definition is hardly a coincidence. To extend the critique, Dick emphasizes the capitalist marketing of animals, an industry fueled by the religious significance that owning an animal has under Mercerism. Like certain forms of Puritanism, Mercerism joins with capitalism to create a system in which the financially privileged merge seamlessly with the religiously sanctified.

Despite this pointed satire, Dick's treatment of Mercerism remains complexly ambiguous. The text refuses an either/or choice and implies that Mercerism is both political hucksterism and a genuinely meaningful experience. In an exposé by Buster Friendly, a radio talk-show host later revealed to be an android, Mercer is proved to be a fake, a drunk hired by unknown parties to act out a few cheesy scenes of humiliation and atonement on cheap sets. Yet Mercer is also an inspiring figure who mysteriously appears to Deckard to tell him that killing androids is both wrong and necessary, just as Mercer acknowledges that he is at once fake and genuine.

These multiple confusions are reenacted when J. R., already operating in the borderland between hallucination and reality, rushes to the empathy box after the tidal wave of kipple engulfs his apartment. As he grabs the handles, he plays out the scenario that Mercer endlessly repeats in a landscape hovering ambiguously between the internal and external worlds.

Like Sisyphus, Mercer is doomed to climb up a dusty hill, while unseen tormentors throw rocks at him, only to slide back down when he reaches the top. But he does not merely regress to the bottom. He plunges all the way down into the tomb world, a world where nothing but him lives, a world populated by the decaying skeletons and rotting carcasses of animals. In the tomb world, time has either stopped or moves so slowly that its passage cannot be perceived. All one can do is wait passively for what seems like eons in the utter desolation, surrounded by death and decay, until very slowly things begin to come alive again and it is possible to climb out.

A clue to the psychological significance of the tomb world comes in Dick's analysis of schizophrenia in "Schizophrenia and *The Book of Changes.*"[21] In high school, Dick experienced an agoraphobia so acute that he had to be tutored at home. As a young man, he compulsively engaged in various neurotic behaviors, including his eating disorder. He had three nervous breakdowns and attempted suicide several times. When he talks about the experience of mental illness, therefore, we may suppose he knows whereof he speaks. Writing at a time when R. D. Laing was calling for a reassessment of schizophrenia, Dick echoed Laing in viewing schizophrenia with sympathy and even admiration.[22] In a letter to Patricia Warrick, he wrote that he wanted to draw a "sharp line" between the neurotic schizoid, whom he saw as an essentially cold person seeking power over others, and the psychotic schizophrenic, who is too "nuts" to be much of a threat to anyone but himself or herself.[23] In contrast to his scathing indictment of the schizoid, who withdraws emotions from the world, Dick saw the schizophrenic as someone who suffers because of projecting emotions too much into the world.

His sympathy is evident in "The Android and the Human" when he observes that for the schizophrenic, time stops because nothing new can happen.[24] The ego has become so distended, so inflated, that it blocks out everything else. Since the self is perceived as responsible for explaining everything and putting it into order, there can be no surprises. The new, the inexplicable, the mysterious, and the unexplained do not exist for the paranoid schizophrenic. The tomb world is a literary and fictional representation of this state seen from the point of view of the person who experiences it. The dreariness, the hopelessness, the feeling that time has stopped and there is nothing to do but wait, the deadness inside projected onto an exterior landscape—these are the markers of extreme mental distress as Dick describes them. The tomb world appears in several of Dick's fictions from the mid-sixties, and it is always associated with a deep confusion of inside/outside boundaries.

The inside/outside confusion links the schizophrenic to the android. Like the schizophrenic, the android is a hybrid figure—part human, part machine—whose very existence calls boundaries into question. Whereas the android's actions are always predictable—the android is characterized by predictability "most of all," Dick wrote in "The Android and the Human" (p. 191)—for the paranoid schizophrenic, the world's actions are always predictable. In the first case, the predictability is understood as coming from the android's internal programming; in the second case, the predictability is perceived as originating in the external world. This distinction is moot in the tomb world, however, for inside and outside merge in its ambiguous landscape. Surely it is no accident that Dick's mid-sixties novels in which the tomb world prominently appears also feature android characters.

The android that Dick writes about in his essays represents the loss of free play, creativity, and most of all, vitality—in short, the triumph of obsession over the flexibility and empathy that a writer needs to create the new. Yet as we have seen, Dick's fictional androids are considerably more complex than this portrayal. We can understand this contrast through a paradox: the simple version of the android represents a loss of vitality that would make writing impossible, yet this view of androidism is precisely what Dick makes into the occasion for writing. Androidism both annihilates writing and makes it possible. The paradox is written into Mercerism through the ambiguities it generates between self and other. The moment a human grasps the empathy box, his consciousness fuses with that of unknown and unnamed others. He is both alone and in company, cut off from his surroundings and in emotional communication with other human beings. In short, he partakes of the instabilities that male subjects feel when they come into close proximity with female characters who participate in the schizoid woman/android/dark-haired girl configuration. He experiences an ego expansion that, although it can be extremely dangerous if it progresses into megalomania, in the empathy box remains relatively contained and so is relatively benign.

Even so, the downside is hardly negligible. The empathy box interpolates the private delusions of the subject into a shared ideology that inscribes his characteristic experiences into scripts invested with religious, political, and social significance. As Jill Galvin points out, there are hints that the government, faced with a declining population in a world rapidly becoming uninhabitable by humans, encourages the use of the empathy box to keep citizens quiescent and tractable.[25] In the empathy box, citizens *feel* empowered, but the endless scenarios they play out make clear that they are in fact powerless, a paradox that is a more insidious version of Dick's empowerment by writing about the hopelessness and androidism of the tomb world. Buster Friendly, his ulterior motives notwithstanding, makes a good point when he asks his listeners to question "what it is that Mercerism does. Well, if we're to believe its many practitioners, the experience fuses . . . men and women throughout the Sol System into a single entity. But an entity which is manageable by the so-called telepathic voice of 'Mercer.' An ambitious politically-minded would-be Hitler could—."[26]

Despite the obvious exploitative potential of Mercerism, Dick also insists that alongside its fakery there exists a possibility for genuine atonement and redemption, a possibility written into *Do Androids Dream* when Mercer's intervention saves Deckard's life. If Mercer is in some sense real as well as fake, then the tomb world must also at once be a delusion and a necessary purgation. The key to understanding its mysterious double nature lies in the schizoid woman/dark-haired girl configuration. As we have seen, the oscillation between the dark-haired girl and the schizoid woman becomes more pronounced the closer the male character draws to her. The male character nurses a fierce ambivalence toward her, both desiring intimacy and fearing it. Because of the multiplicity of her nature as Dick constructs it, she is the perfect screen on which to project this ambivalence. On the one hand, she represents a rejection all the more inevitable and absolute because it springs not from deficiencies in the male, which he could presumably correct, but from her own inability to relate to anyone. On the other hand, she represents all that the male finds desirable and vital, so for him to be cut off from her is tantamount to not living. If she rejects him, it means he is not really alive and thus is an android. If she accepts him, it means he will be tied to her and thus exposed to the coldness he most fears. Either way, exposure to her compromises his humanity with a touch of androidism, a possibility brilliantly realized in *Do Androids Dream* through the intimation that Deckard himself may be an android. The tomb world acknowledges this pollution and tries to atone for it.

Occasionally male characters, who are constantly in danger of being sucked into the maelstrom of conflicting impulses that the schizoid android evokes, take revenge on the woman who attracts them, presumably with authorial sanction. The violence that the narrator (and, behind him, Dick) can visit on the schizoid woman is revealed in the scene from *The Three Stigmata* in which Leo Bulero propositions his attractive assistant, Roni Fugate, a character who bears more than a passing resemblance to the schizoid woman. When she turns him down, Leo spitefully wishes she were old, over one hundred years old. Leo has consumed (and been consumed by) Chew-Z and so has never really left the space where Palmer Eldritch makes the rules. Too late he remembers that in this delusional world, his thoughts are coextensive with the apparently real landscape. He turns around to see a "spider web, gray fungoid strands wrapped one around another to form a brittle column that swayed . . . he saw the head, sunken at the cheeks, with eyes like dead spots of soft, inert white slime that leaked out gummy, slow-moving tears." Aghast, he wishes her alive again. The gray mass becomes a puddle that "flowed gradually outward, then shuddered, and retracted into itself; in the center the fragments of hard gray matter swam together, and cohered into a roughly shaped ball with tangled, matted strands of hair floating at its crown. Vague eyesockets, empty, formed; it was becoming a skull, but of some life-formation to come: his unconscious desire for her to experience evolution in its horrific aspect had conjured this monstrosity into being."[27]

Two lessons emerge from this episode. First is the connection between the schizoid woman and the tomb world, here made explicit when the male who perceives himself insulted by her takes revenge by making *her* the victim of the tomb world dynamics. Although he may be punished by perceiving the world as a dust heap, the punishment he visits upon her is the more extreme violence of being made to incorporate within her own body the preternatually accelerated entropic decay of the tomb world. The second lesson is the inescapability of a landscape in which the subject's "inside" has merged with the world's "outside." Time can go forward again into life, but it can't be reversed back to where it was before the psychotic episode happened, for Leo or, more drastically, for Roni either. Leo learns that one can climb out of the tomb world, as Mercer eventually does, but once one has fallen into it, the tomb world itself never goes away. It stays there, waiting, until the boundaries separating inside and outside again become so unstable that the subject slides down into it once more, as Jack Bohlen does in Dick's novel *Martian Time-Slip*.

Given these complexities, how does resolution occur in Dick's novels? How are the endlessly complex splittings and recombinations of the schizoid android stabilized enough to achieve closure? To answer this question, I turn now to the staging of the schizoid android in his novel *Dr. Bloodmoney*.

Turning Right-Side Out in *Dr. Bloodmoney*

In a brilliant article entitled "After Armageddon: Character Systems in *Dr. Bloodmoney*," Fredric Jameson uses the semiotic square to elucidate the relationships between characters in *Dr. Bloodmoney*.[28] His analysis posits a primary axis between "organic" and "mechanical," where the terms exist in an oppositional relation to each other. The appearance of this axis is scarcely a surprise, since the opposition between human and mechanical is a prominent theme in Dick's fiction. Less obvious is the axis labeled contradictory, in which the terms are not opposites to the terms of the first axis but exist in a subtler relation of including realms that the first terms exclude. The lower-right position is occupied by the "not-organic," which Jameson interprets as lacking organs, that is, as the dead, and the lower-left position is held by "neither mechanical nor organic," to which he assigns those characters who possess extraordinary or spiritual powers, such as talking animals and humans gifted with preternatural abilities. Most of the major characters exist in positions synthesizing these four primary terms. From this analysis, it becomes clear that the characters are arranged according to their power over words or things, a conclusion with important implications in these fictions where words are used to reveal the unreality of things and where things are used to reveal the instabilities in words.

In my view, the only unconvincing part of Jameson's argument is the supposition he uses to launch his analysis. He points out that *Dr. Bloodmoney* shows the event that the other mid-sixties novels presume but do not depict: the nuclear holocaust that destroys the environment and permanently changes the relation of the human race to Earth. He argues that the bomb demands a "flat *yes* or *no*," thus defeating Dick's aesthetic of creating an indeterminate reality and requiring a new technique.[29] The character system, he suggests, is this new technique, although how it works to solve the problem he posits is never made clear. But Dick has no problem rendering the reality of the bomb problematic (as we shall see shortly), so from this point of view, the character system is overkill. I want to suggest, on the contrary, that the character system is more aptly understood in relation to a different interpretive problem, one that I believe is central to the narrative dynamics of *Dr. Bloodmoney:* the boundary work of turning characters who are inside out right-side out.

The first time the bombs come, the narrative makes clear that the event is "actually" taking place, although significantly it is rendered through the eyes of Bruno Bluthgeld (the "Dr. Bloodmoney" of the title), a theoretical physicist who convinced people that high-altitude nuclear tests would be safe. His calculations proved tragically off-base, and a generation of malformed children were born as a result. Now, nine years later, he goes by the name "Mr. Tree," fantasizes that large disfiguring blotches under his skin mark him for those in the know, and believes that there are massive conspiracies afoot to kill him. Leaving the psychiatrist he has consulted to get rid of his "infection" (which he construes physiologically but which the reader has no trouble recognizing as paranoid schizophrenia), he sees the San Francisco street suddenly sink and tilt to the left. He attributes the phenomenon to his astigmatism. "Sense-data so vital, he thought. Not merely what you perceive but how. . . . Perhaps I have picked up a mild labyrinthitis, a virus infection of the inner ear."[30] The monstrous incommensurability between a minor ear infection and a nuclear holocaust reveals how out of whack are Bluthgeld's perceptions. Like other schizophrenics in Dick's fiction, he has experienced an inflation of ego so extreme that he believes he alone is responsible for everything that happens. In his delirium, he interprets the holocaust as a defensive measure he was "forced" to take to punish those conspiring against him.

In contrast to this first scene is the second holocaust, more shadowy than real. Bluthgeld, now living as "Mr. Tree" in a postnuclear society on the Marin headlands, grotesquely decides that the conspiracy against him has been reactivated when Stuart McConchie arrives in the community. (In the afterword to *Dr. Bloodmoney,* Dick identified McConchie, an African American, as representing his own viewpoint within the novel.) Alarmed by this fantasized conspiracy, Bluthgeld/Mr. Tree concentrates on reactivating the destructive forces that will drop the bombs. The second time around, it is not so easy for us to decide the ontological status of the bomb. Since many of Mr. Tree's other fantasies obviously don't work, the reader may be tempted to dismiss his calling forth the bomb as a private delusion. But other characters corroborate that some version of a nuclear hit is taking place, although exactly in what sense is not clear. The most important of these corroborations is the viewpoint of Walt Dangerfield, a would-be Martian émigré trapped in a satellite after war broke out. From high over Earth, Dangerfield sees on the horizon a flash that he thinks he recognizes from the holocaust nine years earlier. "Seconds passed and there were no further explosions. And the one he had seen; it had been peculiarly vague and shadowy, with a diffuseness that had made it seem somehow unreal, as if it was only imagined" (p. 230). Thus Dick constructs the event so it hovers suspended between internal fantasy and exterior corroboration, using much the same multifocal narrative techniques that operate in his other novels. Whatever the purpose of the character system, its object cannot be to infect the holocaust with ontological uncertainty, for Dick accomplishes this by other means.

So what does the character system accomplish? As I suggested earlier, I believe it is directed toward the different problem of how to escape becoming trapped in the "inside" of a power-mad fantasy and how to turn the world right-side out again. Jameson rightly notices that Dick runs into a problem when he tries to defeat violent characters by using characters still more violent, for the cure quickly becomes worse than the disease. Exactly this problem arises in *Dr. Bloodmoney*. When Mr. Tree goes crazy and starts setting off bombs again, Hoppy, a "phocomelus" born with no legs, with flippers for arms, and with fearsome telekinetic powers, saves himself (and incidentally the town) by killing Mr. Tree, tossing Mr. Tree high into the air by using his telekinetic powers. But Hoppy has a growing megalomania akin to that of Bluthgeld. Drunk with his powers, he has no empathy for anyone else and has such an inflated ego that only his own needs and desires are real to him. Moreover, his reach extends beyond the town. He has been preparing himself to become the world's dictator by using telekinesis to kill Walt Dangerfield and take over his satellite broadcasts, which have hitherto provided the one bright spot in an otherwise dark world. Like Mr. Tree, Hoppy wants to locate others "inside" his fantasy and arrange matters so that others are forced to live there on his terms.

The resolution comes from another direction entirely. Opposing Hoppy *and* Mr. Tree is Edie Keller, a young girl who carries the homunculus of her twin brother, Bill, who formed inside her body when they were in the womb and communicates with her telepathically. No doubt her characterization reflects Dick's belief that he carried the spirit of Jane inside his body; Edie and Bill are Dick and Jane turned inside out. Whereas Mr. Tree and Hoppy are completely narcissistic, the confusion of boundaries that Bill and Edie experience includes genuine concern for each other. Bill's greatest hope is that he can exit Edie's body and live on his own, rather than vicariously through her reports to him. Although Edie can be thoughtless and cruel, she also tries earnestly to find a suitable home for him. When she hears that Mr. Tree has gone crazy, she hurries to him, reasoning that Bill will make better use of Mr. Tree's body. But Hoppy kills Mr. Tree before she can reach him. Her next plan is to trick Hoppy into allowing her to get close to him, so that Bill can expropriate Hoppy's body. But Hoppy, telepathic as well as telekinetic, takes Bill out of her body before she can reach the phocomelus; he tosses Bill's tiny, malformed body into the air, as he did with Bluthgeld.

Bill, however, has authorial resources that Bluthgeld lacked. Tongue in cheek, Dick uses an *avis ex machina* to rescue Bill. Bill succeeds in exchanging bodies with Hoppy seconds before the homunculus body dies. In contrast to Hoppy's megalomania, Bill's aspirations are modest. Although Hoppy's body is severely deformed, it is so superior to Bill's previous body that he is delighted with it, for now he can see and hear on his own. Thus the inside/outside confusion is resolved in two ways. Bluthgeld and Hoppy, the characters who threatened to expand until others were condemned to live in the "inside" of their horrific mental worlds, are killed; and Bill and Edie, innocents who through no fault of their own were caught in a tragic encapsulation that threatened both their lives, are turned right-side out. The happy conclusion reverses the tragic end of Dick and Jane's twinship. Instead of two children becoming one when the girl twin dies, here one child becomes two when the boy twin succeeds in leaving his sister's body and living on his own.

Why can this accommodation be achieved in *Dr. Bloodmoney* when it eludes the other mid-sixties novels, with their darker endings? Central to this "extremely hopeful novel" (Dick's phrase in the afterword, p. 300) is the characterization of Bonny Keller, who remains largely outside Jameson's semiotic square. Depicted as a beautiful woman, Bonny is one of the very few attractive females who is neither fetishized as the dark-haired girl nor feared as the schizoid woman in Dick's fiction. Her deepest allegiance seems to be toward living life as fully, honestly, and joyfully as possible. The day the bombs fall, her immediate reaction is to make love with the first man who comes along, Andrew Gill, as if to affirm that life can still go on. As a result, Edie and Bill are conceived. As if to confirm that she is emphatic as well as life-affirming, her next reaction is to weep for all the people who have died in the city.

Bonny's moment of truth comes, significantly, when she refuses to take responsibility any longer for Bluthgeld's madness. Throughout the years she has tried to protect him and even to reason with him, but when his madness breaks through again, she leaves him to his folly. She has a similar reaction when Hoppy starts terrorizing the town. Rather than fight him, she intends to get as far away as possible. Somehow her shrugging off these self-assigned responsibilities comes across as the right thing to do, even though it means leaving her children behind. Talking with Dr. Stockwell, Bill wonders what his mother will think when she realizes she has twins rather than a single child. But he and his sister never find out, for Bonny has already run off with Andrew Gill. The Gordian knot formed by the tangled complexities of mothers who cannot properly care for their children, of twins whose lives enmesh with disastrous results, and of vicious circles that form when a male character both desires and fears the dark-haired girl/schizoid woman is simply cut through by a knifelike clarity that says in effect: "This mess is not my responsibility. I have my own life to lead."

Jameson remarks that one of the purposes the character system serves is a "freshening" of the world, whereby commodified products that used to be taken for granted become homemade luxuries that bring delight and pleasure to the lives of those who consume them. This redemption of capitalism, the refusal of the double bind of the dark-haired girl/schizoid woman, the destruction of those who would engorge themselves so that they can forcibly encapsulate others in their "inside," the turning right-side out of a tragically enfolded twin boy and girl, the choice for life over futile self-assigned and self-defeating responsibilities—these are the entwined complexities that the elaborate textual machinery of the character system is designed to straighten out. More so than other texts from this period (with the possible exception of *Martian Time-Slip*), *Dr. Bloodmoney* succeeds in cutting through the entanglements figured by schizoid android.

As Dick moved out of the mid-sixties era, the frequency with which the dark-haired girl/schizoid woman appears in his fiction diminishes. *Ubik,* written at the outer range of this period, in 1966, functions as an interesting transitional text, for it suggests that Dick was able to resolve the deep ambiguities of the dark-haired girl/schizoid woman configuration through his writing. Structured as a series of revelations, each of which exposes its predecessor as a facade rather than an authentic reality, *Ubik* uses this favorite Dickian technique to suggest that the dark-haired girl/schizoid woman configuration is itself merely a facade underlaid by a deeper reality. To see how Dick achieves this psychological resolution, let us turn now to this complexly reflexive text.

Dark-Haired Girl/Schizoid Woman as Facade: The Reality Underneath

In *Ubik,* the struggle to occupy an "outside" relative to someone else's "inside" takes place on multiple levels. The little guy of the novel, Joe Chip, works for Glen Runciter, head of a "prudence organization" that specializes in "anti-psis" who can neutralize the psionic talents of telepaths, pre-cogs, and the like. Lured to Luna by a business rival, Glen Runciter, Joe Chip, and a group of their "anti-psis" are hit by a bomb. What happens afterward is notoriously unclear. For a while Joe believes that Glen has been fatally injured, and he and his team rush back to Earth to put Glen in cryogenic suspension, where the little life force that remains can be stretched out into several years of "half-life" in a moratorium, a neologism that points toward the liminal state that half-lifers occupy as they hover suspended between life and death. No sooner does Joe return to Earth, however, than he finds the world around him manifesting preternaturally rapid decay. Unlike earlier protagonists, who accepted the tomb world as reality, Joe puzzles over where the boundaries between inside and outside should be drawn. "It is-n't the universe which is being entombed," he thinks. "All this is going on within me, and yet I seem to see it outside. . . . Is the whole world inside me? Engulfed by my body?"[31] The mystery seems to clear when he re-ceives messages inscribed on various media—recorded telephone calls, matchbook covers, parking citations, bathroom graffiti—implying that it was he and his team who were fatally injured in the explosion. In this ver-sion of reality, Runciter is "outside" in a moratorium trying to communicate with those "inside" the world of dreaming half-lifers.

Although this explanation may account for the messages, Joe does not understand how it relates to the decay and regression, which soon attack people as well as objects. The woman he aspires to marry, Wendy Wright, who at first looks so "durable" that he can't imagine her aging—"she had too much control over herself and outside reality for that" (p. 55)—he later finds desiccated in his closet, "a huddled heap, dehydrated, almost mum-mified" (p. 90). He suspects that the culprit causing this disastrous intrusion of the entropic tomb world into his "reality" is Pat Conley, a dev-astatingly attractive woman with a lithe body, tell-tale dark eyes, and long black hair (p. 31). Pat is particularly dangerous because she can change what happened in the past, thus creating a different present—and, more-over, a present that other people do not realize is not the same present as the one they were living in a moment ago. After several members of Joe's party turn into desiccated heaps, the decay attacks him. As he crawls up the stairs to his room, heroically struggling so that he can die in decent privacy, Pat taunts him and gloats over his imminent demise, revealing herself as a particularly vicious instantiation of the schizoid woman. "The thing we call Pat," Joe significantly names her in this moment of confrontation (p. 159). Yet this signing of the dark-haired girl as an android, a spectacle that in ear-lier texts had proved so obsessively engrossing that it derailed the plot (wit-ness *We Can Build You*), here is revealed as a mere facade, behind which stands a more "authentic" reality. The real culprit responsible for the desic-cation, Joe discovers, is not Pat but the teenage Jory, characterized princi-pally by his *voracious appetite for life*. Condemned to half-life while still an adolescent, Jory maintains his crude but vibrant vitality by feeding on the life force of half-lifers weaker than himself. "I eat their life, what remains of it," Jory tells Joe Chip when he is exposed. "I need a lot of them. I used to wait until they had been in half-life awhile, but now I have to have them im-mediately" (p. 174).

If part of Dick's fascination with the schizoid woman/dark-haired girl comes from his guilt over Jane's death, Jory's unveiling moves toward untangling the complexities of the dark-haired girl/schizoid android configuration, for it brings into view the child who performs the fantasized act of eating what others need to survive. The flat horror that attends Jory's cannibalistic appetite recalls Dick's teenage phobia of eating in public. When Joe hears that Jory consumes his victims, he recoils in horror. "'How do you mean "ate"?' Literally? he wondered, his flesh undulating with aversion; the gross physical motion rolled through him, engulfing him, as if his body wanted to shrink away" (p. 173). He learns firsthand what Jory does when the youngster leaps at him. Even though the perceived actions must be happening symbolically rather than literally (in half-life each person is cryogenically suspended in his or her individual container, so physical assault is impossible), the description of the attack is vividly animalistic and horrific. "Snarling, Jory bit him. The great shovel teeth fastened into Joe's right hand. They hung on as, meanwhile, Jory raised his head, lifting Joe's hand with his jaw; Jory stared at him with unwinking eyes, snoring wetly as he tried to close his jaws" (pp. 175–76).

The intensity of Dick's revulsion here is unmistakable. Nevertheless, Jory finds a measure of absolution, as if Dick recognized that Jory is only doing what he must to survive. Ella Runciter, Glen's wife who entered half-life at age twenty-two and who is described as "pretty," "light-skinned," and blue-eyed in the tradition of Bonny Keller, acts like the mother that Dick might have wished he had. She insists, "God knows I detest Jory." But she also accepts Jory and his kind as a condition of life, seeing his preying on others as "a verity, a rule, of our kind of life." She urges Joe to come to terms with Jory's predation, insisting that moving to a new site won't help because there are "Jorys in every moratorium." When Joe insists that he wants to "defeat" Jory, she cautions, "I doubt if you can truly destroy him—in other words consume him—as he does to half-lifers placed near him in the moratorium" (p. 183). The language makes clear that consuming the predator cannot be a final solution, for the only way to achieve this illusory closure is by taking the predator "inside," thus symbolically and literally becoming a predator oneself.

As in Dick's other novels from this period, the psychological interpenetrates the social and the economic. It comes as no surprise, then, that Jory's appetite is also linked with the relentless capitalism that preempts the beginning of each chapter with a commercial for Ubik. When Joe asks Ella why Jory can't simply be physically removed from proximity with the half-lifers he preys on, she replies, "Herbert [the moratorium owner] is paid a great deal of money annually, by Jory's family, to keep him with the others and to think up plausible reasons for doing so" (p. 183). In the commercials that serve as epigraphs for the chapters, Ubik signifies all manner of capitalist predation, from used cars to foods, accompanied by the ominous disclaimer "Safe when used as directed." As in *Dr. Bloodmoney,* resolution of a personal crisis is mysteriously linked with the redemption of capitalism, even though resolution in the economic sphere is not logically motivated in the plot.

Thus when Jory is revealed and accepted as an inevitability, the meaning of Ubik is mysteriously changed as well. In the final epigraph, we find the following proclamation: "I am Ubik. Before the universe was, I am . . . I created the lives and the places they inhabit; I move them here, I put them there. They go as I say, they do as I tell them. . . . I am called Ubik, but that is not my name. I am. I shall always be" (p. 190). When Joe Chip discovers Ella as the force opposing Jory and muses, "I've reached the last entities involved," she replies caustically: "I don't think of myself as an 'entity.' I usually think of myself as Ella Runciter" (p. 182). Now, the unlocatable voice speaking the epigraph seems to reveal itself as the "final entity." To my knowledge, no one has attempted to explain why Ubik changes from signifying the worst excesses of capitalism to standing for a ubiquitous God. Many critics even suggest that Ubik has somehow "really" been God all along. I want to suggest that on the contrary, Ubik undergoes a sudden transformation and that this transformation cannot be understood except in relation to the revelation that behind Pat stands Jory and behind Jory stands his animalistic appetite. Only after acknowledging this appetite (which must be understood as operating on the multiple levels signified by "consuming") can the author discern, among the trashy surfaces of capitalist excess, the divine within the world—and by implication, within himself.

For if Ubik is intended to signify an ultimate "authentic" reality, it can do so only from a perspective *inside the text.* Outside the text (let us suppose, Derrida notwithstanding, that we can imagine such a vantage), Ubik must be none other than Philip K. Dick. It is ultimately Dick who "created the lives and the places [the characters] inhabit," who "put them there" in this text. Confused about where Ubik comes from, Joe at first assumes that Runciter has smuggled it to him, but Jory insists that no *objects* can come into the half-life world from the outside, only *words.* The distinction between words and things encoded into the character system of *Dr. Bloodmoney* is here invoked to remind us of the difference between resolution in fictional worlds, where writing has efficacy only in the performative realm of symbolic action, and resolution in the real world, where the materiality of things is often stubbornly resistant to verbal interventions. This split between words and things maps onto the split in Dick's life between past and present, sedimented psychological formation and present active intention. If he is writing his way toward resolving deep conflicts in his life, he can do so through words that have only a mediated and uncertain relation to the ghosts inhabiting his psyche. As a writer, he passes messages through his fiction into his own heart of darkness, hoping that somehow they might prove efficacious. Within the world of the text, the murmurs the half-lifers hear from the world "outside" trope this situation, for no *things* can pass between "inside" and "outside," only *words*. Joe Chip seems to comment on this aspect of Dick's writing when he remarks, from his perspective in half-life: "We are served by organic ghosts . . . who, speaking and writing, pass through this our new environment. Watching, wise, physical ghosts from the full-life world, elements of which have become for us invading but agreeable splinters of a substance that pulsates like a former heart" (pp. 188–89).

Ubik's distinctive achievement is to represent *simultaneously* the performative power of language and the mediated, uncertain relation of language to the material world while also mapping this difference onto an "inside"-"outside" boundary that hints at the complexity of communication between self and other, conscious and unconscious. The hope *Ubik* holds out is that although boundary disputes will never disappear, inside and outside can be made to touch each other through the medium of a writing that is no less valuable for infecting our world with all manner of epistemological and ontological instabilities.

Punctuating the Endless Regress of Reflexivity

Like Maturana and Varela, Dick is a system builder. He takes seriously an idea they also propose (following Spencer-Brown): that the observer creates a system by drawing a distinction between inside and outside. For Maturana and Varela, this move introduced certain instabilities into the foundation of their system, instabilities that they sought to fix by locating the formation of the system boundaries in reality as such. For Dick, such instabilities have potentially deadly consequences as subjects struggle to define boundaries that are "outside" relative to others' "inside." Consistent with their base in the biological sciences, Maturana and Varela tend to assume rational observers. The observer they posit is the kind of observer who sits in a laboratory watching an instrument dial. Granting constructive power to the observer may be epistemologically radical, but it is not necessarily politically or psychologically radical, for the rational observer can be assumed to exercise restraint. Dick makes no such assumption. His conclusions are darker and more psychologically complex because he is acutely aware of cases in which the stability of the observing mind cannot be assumed, especially when the act of creating a world may stimulate an already insatiable appetite for power and self-expansion. Thus Dick uses the inclusion of the observer to opposite effect. Whereas Maturana and Varela use the "domain of the observer" to recuperate everyday notions like cause and effect, Dick uses it to estrange further consensus reality.

Similarly, Dick shares with second-wave cybernetics an emphasis on reflexivity, though he changes its use and intent. Maturana sought to rescue the reflexivity inherent in autopoiesis from an infinite regress by asserting, "We do not see what we do not see, and what we do not see does not exist."[32] The infinite regress of observers watching other observers is thus contained, for the reflexive spiral does not continue past the boundaries of the observer. Sharing Maturana's passion for systems, Dick used a different strategy. Instead of bracketing "reality," he turned to the creation of ever more inclusive systems. His most ambitious attempt at system creation sprang from a series of visionary experiences he had in February and March of 1974, dates that he abbreviated as 2-3-74. To explain these experiences, he wrote a vast tract, entitled *Exegesis,* that ran to more than three thousand pages in length, selections of which have been published.

There is a possible physiological explanation, skeptically entertained by Dick's biographer Lawrence Sutin and embraced by others, for the visions of 2-3-74.[33] Within eight years Dick would die of a massive stroke, and throughout this period, he had extreme hypertension. The visions he experienced are consistent with symptoms experienced by people who have had small strokes in the brain, which can stimulate the auditory centers and cause hallucinations. Dick himself was skeptical of the visions and entertained numerous hypotheses about them. As if to make good Maturana's assertion that there can be no difference between a hallucination and reality for one who experiences the hallucination,[34] Dick finally concluded that the most likely explanation was that he had been contacted by an extraterrestrial intelligence he called the Vast Active Living Intelligent System, or VALIS, the subject of a final trilogy of novels that are among the best of his fiction. On November 17, 1980, he had another visionary experience, in which he believed God contacted him directly. For Dick, the contact solved the problem of infinite regress that inevitably haunts reflexive constructions. Wherever a regress appeared, the voice claimed that Dick had encountered the infinite, therefore the divine, therefore God. Here is Dick's account of the experience, as it is recorded in *Exegesis.*

[God] said, "I am the infinite. I will show you. Where I am, infinity is; where infinity is, there I am. Construct lines of reasoning by which to understand your experience in 1974. I will enter the field against their shifting nature. You think they are logical but they are not; they are infinitely creative."

I thought a thought and then an infinite regress of theses and countertheses came into being. God said, "Here I am; here is infinity." I thought another explanation; again an infinite series of thoughts split off in dialectical antithetical interaction. God said, "Here is infinity; here I am."[35]

For Dick, the construction of the observer cannot finally be separated from the construction of reality. Both end at the same point, in infinite regresses that, for mystical reasons, he chose to call God rather than a Maturanian reality that remains outside the compass of human knowing. In this way, Dick constructs an outside, authorized with the name of God and made invulnerable by continuing to infinity, an outside that is safe from being co-opted and forced to become an "inside." The irony, of course, is that this very construction may itself have been precipitated by a physical event inside his brain.

In contrast to the ambitious system building that Dick undertook in response to the visions of 2-3-74, his fiction of the mid-sixties tends toward a different kind of affirmation, one that I find more appealing. It can be illustrated by the ending of *Do Androids Dream.* After Deckard succeeds in killing the last three androids on his list and returns home to find out that Rachael has pushed his beloved goat off the roof, he is so exhausted and demoralized that he heads for the desolate country north of San Francisco, "where no living thing would go. Not unless it felt that the end had come."[36] There he has a visionary experience. Hundreds of miles from his government-sanctioned empathy box, he feels a rock hit him as he stands on top of a dead hill, as if he were enacting Mercer's endless scenario. Panicked, he calls the office on the car phone and tells his secretary: "I'm Wilbur Mercer. I've permanently fused with him. And I can't unfuse" (p. 207). Here the expansion of the ego comes not from megalomania but from suffering and inner conflict, a result of the empathy that Deckard has increasingly come to feel for the androids. Psychologically if not intellectually, he has refused the distinctions that make androids fair game for bounty hunters. The fusion experience ends only when he spies a toad, an animal sacred to Mercer, half-buried in the lifeless dust. Awed by his discovery, he takes it as a sign that he is meant to go on living.

When he returns home to Iran, his wife, and shows her the toad, she discovers the trap door hidden in its belly. It is electrical, like the ersatz sheep with which he tried to fool his neighbors. The sign is a delusion; the miracle is a fake. But this ironic turn is not quite the end. He tells Iran that the delusion doesn't matter: "The electric things have their lives, too. Paltry as those lives are" (p. 214). When Iran tells him to dial "Long deserved peace" on the mood organ, he agrees but falls asleep without it, finding peace without needing a cybernetic benediction. She then orders artificial flies for the electric toad to eat, showing a tenderness toward her husband that was notably lacking at the beginning of the novel. Although nothing has happened to explain how she moved from bitterness to tenderness, it is as if Deckard's struggle with the schizoid android has somehow resolved tensions in their relationship as well. The symbolic way in which resolution occurs emphasizes that no big problems are solved here. Only a modest accommodation has been reached, infused with multiple ironies, that emphasizes survival and the mixed condition of humans who are at their best when they show tolerance and affection for the creatures, biological and mechanical, with whom they share the planet. One could do worse than to accept this as a fitting conclusion to the deep epistemological and ethical problems that second-wave cybernetics raised but did not conclusively solve.

Chapter Eight

THE MATERIALITY OF INFORMATICS

Every epoch has beliefs, widely accepted by contemporaries, that appear fantastic to later generations. Of such are New Historical studies made—with good reason, for understanding the constellation of practices, metaphors, and presuppositions that underlie apparently bizarre beliefs opens a window onto a culture's ideology. One contemporary belief likely to stupefy future generations is the postmodern orthodoxy that the body is primarily, if not entirely, a linguistic and discursive construction. Coincident with cybernetic developments that stripped information of its body were discursive analyses within the humanities, especially the archaeology of knowledge pioneered by Michel Foucault, that saw the body as a play of discourse systems. Although researchers in the physical and human sciences acknowledged the importance of materiality in different ways, they nevertheless collaborated in creating the postmodern ideology that the body's materiality is secondary to the logical or semiotic structures it encodes.

It is not difficult to find pronouncements supporting an ideology of disembodiment in cultural theory, no less than in cybernetics. Consider, for example, the following claims. "The human body, our body, seems superfluous in its proper expanse, in the complexity and multiplicity of its organs, of its tissue and functions, because today everything is concentrated in the brain and the genetic code, which alone sum up the operational definition of being," Jean Baudrillard wrote in *The Ecstasy of Communication*.[1] Arthur Kroker and Marilouise Kroker out-Baudrillard Baudrillard in *Body Invaders: Panic Sex in America,* imagining "second-order simulacra" and "floating body parts" that herald the disappearance of the body into a fluid and changing display of signs. "If, today, there can be such an intense fascination with the fate of the body, might this not be because the body no longer exists?" they ask in what they evidently believe is a rhetorical question. They count the ways the body is disappearing: ideologically, into the signs of fashion; epistemologically, as the Cartesian consciousness guaranteeing its existence falls apart (that "grisly and false sense of subjectivity"); semiotically, into tattoos and floating signs; and technologically, into "ultra refuse" and "hyper-functionality."[2] O. B. Hardison concludes his disappearing act by writing the body into computers. He pensively observes, "No matter what precautions are taken, no matter how lucky the body is, in the end it betrays itself." Echoing Hans Moravec, he imagines "the relation between carbon man and the silicon devices he is creating" to be like "the relation between the caterpillar and the iridescent, winged creature that the caterpillar unconsciously prepares to become."[3]

How can we account for these ecstatic pronouncements and delirious dreams? As I suggested in chapters 1 and 2, I believe they should be taken as evidence not that the body has disappeared but that a certain kind of subjectivity has emerged. This subjectivity is constituted by the crossing of the materiality of informatics with the immateriality of information.[4] The very theorists who most emphatically claim that the body is disappearing also operate within material and cultural circumstances that make the claim for the body's disappearance seem plausible. The body's dematerialization depends in complex and highly specific ways on the *embodied* circumstances that an ideology of dematerialization would obscure. Excavating these connections requires a way of talking about the body responsive to its construction as discourse/information and yet not trapped within it. This chapter suggests a new, more flexible framework in which to think about embodiment in an age of virtuality. This framework comprises two dynamically interacting polarities. The first polarity unfolds as an interplay between the body as a cultural construct and the experiences of embodiment that individual people within a culture feel and articulate. The second polarity can be understood as a dance between inscribing and incorporating practices. Since the body and embodiment, inscription and incorporation, are in constant interaction, the distinctions forming these polarities are heuristic rather than absolute. They nevertheless play an important role in understanding the connections between an ideology of immateriality and the material conditions that produce the ideology.

Thus one purpose of this chapter is to develop a theoretical framework to integrate the two camps of abstraction and embodiment that have been sitting uneasily side by side since my discussion of the Macy Conferences. A second purpose is to demonstrate the usefulness of the framework for reading texts. William Burroughs's *The Ticket That Exploded* serves as my example, in part because Burroughs turns the tables on those who advocate disembodiment. Instead of discourse dematerializing the body, in *Ticket* the body materializes discourse. Situating this text in the high technology (for its time) of magnetic tape-recording, I demonstrate how the theoretical framework can be used to foreground embodiment while still being attentive to the complexities of representational codes. To prepare the ground for these discussions, let us turn now to a brief consideration of Foucault's archaeology and its treatment of embodiment.

Foucault's Archaeology and the Erasure of Embodiment

Acknowledging that the Panopticon was never built, Foucault nevertheless argues that it "must not be understood as a dream building; it is the diagram of a mechanism of power reduced to its ideal form; its functioning, abstracted from any obstacle, resistance or friction, must be represented as a pure architectural and optical system; it is in fact a figure of political technology that may and must be detached from any specific use."[5] On the one hand, the abstraction of the Panopticon beyond "any obstacle, resistance or friction" into a system of disciplines dispersed throughout society gives Foucault's analysis its power and universality. On the other hand, it diverts attention away from how actual bodies, in their cultural and physical specificities, impose, incorporate, and resist incorporation of the material practices he describes.

It is not coincidental that the Panopticon abstracts power out of the bodies of disciplinarians into a universal, disembodied gaze. On the contrary, it is precisely this move that gives the Panopticon its force, for when the bodies of the disciplinarians seem to disappear into the technology, the limitations of corporeality are hidden. Although the bodies of the disciplined do not disappear in Foucault's account, the specificities of their corporealities fade into the technology as well, becoming a universalized body worked upon in a uniform way by surveillance techniques and practices. When actual situations involving embodied agents are considered, limits appear that are obscured when the Panopticon is considered only as an abstract mechanism. Failing to recognize these limits, Foucault's analysis reinscribes as well as challenges the presuppositions of the Panoptic society. Foucault thus participates in, as well as deconstructs, the Panoptic move of disembodiment. Exposing the assumptions underlying Panoptic society, his analysis also fetishistically reconstructs them by positing a body constituted through discursive formations and material practices that erase the contextual enactments embodiment always entails.[6] A useful antidote to this view is Elaine Scarry's study of torture in *The Body in Pain: The Making and Unmaking of the World.*[7] Like Foucault, Scarry interrogates the cultural assumptions and political purposes that underlie the use of torture; also like Foucault, when she talks about assaults on the body, she uses representations to bring them into the realm of discourse (what other choice could there be?). But unlike Foucault's discussions, her representations are crafted to emphasize that bodily practices have a physical reality that can never be fully assimilated into discourse.

Although the absorption of embodiment into discourse imparts interpretive power to Foucault, it also limits his analysis in significant ways. Many commentators have criticized the universality of the Foucaultian body; this universality is a direct result of concentrating on discourse rather than embodiment.[8] Fissuring along lines of class, gender, race, and privilege, embodied practices create heterogeneous spaces even when the discursive formations describing those practices seem uniformly dispersed throughout the society. The assimilation of embodiment into discourse has the additional disadvantage of making it difficult to understand exactly how certain practices spread through a society. Foucault delineates the transformations that occurred when corporeal punishment gave way to surveillance, but the engine driving these changes remains obscure. Focusing on embodiment would help to clarify the mechanisms of change, for it links a changing technological landscape with the instantiated enactments that create feedback loops between materiality and discourse. Building on Foucault's work while going beyond it requires understanding how embodiment moves in conjunction with inscription, technology, and ideology. Attentive to discursive constructions, such an analysis would also examine how embodied humans interact with the material conditions in which they are placed.

Elizabeth Grosz makes a good start in her valuable study, *Volatile Bodies: Toward a Corporeal Feminism.*[9] She argues that the mind/body split, pervasive in the Western tradition, is so bound up with philosophical thinking that philosophy literally cannot conceive of itself as having a body. "Philosophy has always considered itself a discipline concerned primarily or exclusively with ideas, concepts, reason, judgment—that is, with terms clearly framed by the concept of mind, terms which marginalize or exclude considerations of the body" (p. 4). Even those philosophers who do take embodiment seriously tend unreflectingly to take the male body as the norm, as Grosz shows in discussing a range of theorists, including Merleau-Ponty, Freud, Lacan, Nietzsche, Foucault, and Deleuze and Guattari. Reading these male writers to find resources for a feminist understanding of embodiment, she offers as a model a Möbius strip in which outside becomes inside becomes outside. The attraction of the model for her is that it undercuts dichotomies by having one turn into the other. So she structures her book by first showing how models of the psyche produce the body, the "inside out," and then how the body produces the psyche, the "outside in."

En route to this analysis, she makes an important observation: "Indeed, there is no body as such; there are only *bodies*—male or female, black, brown, white, large or small—and the gradations in between. Bodies can be represented or understood not as entities in themselves or simply on a linear continuum with its polar extremes occupied by male and female bodies . . . but as a field, a two-dimensional continuum in which race (and possibly even class, caste, or religion) form body specifications" (p. 19). Although I am fully sympathetic with Grosz's project, the Möbius strip model has limitations, as she recognizes (pp. 209–10). In particular, the imperceptible transformations of inside/outside make it difficult to chart gradations within the continuum. It seems to me that the "field" in which bodies take shape may profitably be represented as an interplay between two intersecting axes. The polarities defining the end points of the axes acknowledge the historical importance of dichotomies, but the field itself is generated by the interplay *between* these end points.

To delineate this field, let me begin by clarifying what I mean by *embodiment*, an understanding that aligns itself with Grosz's comment that "there is no body as such; there are only *bodies*." Embodiment differs from the concept of the body in that the body is always normative relative to some set of criteria. To explore how the body is constructed within Renaissance medical discourse, for example, is to investigate the normative assumptions used to constitute a particular kind of social and discursive concept. Normalization can also take place with someone's particular experiences of embodiment, converting the heterogeneous flux of perception into a reified stable object. In contemporary scientific visualization technologies such as positron-emission tomography (PET), for example, embodiment is converted into a body through imaging technologies that create a normalized construct averaged over many data points to give an idealized version of the object in question.[10] In contrast to the body, embodiment is contextual, enmeshed within the specifics of place, time, physiology, and culture, which together compose enactment. Embodiment never coincides exactly with "the body," however that normalized concept is understood. Whereas the body is an idealized form that gestures toward a Platonic reality, embodiment is the specific instantiation generated from the noise of difference. Relative to the body, embodiment is other and elsewhere, at once excessive and deficient in its infinite variations, particularities, and abnormalities.

During any given period, experiences of embodiment are in continual interaction with constructions of the body. Consider, for example, the stress put on the vaginal orgasm during the early part of the twentieth century across a range of cultural sites, from Freudian psychoanalysis to the novels of D. H. Lawrence. Women's experiences of embodiment interacted with this concept in a variety of ways. Some women disciplined their experiences to bring them into line with the concept; others registered their experiences as defective because they were other than the concept; still others were skeptical about the concept because it did not match their experiences. Experiences of embodiment, far from existing apart from culture, are always already imbricated within it. Yet because embodiment is individually articulated, there is also at least an incipient tension between it and hegemonic cultural constructs. Embodiment is thus inherently destabilizing with respect to the body, for at any time this tension can widen into a perceived disparity.

Foucault is not exceptional in focusing on the body rather than embodiment. Most theorists who write on corporeality make the same choice, for theory by its nature seeks to articulate general patterns and overall trends rather than individual instantiations. Theories, like numbers, require a certain level of abstraction and generality to work. A theory that did not generalize would be like the number scheme that Jorge Luis Borges imagines in "Funes the Memorious."[11] Funes, blessed or cursed by a head injury that enables him to remember each sensation and thought in all its particularity and uniqueness, proposes that each number be assigned a unique, nonsystematic name bearing no relation to the numbers that come before and after it. If embodiment could be articulated separate from the body—an impossibility for several reasons, not least because articulation systematizes and normalizes experiences in the act of naming them—it would be like Funes's numbers, a froth of discrete utterances registering the continuous and infinite play of difference.

Yet there are theories, like this one, that abstractly and generally insist on the importance of the particular. Michel de Certeau, for example, provides a useful corrective to Foucault in pointing to the importance of individual articulations of cultural appropriations.[12] Embodiment is akin to articulation in that it is inherently performative, subject to individual enactments, and therefore always to some extent improvisational. Whereas the body can disappear into information with scarcely a murmur of protest, embodiment cannot, for it is tied to the circumstances of the occasion and the person. As soon as embodiment is acknowledged, the abstractions of the Panopticon disintegrate into the particularities of specific people embedded in specific contexts. Along with these particularities come concomitant strategies for resistances and subversions, excesses and deviations.

It is primarily the body that is naturalized within a culture; embodiment becomes naturalized only secondarily through its interactions with concepts of the body. Consequently, when theorists uncover the ideological underpinnings of naturalization, they denaturalize the body rather than embodiment. As the example of Foucault illustrates, it is possible to deconstruct the *content* of the abstraction while still leaving the *mechanism* of abstraction intact. Moving out of the frictionless and disembodied realm of abstraction requires articulating embodiment and the body together. How can this articulation be accomplished without simply absorbing embodiment back into the body?

One possibility is to complicate and enrich the tension between embodiment and the body by juxtaposing this tension with another binary distinction—inscription and incorporation—that partly converges and partly diverges from it. I envision these two bimodalities acting in complex syncopation with each other, like two sine waves moving at different frequencies and with different periods of repetition. How does the inscription/incorporation coupling relate to body/embodiment? Like the body, inscription is normalized and abstract, in the sense that it is usually considered as a system of signs operating independently of any particular manifestation. In Foucault's analysis of Linnaeus's biological taxonomies, it does not matter whether the taxonomies were originally printed in Gothic or Roman type; their significance derives from the concepts they express, not from the medium in which they appear. When the concepts are transported from one medium to another, for instance by being cited in Foucault's text and thus printed in a different typeface, the original medium disappears from sight. Moreover, even the awareness that the original medium has disappeared is erased by the implicit assumption that Linnaeus's words have been exactly reproduced. Such writing practices are so common that we do not normally attend to them. I foreground them now to point out that they constitute inscription as a conceptual abstraction rather than as an instantiated materiality.

In contrast to inscription is incorporation. An incorporating practice such as a good-bye wave cannot be separated from its embodied medium, for it exists as such only when it is instantiated in a particular hand making a particular kind of gesture. It is possible, of course, to abstract a sign from the embodied gesture by representing it in a different medium, for example by drawing on a page the outline of a stylized hand, with wavy lines indicating motion. In this case, however, the gesture is no longer an incorporating practice. Rather, it has been transformed precisely into an inscription that functions as if it were independent of any particular instantiation.

This line of thought leads to the following homology: as the body is to embodiment, so inscription is to incorporation. Just as embodiment is in constant interplay with the body, so incorporating practices are in constant interplay with inscriptions that abstract the practices into signs. When the focus is on the body, the particularities of embodiment tend to fade from view; similarly, when the focus is on inscription, the particularities of incorporation tend to fade from view. Conversely, when the focus shifts to embodiment, a specific material experience emerges out of the abstraction of the body, just as the particularities of an incorporating practice emerge out of the abstraction of inscription. Embodiment cannot exist without a material structure that always deviates in some measure from its abstract representations; an incorporating practice cannot exist without an embodied creature to enact it, a creature who always deviates in some measure from the norms. One path into further understanding the articulation between embodiment and the body, then, is to explore the connection between inscribing and incorporating practices.

Incorporating Practices and Embodied Knowledge

The distinction between incorporating and inscribing practices, a distinction implicit in Maurice Merleau-Ponty's *Phenomenology of Perception,*[13] has been developed further by Paul Connerton in *How Societies Remember.* Following Connerton, I mean by an *incorporating practice* an action that is encoded into bodily memory by repeated performances until it becomes habitual. Learning to type is an incorporating practice, as both Connerton and Merleau-Ponty observe. When we say that someone knows how to type, we do not mean that the person can cognitively map the location of the keys or can understand the mechanism producing the marks. Rather, we mean that this person has repeatedly performed certain actions until the keys seem to be extensions of his or her fingers. Someone can know how to type but not know how to read the words produced, such as when a typist reproduces script in a language that the typist does not speak; conversely, just as someone can be able to read a typescript without knowing how to type. The body's competencies and skills are distinct from discourse, although in some contexts they can produce discourse or can be read discursively. This is Connerton's point when he notes that the meaning of a bodily practice "cannot be reduced to a sign which exists on a separate 'level' outside the immediate sphere of the body's acts. Habit is a knowledge and a remembering in the hands and in the body; and in the cultivation of habit it is our body which 'understands.'"[14]

In distinguishing between inscribing practices and incorporating practices, I do not mean to imply that incorporating practices are in any sense more "natural," more universal, or less expressive of culture than inscribing practices. The body is enculturated through both kinds of practices. Characteristic ways of sitting, gesturing, walking, and moving are culturally specific, just as are characteristic ways of talking and writing. Moreover, culture not only flows from the environment into the body but also emanates from the body into the environment. The body produces culture at the same time that culture produces the body. Posture and the extension of limbs in the space around the body, for example, convey to children the gendered ways in which men and women occupy space. These nonverbal lessons are frequently reinforced verbally; "boys don't walk like that," or "girls don't sit with their legs open." It is significant that verbal injunctions often take a negative form, as in these illustrations, for the positive content is much more effectively conveyed through incorporating rather than inscribing practices. Showing someone how to stand is easy, but describing in words all the nuances of the desired posture is difficult. Incorporating practices perform the bodily content; inscribing practices correct and modulate the performance. Thus incorporating and inscribing practices work together to create cultural constructs. Gender, the focus of these examples, is produced and maintained not only by gendered languages but also by gendered body practices that serve to discipline and incorporate bodies into the complex significations and performances that constitute gender within a given culture.[15]

Because incorporating practices are always performative and instantiated, they necessarily contain improvisational elements that are context-specific. Postures are generalizable to some extent, but their enactments also depend on the specifics of the embodied individual: the precise length of limbs and torso, the exact musculature connecting tendon and joint, the sedimented history of body experience shaping the muscle tension and strength. Incorporation emerges from the collaboration between the body and embodiment, between the abstract model and the specific contexts in which the model is instantiated. In contrast to inscription, which can be transported from context to context once it has been performed, incorporation can never be cut entirely free from its context. As we shall see, the contextual components of incorporation give it qualities that are distinctively different from those of inscription.

Just as incorporating practices are not necessarily more "natural" than inscribing practices, so embodiment is not more essentialist than the body. Indeed, it is difficult to see what essentialism would mean in the context of embodiment. Essentialism is normative in its impulse, denoting qualities or attributes shared by all human beings. Though it is true that all humans share embodiment, embodied experience is dispersed along a spectrum of possibilities. Which possibilities are activated depends on the contexts of enactment, so that no one position is more essential than any other. For similar reasons, embodiment does not imply an essentialist self. As Francisco Varela, Evan Thompson, and Eleanor Rosch argue in *The Embodied Mind,* a coherent, continuous, essential self is neither necessary nor sufficient to explain embodied experience.[16] The closer one comes to the flux of embodiment, Varela and his coauthors believe, the more one is aware that the coherent self is a fiction invented out of panic and fear. In this view, embodiment subversively undercuts essentialism rather than reinforces it.

If embodiment is not essentialist, it is also not algorithmic. This conclusion has important implications for debates over what difference embodiment makes to thinking and learning. In *What Computers Can't Do,* Hubert Dreyfus argues that many human behaviors cannot be formalized in a heuristic program for a digital computer because these behaviors are embodied. For Dreyfus, embodiment means that humans have available to them a mode of learning, and hence of intellection, different from that deriving from cogitation alone. He gives the example of a child learning to pick up a cup. The child need not have an analytic understanding of the motor responses and dynamics involved in this action; the child need only flail around until managing to connect. Then, to learn the action to be able to perform it at will, the child only has to repeat what was done before. At no point does the child have to break down the action into analytical components or explicit instructions.

The advantage of this kind of learning is that everything does not need to be specified in advance. Moreover, the learning can be structured into complex relations without the necessity of a formal recognition that the relations exist. Drawing from Maurice Merleau-Ponty, Karl Polanyi, Jean Piaget, and other phenomenologists, Dreyfus delineates three functions that are characteristic of embodied learning and are not present in computer programs: an "inner horizon" that consists of a partly determined, partly open context of anticipation; the global character of the anticipation, which relates it to other pertinent contexts in fluid, shifting patterns of connection; and the transferability of such anticipation from one sense modality to another.[17] One implication of this view of embodied learning is that humans know much more than they consciously realize they know. Another is that this embodied knowledge may not be completely formalizable, since the openness of the horizon allows for ambiguities and new permutations that cannot be programmed into explicit decision procedures. As we shall see in chapter 9, this provides researchers in mobile robotics such as Rodney Brooks with an argument for the superiority of mobile (embodied) robots over computer programs, which have no capacity to move about and explore the environment. In ways Dreyfus did not anticipate, artificial-life researchers have moved closer to his position than to the artificial-intelligence research programs that he argues against.

A further implication of embodied interaction with the environment is developed by Pierre Bourdieu. He argues that even if one is successful in reducing some area of embodied knowledge to analytical categories and explicit procedures, one has in the process changed the kind of knowledge it is, for the fluid, contextual interconnections that define the open horizons of embodied interactions would have solidified into discrete entities and sequential instructions. He makes this point—that largely unnoticed and unacknowledged changes occur when embodied knowledge is expressed through analytical schema—in his discussion of the seasonal rituals of the Kabyle, a group of Berber tribes living in Algeria and Tunisia. The calendar that the Kabyle enact through improvisational embodied practices is not the same calendar that the anthropologist extracts in schematic form from data provided by informants. Whereas the anthropologist's schema will show fields, houses, and calendars arranged according to such dualities as hot and cold, male and female, for the Kabyle this knowledge exists not as abstractions but as patterns of daily life learned by practicing actions until they become habitual. Abstraction thus not only affects how one describes learning but also changes the account of what is learned.

Bourdieu's work illustrates how embodied knowledge can be structurally elaborate, conceptually coherent, and durably installed without ever having to be cognitively recognized as such. Through observation and repetition, the child attains "a practical mastery of the classificatory schemes which in no way implies symbolic mastery." By transposing terms of symmetry relations, the child is able to grasp the rationale of what Bourdieu calls the "habitus" (a word coined to recall the habitual nature of embodied actions), defined as the "durably installed generative principle of regulated improvisations."[18] The habitus, which is learned, perpetuated, and changed through embodied practices, should not be thought of as a collection of rules but as a series of dispositions and inclinations that are both subject to circumstances and durable enough to pass down through generations. The habitus is conveyed through the orientation and movement of the body as it traverses cultural spaces and experiences temporal rhythms. For the Kabyle, the spatial arrangements of home, village, and field instantiate the dichotomies that serve as generative principles stimulating improvisation within the regulated exchanges defined by the habitus. Living in these spaces and participating in their organization form the body in characteristic ways, which in turn provides a matrix of permutations for thought and action.

To look at thought in this way is to turn Descartes upside down. The central premise is not that the cogitating mind can be certain only of its ability to be present to itself but rather that the body exists in space and time and that, through its interaction with the environment, it defines the parameters within which the cogitating mind can arrive at "certainties," which not coincidentally almost never include the fundamental homologies generating the boundaries of thought. What counts as knowledge is also radically revised, for conscious thought becomes an epiphenomenon corresponding to the phenomenal base the body provides.

In "Eye and Mind," Merleau-Ponty articulates a vision similar to Bourdieu's when he states that the body is not "a chunk of space or a bundle of functions" but "an intertwining of vision and movement."[19] Whereas the causal thinking that Descartes admired in geometry and sought to emulate in philosophy erases context by abstracting experience into generalized patterns, embodiment creates context by forging connections between instantiated action and environmental conditions. Marking a turn from foundation to flux, embodiment emphasizes the importance of context to human cognition. Here, in another key and a reverse direction, we see replayed the decontextualization that information underwent when it lost its body. Just as disembodiment required that context be erased, so remembering embodiment means that context be put back into the picture.

When accounts of learning change, so do accounts of cultural transmission. In *How Societies Remember*, Paul Connerton links embodiment with memory. He points out that rituals, commemorative ceremonies, and other bodily practices have a performative aspect that an analysis of the content does not grasp. Like performative language, performative rituals must be enacted to take place. A liturgy, for example, "is an ordering of speech acts which occurs when, and only when, these utterances are performed; if there is no performance there is no ritual." Although liturgies are primarily verbal, they are not exclusively so. Gestures and movements accompany the words, in addition to the sense data created by speaking and hearing. Over and above (or better, below) the verbal aspects is the incorporation enacted through sensory responses, motor control, and proprioception. Because these ceremonies are embodied practices, to perform them is always in some sense to accept them, whatever one's conscious beliefs. "We may suppose the beliefs someone else holds sacred to be merely fantastic," Connerton wrote, "but it can never be a light matter to demand that their actual expression be violated. . . . To make patriots insult their flag or to force pagans to receive baptism is to violate them."[20]

Bodily practices have this power because they sediment into habitual actions and movements, sinking below conscious awareness. At this level they achieve an inertia that can prove surprisingly resistant to conscious intentions to modify or change them. By their nature, habits do not occupy conscious thought; they are habitual precisely because they are done more or less automatically, as if the knowledge of how to perform the actions resided in one's fingers or physical mobility rather than in one's mind. This property of the habitual has political implications. When a new regime takes over, it attacks old habits vigorously, for this is where the most refractory resistance to change will be met. Bourdieu comments that all societies wanting to make a "new man" should approach the task through processes of "deculturation" and "reculturation" focused on bodily practices. Hence revolutionaries place great emphasis "on the seemingly most insignificant details of *dress, bearing,* physical and verbal *manners,*" because "they entrust to [the body] in abbreviated and practical, i.e., mnemonic, form the fundamental principles of the arbitrary content of the culture."[21]

Bourdieu somewhat overstates the case when he asserts that "principles em-bodied in this way are placed beyond the grasp of consciousness and hence cannot be touched by voluntary, deliberate transformation" (p. 94), but he is correct in emphasizing the resistance of such practices to intellection. He also rightly sees the importance of these practices for education and discipline. "The whole trick of pedagogic reason," he observes, "lies precisely in the way it extorts the essential while seeming to demand the insignificant: in obtaining the respect for form and forms of respect which constitute the most visible and at the same time the best-hidden (because most 'natural') manifestation of submission to the established order. . . . The concession of *politeness* always contains *political* concessions" (pp. 94–95). Along similar lines, Connerton wrote: "Every group will entrust to bodily automatisms the values and categories which they are most anxious to conserve. They will know how well the past can be kept in mind by a habitual memory sedimented in the body."[22]

To summarize: four distinguishing characteristics of knowledge gained through incorporating practices have emerged from the discussion so far. First, incorporated knowledge retains improvisational elements that make it contextual rather than abstract, that keep it tied to the circumstances of its instantiation. Second, it is deeply sedimented into the body and is highly resistant to change. Third, incorporated knowledge is partly screened from conscious view because it is habitual. Fourth, because it is contextual, resistant to change, and obscure to the cogitating mind, it has the power to define the boundaries within which conscious thought takes place. To these four characteristics I want to add a fifth. When changes in incorporating practices take place, they are often linked with new technologies that affect how people use their bodies and experience space and time. Formed by technology at the same time that it creates technology, embodiment mediates between technology and discourse by creating new experiential frameworks that serve as boundary markers for the creation of corresponding discursive systems. In the feedback loop between technological innovations and discursive practices, incorporation is a crucial link.

Having distinguished between incorporating and inscribing practices, I want to explore the connections between them. To complete the model I have been constructing, I turn now to Mark Johnson's *The Body in the Mind*.[23] It is a truism in contemporary theory that discourse writes the body; Johnson illustrates how the body writes discourse. He shows that the body's orientation in time and space, deriving from such common experiences as walking upright and finding a vertical stance more conducive to mobility than a horizontal position, creates a repository of experiences that are encoded into language through pervasive metaphoric networks. Consider, for example, metaphors having to do with verticality. We speak of someone being "upright" in a moral or ethical sense, of people "at the top," and of "upscale" lifestyles. Depressing events are a "downer," in a recession people are "down on their luck," and entry-level people start at the "bottom of the ladder." The hierarchical structures expressed and constituted through these metaphors, Johnson argues, have a basis in bodily experience that reinforces and reinscribes their social and linguistic implications. Other common body experiences giving rise to extensive metaphoric networks include in/out, front/back, and contained/uncontained. Johnson characterizes such schema as prepropositional. The point of his inquiry is to show that these encoded experiences bubble up into language in propositional statements, such as "He got high," and in metapropositional statements having to do with the truth or goodness of propositions, such as "That statement expresses a higher truth." An obvious implication is that if we had bodies with significantly different physiological structures, for example exoskeletons rather than endoskeletons or unilateral rather than bilateral symmetries, the schema underlying pervasive metaphoric networks would also be radically altered.

Of the theorists discussed here, Johnson launches perhaps the most severe attack on objectivism. Thus it is ironic that he reinscribes objectivist presuppositions in positing a universal body unmarked by gender, ethnicity, physical disability, or culture. Insisting that the body is an important part of the context from which language emerges, he erases the specific contexts provided by embodiment. The consequences of this erasure can be seen in his discussion of a passage from *Men on Rape* in which a law clerk tells why rape is, in his view, sometimes justified. Johnson shows that the clerk's reasoning is based on a series of interrelated propositions that begin with the idea that "physical attraction is a physical force." The clerk constructs a woman's physical attractiveness as an aggression that she practices on men and to which they sometimes respond with (allegedly retaliatory) violence.

In some ways Johnson's analysis is remarkably astute, for it reveals how gendered experiences of embodiment get encoded into implicit propositions. Yet with stunning reticence, he never remarks on the gender politics so obviously foregrounded by this series of propositions, treating the example as if it were sexually and culturally neutral. More than one graduate student whom I have asked to read Johnson's book has thrown it down in disgust at this point, assuming that any analysis so gender-blind could have nothing significant to say to her. Although I sympathize with the reaction, it is a mistake, for the general point that embodiment is encoded into language through metaphoric networks is strengthened rather than undercut by insisting on the specificities of physically diverse and differentially marked bodies. Just as I can imagine that schema would vary for different physiologies, so I can envision that metaphors would vary in response to different experiences of embodiment created by historically positioned and culturally constructed bodies. From such considerations emerges an enriched appreciation of how inscribing and incorporating practices work together to create the heterogeneous spaces of postmodern technologies and cultures.

Although Johnson does not develop this implication, his analysis suggests that when people begin using their bodies in significantly different ways, either because of technological innovations or other cultural shifts, changing experiences of embodiment bubble up into language, affecting the metaphoric networks at play within the culture. At the same time, discursive constructions affect how bodies move through space and time, influence what technologies are developed, and help to structure the interfaces between bodies and technologies. By concentrating on a period when a new technology comes into being and is diffusing throughout the culture, one should be able to triangulate between incorporation, inscription, and technological materiality to arrive at a fuller description of these feedback loops than discursive analysis alone would yield. To develop such an analysis, I concentrate in the following section on an information technology appropriate to the era under discussion, specifically the use of magnetic tape-recording from the early 1900s to 1962, the year Burroughs published *The Ticket That Exploded*. Chapters 9 and 10 continue the analysis into an array of contemporary virtual technologies. Let us now return to an earlier period, when it was a startling discovery to learn that one's voice could be taken out of the body and put into a machine, where it could be manipulated to say something that the speaker had never heard before.

Audiotape and Its Cultural Niche

In his groundbreaking work *Reading Voices: Literature and the Phonotext,* Garrett Stewart asks not how we read, or why we read, but where we read.[24] He decides we read in the body, particularly in the vocal apparatus that produces subvocalization during silent reading. This subvocalization is essential, he argues, to the production of literary language. Language becomes literary for Stewart when it cannot be adequately replaced by other words, when that particular language is essential to achieving its effects. Literary language works by surrounding its utterances with a shimmer of virtual sounds, homophonic variants that suggest alternative readings to the words actually printed on the page. Subvocalization actualizes these possibilities in the body and makes them available for interpretation. Several interesting consequences flow from this argument. First, the bodily enactment of suppressed sound plays a central role in the reading process. Second, reading is akin to the interior monologue that we all engage in, except that it supplies us with another story, usually a more interesting one than that provided by the stream of subvocalized sound coming out of our own consciousness. Third, the production of subvocalized sound may be as important to subjectivity as it is to literary language.[25]

We are now in a position to think about what tape-recording means for certain literary texts. Audiotape opens the possibility that the voice can be taken out of the body and placed into a machine. If the production of sub-vocalized sound is essential to reading literary texts, what happens to the stories we tell ourselves if this sound is no longer situated in the body's sub-vocalizations but is in the machine? Often histories of technology and literature treat technology as a theme or subject to be represented within the world of the text. I want to take a different approach, focusing on the technical qualities of audiotape that changed the relation of voice and body, a change Burroughs associates in *Ticket* with the production of a new kind of subjectivity. In the mutating and metamorphosing bodies of *Ticket,* we can see a harbinger of the posthuman body that will be fully articulated in the following chapters. These mutations are intimately bound up with internal monologues that, in Burroughs's view, parasitically inhabit the body. But I am getting ahead of my story. First we need to trace the development of the audio technology that he uses to effect this startling view of discourse as a bodily infestation.

Born in the early 1900s and coming of age after World War II, audiotape may already be reaching old age, fading from the marketplace as it is re-placed by compact disks, computer hypermedia, and the like. The period when audiotape played an important role in U.S. and European consumer culture may well be limited to the four decades of 1950–90. Writing his cybernetic trilogy—*The Ticket That Exploded, The Soft Machine,* and *Nova Express*—in the late 1950s and 1960s, William Burroughs was close enough to the beginnings of audiotape to regard it as a technology of revolutionary power. Long after writing dissociated presence from inscription, voice continued to imply a subject who was present in the moment and in the flesh. Audiotape was of course not the first technology to challenge this assumption, and the cultural work it did can best be understood in the context of related audio technologies, particularly telephone, radio, and phonograph.

Telephone and radio broke the link between presence and voice by making it possible to transport voice over distance.[26] Before audiotape and phonograph, however, telephone and radio happened in the present. Speaker and listener, although physically separated, had to share the same moment in time. Telephone and radio thus continued to participate in the phenomenology of presence through the simultaneity that they produced and that produced them. In this sense they were more like each other than either was like the phonograph. By contrast, the phonograph functioned primarily as a technology of inscription, reproducing sound through a rigid disk that allowed neither the interactive spontaneity of telephone nor the ephemerality of radio.

The niche that audiotape filled was configured through the interlocking qualities of the audio technologies that preceded it, in a process Friedrich Kittler has aptly called "medial ecology" (discussed briefly in chapter 2).[27] Like the phonograph, audiotape was a technology of inscription, but with the crucial difference that it permitted erasure and rewriting. As early as 1888, Oberlin Smith, at one time president of the American Society of Mechanical Engineers, proposed that sound could be recorded by magnetizing iron particles that adhered to a carrier.[28] He was too busy to implement his idea, however, and the ball passed to Valdemar Poulsen, a young Danish engineer who accidentally discovered that patterns traced on the side of a magnetized tuning fork became visible when the fork was dipped in iron powder. When the fork was demagnetized, the patterns were erased. He saw in the imprinting and erasure of these patterns the possibility of a recording device for sound, using iron wire as the carrier. Its immediate commercial use, he imagined, would lie in providing tangible records of telephone conversations. He called the device a "Telegraphone," which he understood to signify "writing the voice at a distance." At the 1900 Paris Exposition, he won the Grand Prix for his invention.[29] Despite extensive publicity, however, he was not able to raise the necessary capital in Europe for its development. By 1903 the patents had passed to the American Telegraphone Company (ATC), which raised a huge amount of money ($5,000,000) by selling stock. Five years later the owners of ATC had still not built a single machine. Their main business, in fact, turned out to be raising money for the machines rather than actually producing the machines. When they did finally turn out a few operational devices in 1911, using the famous model Phoebe Snow to advertise them as dictation machines, the sound quality was so bad that the Dupont Company, after installing them in a central dictation system, ended up suing. The questionable status of the machines was exacerbated during World War I, when the Telefunken Company of America was accused of using them to encode and transmit secret messages to Germany. From the beginning, audiotape was marked with the imprint of international capitalism and politics as surely as it was with the imprint of voices.

By 1932, steel tape had become the carrier of choice in high-end machines, and the British Broadcasting Corporation (BBC) became actively interested in the development of steel tape, using it to carry the Christmas address of King George V in that year. Film tape, created by coating paper or plastic tape with iron oxide and feeding it through a ring-shaped head, appeared on the scene by 1935.[30] The great advantage of film tape was that it could be easily spliced, but originally it had such poor sound quality that it could not compete with steel tape. The problem of establishing good correspondence between sound frequency and the pattern on film tape (that is, controlling hysteresis) was partly solved by the introduction of high-frequency bias in 1938.[31] By 1941 the sound quality of film tape had so improved that it was competitive with steel tape in studio work. On the consumer market, machines with wire were still common. It was not until after World War II that systematic research was carried out to find the optimum coating material for film tape, and only in 1948 was the first American patent issued for a magnetic recording machine using film tape and a ring head. The use of film tape then expanded rapidly, and within a decade it had rendered steel tape obsolete, with film tape being used in the consumer market as well as the professional studios.

By the late 1950s, then, magnetic tape had acquired the qualities that, within the existing cultural formation, gave it the force of paradox. It was a mode of voice inscription at once permanent and mutable, repeating past moments exactly yet also permitting present interventions that radically altered its form and meaning. These interventions could, moreover, be done at home by anyone who had the appropriate equipment. Whereas the phonograph produced objects that could be consumed only in their manufactured form, magnetic tape allowed the consumer to be a producer as well. The switches activating the powerful and paradoxical technoconceptual actors of repetition and mutation, presence and absence, were in the hands of the masses, at least the masses who could afford the equipment.

The paradoxical qualities that magnetic tape was perceived to have in the late 1950s were forcefully expressed by Roy Walker, involved in making tape-recordings for the BBC during this period. "Anyone who has made a BBC recording and been in on the editing session may emerge feeling that he can no longer call himself his own. Cuts and transpositions can be and are made. Halves of sentences spoken at different times can be amalgamated to let a speaker hear himself say the opposite of what he knows he said. Hearing oneself say something and continue with something else said half an hour earlier can be peculiarly disconcerting. You might have the feeling that if you went quickly out of the studio you might catch yourself coming in."[32] His language locates the disconcerting effect both in the time delay ("sentences spoken at different times can be amalgamated") and in the disjunction between voice and presence ("he can no longer call himself his own"). When these qualities of audiotape were enacted within literary productions, a complex interplay was set up between representational codes and the specificities of the technology. When voice was displaced onto tape, the body metonymically participated in the transformations that voice underwent in this medium. For certain texts after 1950, the body became a tape-recorder.

When Burroughs wrote *The Ticket That Exploded,* he took seriously the possibilities for the metonymic equation between tape-recorder and body.[33] He reasoned that if the body can become a tape-recorder, the voice can be understood not as a naturalized union of voice and presence but as a mechanical production with the frightening ability to appropriate the body's vocal apparatus and use it for ends alien to the self. "The word is now a virus" (*TTE,* p. 49), the narrator says in a phrase indebted to the Buddhist-inspired idea that one's sense of selfhood is maintained through an internal monologue, which is nothing other than the story the self tells to assure itself that it exists.[34] Woven into this monologue are the fictions that society wants its members to believe; the monologue enacts self-discipline as well as self-creation. Burroughs proposes to stop the interior monologue by making it external and mechanical, recording it on tape and subjecting the recording to various manipulations. "Communication must become total and conscious before we can stop it," the narrator asserts (*TTE,* p. 51). Yet splicing tape is far from innocuous. Once someone's vocalizations and body sounds are spliced into someone else's, the effects can feed back into the bodies, setting off a riot of mutations. The tape-recorder acts both as a metaphor for these mutations and as the instrumentality that brings them about. The taped body can separate at the vertical "divide line," grotesquely becoming half one person and half another, as if it were tape spliced lengthwise. In a disturbingly literal sense, the tape-recorder becomes a two-edged sword, cutting through bodies as well as through the programs that control and discipline them.

In *The Ticket That Exploded,* the body is a site for contestation and resistance on many levels, as metaphor, as physical reality, as linguistic construct, and last but hardly least, as tape-recorder. The tape-recorder is central to understanding Burroughs's vision of how the politics of co-optation work. Entwined into human flesh are "pre-recordings" that function as parasites ready to take over the organism. These "pre-recordings" may be thought of as social conditioning, for example an "American upper middle-class upbringing with maximum sexual frustration and humiliations imposed by Middle-Western matriarchs" (*TTE,* p. 139), which not coincidentally matches Burroughs's own experience. A strong sense of sexual nausea pervades the text, and sexuality is another manifestation of prerecording. Parodically rewriting the fable in Plato's *Symposium* about the spherical beings who were cut in half to make humans, the narrator asserts: "All human sex is this unsanitary arrangement whereby two entities attempt to occupy the same three-dimensional coordinate points giving rise to the sordid latrine brawls. . . . It will be readily understandable that a program of systematic frustration was necessary in order to sell this crock of sewage as Immortality, and Garden of Delights, and *love*" (*TTE*, p. 52).

The idea of two entities trying to occupy the same space is further reinforced by the vertical "divide line" crossing the body, the physically marked line in bone, muscle, and skin where the neural canal of the month-old fetus closes to create the beginnings of the torso.[35] The early point at which the "divide line" is imprinted on human flesh suggests how deeply implicated into the organism are the prerecordings that socialize it into community norms. In one scene the narrator sees his body "on the operating table split down the middle," while a "doctor with forceps was extracting crab parasites from his brain and spine—and squeezing green fish parasites from the separated flesh." "My God what a mess," the doctor exclaims. "The difficulty is with two halves—other parasites will invade sooner or later. . . . Sew him up nurse" (*TTE*, p. 85). As the doctor intimates, the body is always already fallen. Divided within itself rather than an organic unity, it is subject to occupation and expropriation by a variety of parasitic forms, both cultural and physical.

Chief among these parasitic forms is "the word." It is a truism in contemporary theory that discursive formations can have material effects in the physical world. Without having read Foucault and Derrida, Burroughs came to similar conclusions a decade earlier, imagining "the word" as the body's "Other Half." The narrator stated: "Word is an organism. The presence of the 'Other Half' a separate organism attached to your nervous system on an air line of words can now be demonstrated experimentally." The experiments to which the narrator alludes were performed by, among others, John Cunningham Lilly, who in the late 1950s and 1960s used isolation tanks to test the malleability of human perception.[36] The experiments required subjects to enter a dark tank and to float, cut off from all sensory input, in water kept at body temperature. The narrator mentions that a common "hallucination" of subjects in sense withdrawal was "the feeling of another body sprawled through the subject's body at an angle" (*TTE* p. 49). "Yes quite an angle," the narrator ironically remarks, identifying the sensation as the subject's perception of his "other half,'" the word virus that invades the organism until it seems as intrinsic to the body as flesh and bone.

For the narrator, the proof of this parasitic invasion and infection is the interior monologue we all experience. "Modern man has lost the option of silence," he asserts. "Try to achieve even ten seconds of inner silence. You will encounter a resisting organism that *forces you to talk*. That organism is the word" (*TTE*, pp. 49–50). Burroughs's project is to offer the reader as many ways as he can imagine to stop the monologue, to rewrite or erase the "pre-recordings," and to extricate the subject from the parasitic invasion of the "Other Half." Tape-recorders are central to this project; "it's all done with tape recorders," the narrator comments (*TTE*, p. 162). One strategy is to "externalize dialogue" by getting "it out of your head and into the machines" (*TTE*, p. 163). He suggests that the reader record the last argument the reader had with a boyfriend or girlfriend, putting the reader's side of the argument on one recorder and the friend's side on the other. Then the two recorders can argue with each other, leaving the human participants free to stop replaying the conversation in their heads and get on with their lives. The narrator also suggests recording random sounds on a third machine—snippets from a news broadcast, say—and mixing them in too. The intrusion of the random element is significant; it aims to break the reader not only out of personal obsessions but also out of the surrounding, culturally constructed envelope of sounds and words. "Wittgenstein said: 'No proposition can contain itself as an argument,'" the narrator remarks, interpreting this as follows: "The only thing *not* prerecorded in a prerecorded universe is the prerecording itself which is to say *any* recording that contains a random factor" (*TTE*, p. 166).

The intrusion of randomness is important in another way as well, for Burroughs is acutely aware of the danger that he might, through his words, spread the viral infection he is trying to combat. It is important, therefore, that disruptive techniques be instantiated within the text's own language. These techniques range from his famous use of the "cut-up," where he physically cuts up previously written narratives and arbitrarily splices them together, to more subtle methods such as shifting between different linguistic registers without transition or explanation.[37] Perhaps the single most important device is the insistent pressure to take metaphors literally—or put another way, to erase the distinction between words and things. Language is not merely like a virus; it *is* a virus, replicating through the host to become visible as green fish in the flesh and crab parasites tearing at the base of the spine and brain. In Burroughs, the material effects of language do not need to be mediated through physical discipline to re-form the body, for example through the prescribed postures and gestures used to teach penmanship in the eighteenth and nineteenth centuries. With a writer's license, he makes language erupt directly into the body. The body itself, moreover, is treated as if it physically were a recorder, regulated by the principles that govern magnetic tape in its reproduction, erasure, and reconfiguration. Here, within the represented world of the text, techniques of inscription merge with incorporated practices in a cyborg configuration of explosive potential. The double edge of this potential is not difficult to understand, for the reifying and infective power of words can be defused only through other words, which can always turn against their master and become infectious in turn. Making the word flesh is both how the virus infects and how the vaccine disinfects. In either case, the flesh will not continue unchanged.

The pressure toward literalization can be seen in the narrative sections that use the conventions of science fiction to figure the invasion of the word as a physical operation (early on the narrator announces, "I am reading a science fiction book called *The Ticket That Exploded*" [*TTE*, p. 5]). On this track, Earth has been invaded by the alien Nova mob, so-called because their strategy is to drive the planet to extreme chaos or "nova." The mob includes such creatures as heavy-metal addicts from Uranus, sex addicts from Venus, and other parasitic organisms that can occupy human flesh. "Nova criminals are not three-dimensional organisms—(though they are quite definite organisms as we shall see)—but they need three-dimensional human agents to operate" (*TTE*, p. 57). A single parasitic alien can take over hundreds of humans, stringing together its hosts to form rows of "coordinate points," analogous to lines of print or to phonemes subordinated through grammar and syntax. The reputed leader of the mob is an appropriately bimorphic creature called variously "Mr Bradly Mr Martin," "Mr and Mrs D," or simply "the Ugly spirit." In this instantiation of the "Other Half," the word itself is split down the middle.

A counterinvasion has been staged by the Nova police, whose weapons include radio static, "camera guns" that destabilize images by vibrating them at supersonic speeds, and of course tape-recorders. Recruiting "Mr. Lee" (this pseudonym used often by Burroughs was his mother's maiden name), the district supervisor tells Mr. Lee that he will receive his instructions "from books, street signs, films, in some cases from agents who purport to be and may actually be members of the organization. There is no certainty. Those who need certainty are of no interest to this department. This is in point of fact a *non-organization* the aim of which is to immunize our agents against fear despair and death. We intend to break the birth-death cycle" (*TTE*, p. 10). One of the criminals the department seeks is Johnny Yen, whose name suggests sexual desire. "Death *is* orgasm *is* rebirth *is* death in orgasm *is* their unsanitary Venusian gimmick *is* the whole birth death cycle of action," the narrator explains. He proceeds, apparently exasperated, to make his point even more obvious. "You got it?—Now do you

understand who Johnny Yen is?—The Boy-Girl Other Half strip tease God
of sexual frustration—Errand boy from the death trauma" (*TTE*, p. 53).

On this track, the action can be read as a physical contest between the
Nova mob and the Nova police, as when a police operative from Minraud
blows a mob crab guard into smithereens. But if the word is a parasite with
material effects, the distinction between metaphor and actuality, representation and reality, is moot. Thus another strategy of resistance is the
"Rewrite Room," the space from which comes the exposé of Johnny Yen
cited above. Johnny Yen is not blown away but rather is rewritten to become
a rather enchanting green fish boy, an amphibious life-form (a benign bimorphic creature) living in the canals and mating with Ali the street boy in
a nonhuman life cycle that destabilizes the human sense of what constitutes
the body, life, and death. The crisis of mutation, the recognition that pattern is always already penetrated by randomness, is here associated with a
form of embodiment that moves through a froth of noise as easily as a fish
through water.

For human subjects, however, this destabilization is bound to be threatening rather than simply liberating, for the narrative attempts to put into
play all the boundaries that define human subjectivity. Body boundaries are
often literally disintegrated, for example by the Sex Skin, an organism that
surrounds its victims with a second skin that gives its victims intense sexual
pleasure while dissolving and ingesting them. Positioned against the clear
threat of this kind of sexual delirium are tape-recorders, potentially liberating but also not without danger. Recording one's body sounds and splicing
them into someone else's can free one from the illusion that body sounds
cannot exist apart from the interior monologue. But just as Burroughs's
words can become parasitic if not self-disrupted, so these sounds have the
potential to constitute a parasitic monologue in turn. According to the narrator, the splicing produces a strong erotic reaction. If it is expressed in actual sexual contact "it acts as an aphrodisiac . . . nothing more. . . . But when
a susceptible subject is spliced in with someone *who is not there* then it acts
as a destructive virus," ironically becoming the phenomenon it was meant
to counteract (*TTE*, p. 20).

As well as disrupting words audibly present, Burroughs wants to create—
or expose—new ones from the substrata of the medium itself. He describes experiments based on "inching tape," manually rubbing the tape
back and forth across the head at varying speeds. "Such exercises bring you
a liberation from old association locks . . . you will hear words that were not
in the original recording new words made by the machine different people
will scan out different words of course but some of the words are quite

clearly there." The technique gives new meaning to Marshall McLuhan's aphorism "The medium is the message," for it is "as if the words themselves had been interrogated and forced to reveal their hidden meanings it is interesting to record these words literally made by the machine itself" (*TTE*, p. 206). Here Burroughs envisions incorporating practices that can produce inscriptions without the mediation of consciousness.

He actually performed the tape-recorder experiments he describes from the late 1950s through the late 1970s. He inched tape to create, as he heard it, new words; he recorded radio broadcasts and spliced the tape to achieve an aural "cut-up"; and he held the microphone to the base of his throat and tried to record his own subvocal speech. As if anticipating Christian fundamentalists who hear Satanic messages hidden in records and tapes (people whose sensibilities he would no doubt enjoy outraging), he also read from his books, including *The Nova Express* and *The Ticket That Exploded*, and spliced the readings in with music played backward. The recordings have been preserved, and some of this archival material has been collected in a phonograph album entitled "Nothing Here Now but the Recordings."[38] Late one night I traveled to the music library at the University of California at San Diego to listen to the album. Even though the experience of sitting in the nearly deserted high-tech facility, insulated from exterior sound, was eerily conducive to hearing the words that Burroughs claims are there, some of the passages are clearly of historic interest only. In particular, the section that records subvocal speech is virtually unintelligible as patterned sound. Perhaps paradoxically, I found the recording less forceful as a demonstration of Burroughs's theories than his writing. For me, the aurality of his prose elicits a greater response than the machine productions it describes and instantiates.

The power of that writing is evident in the "writing machine" section of *The Ticket That Exploded.* The narrator describes an "Exhibition," which includes "a room with metal walls magnetic mobiles under flickering blue light and smell of ozone" (*TTE*, p. 62). The room is situated, of course, inside a tape-recorder. Normally, narrative fiction leaps over the technologies (printing press, paper, ink) that produce it and represents the external world as if this act of representation did not require a material basis for its production. Burroughs turns this convention inside out, locating the "external" world *inside* the technological artifact. The move constructs a completely different relation between fiction and the material means of its production, constituting the technology as the ground out of which the narrative action evolves. This technique hints that the technology is not merely a medium to represent thoughts that already exist but is itself capable of dynamic interactions *producing* the thoughts it describes. At issue, then, is the technology not only as a theme but as an articulation capable of producing new kinds of subjectivities.

The tape-recording qualities important in the Exhibition are the twin and somewhat contradictory powers of inscription and mutation. Unlike marks on paper, this writing can easily be erased and rewritten in other forms. As spectators clink through turnstiles of the Exhibition, "great sheets of magnetized print held color and disintegrated in cold mineral silence as word dust falls from demagnetized patterns" (*TTE*, p. 62). The description points to the attraction the recorder has for Burroughs. Sound, unlike print, dies away unless it is constantly renewed. Its ephemerality calls forth a double response that finds material expression in the technology. On the one hand, magnetic tape allows sound to be preserved over time; in this respect it counters the ephemerality of sound by transforming it into inscription. On the other hand, inscriptions can be easily erased and reconfigured; in this sense, it reproduces the impermanence of sound. Burroughs was drawn to both aspects of the technology. The inscription of sound in a durable medium suited his belief that the word is material, whereas the malleability of sound meant that interventions were possible, interventions that could radically change or eradicate the record.

At the Exhibition, language is inscribed through "word dust" that falls from the walls as pervasively as smog particles from the Los Angeles sky. Anticipating videotape, Burroughs imagines that "picture dust" also falls from the walls. "Photomontage fragments backed with iron stuck to patterns and fell in swirls mixing with color dust to form new patterns, shimmering, falling, magnetized, demagnetized to the flicker of blue cylinders pulsing neon tubes and globes" (*TTE,* pp. 62–63). When the Nova police counterinvade the planet, "falling" phrases repeatedly appear, as if they were news bulletins read over and over on the radio: "Word falling—Photo falling—Time falling—Break through in Grey Room"; "Shift linguals— Cut word lines—Vibrate tourists—Free Doorways—Pinball led streets— Word falling—Photo falling—Break through in Grey Room—Towers, open fire" (*TTE*, p. 104); "cut all tape"; "Break through in Grey Room— 'Love' is falling—Sex word is falling—Break photograph—Shift body halves" (*TTE*, p. 105). The "Grey Room" evidently refers to the mob's communication and control center, perhaps the "board room" where, the narrator tells us, multinationals plot to take over outer and inner space.

In opposition to the linear centralized control of the "Grey Room" is the chaotic recursivity of the Exhibition. Here there is no clear line between those who act and those who are acted upon. The traffic flow through the room is structured like a recursive loop. As the spectators pass, they are recorded "by a battery of tape-recorders recording and playing back moving on conveyor belts and tracks and cable cars spilling the talk and metal music fountains and speech as the recorders moved from one exhibit to another." The narrator remarks parenthetically, "Since the recorders and movies of the exhibition are in constant operation it will be readily seen that any spectator appears on the screen sooner or later if not today then yesterday or tomorrow" (*TTE*, p. 64). Thus spectators move along within the room, hearing and watching recordings of themselves as the recordings are played back from machines that are also moving along a conveyor belt. Their reactions as they hear and watch are also recorded in turn by other machines, creating an infinite regress in which body and tape, recording and voice, image and sight, endlessly reproduce each other. Within this world, it makes a weird kind of sense for bodies to mutate as easily as spliced tape, for the distinction between reality and representation has been largely deconstructed. "Characters walk in and out of screen flickering different films on and off" (*TTE*, p. 64); bodies split in half lengthwise; screens show two films simultaneously, half of one on one side, half of the other on the other side; a writing machine "shifts one half one text and half the other through a page frame on conveyor belts" (*TTE*, p. 65). Inscriptions, bodies, sounds, and images all follow the same dynamics and the same logic of splices running lengthwise to create mutated posthuman forms that both express and strive to escape from the conditioning that makes them into split beings.

In a wonderfully oxymoronic phrase, Burroughs calls the place where culture produces its replicating sound and image tracks the "reality studio." "Clearly no portentous exciting events are about to transpire," the narrator says, implicitly mocking the melodrama of his own space-alien track. "You will readily understand why people will go to any lengths to get in the film to cover themselves with any old film scrap . . . anything to avoid the hopeless dead-end horror of being just who and where you all are: dying animals on a doomed planet." Connecting capitalist financing with cultural productions (as if remembering the American Telegraphone Company), he continues: "The film stock issued now isn't worth the celluloid its [*sic*] printed on. There is nothing to back it up. The film bank is empty. To conceal the bankruptcy of the reality studio it is essential that no one should be in position to set up another reality set. . . . Work for the reality studio or else. Or else you will find out how it feels to be *outside the film*" (*TTE*, p. 151).

As the text draws to a close, the narrator directs the reader's attention to the possibility that the reality studio may indeed be closing down and that the reader will therefore shortly be outside the film, off the recording. A similar message is given in a different medium, when at the end of the penultimate section, the print of the text is disrupted by several lines of cursive script, English alternating with Arabic. Each line runs through a permutation of "To say good by silence," with the lines gradually becoming more random and indecipherable as they proceed down the page (*TTE*, p. 203) Perhaps Burroughs is trying to prepare the reader for the panic that sets in when the interior monologue is disrupted and, for the first time in one's life, one hears silence instead of language. For whatever reason, he takes extraordinary care to achieve a feeling of closure unusual in his works from this period. Compared with *Naked Lunch*, the ending here is formally elaborate and thematically conclusive.

Echoing *The Tempest*, the text as it winds down splices in dialogue from Shakespeare's play with visions of contemporary technologies. "i foretold you were all spirits watching TV program—Terminal electric voices end—These our actors cut in—A few seconds later you are melted into air—Rub out promised by our ever-living poet—Mr Bradly Mr Martin, five times our summons—no shelter in setting forth" (*TTE*, p. 174). The splices invite the reader to tease out resonances between the two works. Whereas insect imagery predominates in *Naked Lunch*, in *Ticket* the usual form of nonhuman life is aquatic or amphibian, recalling Caliban's characterization as a "fishy monster." Prospero conjures spirits from the air, and yet his magic has a terrible materiality; he can, we are told, raise bodies from the dead. Most of all he is a supreme technician, blending illusion with reality so skillfully that his art can effect changes in the real world. Burroughs aims for nothing less, using language to disrupt the viral power of the word, creating recordings to stop the playbacks that imprison our future in the sounds of our past. If the tape-recorder is, as Paul Bowles called it, "God's little toy," *The Ticket That Exploded* is the tape that reveals this god-machine's life-transforming possibilities (Bowles quoted in *TTE*, p. 166).

"What we see is dictated by what we hear," the narrator of *Ticket* asserts (*TTE*, p. 168). There is considerable anecdotal evidence to support his claim. Whereas sight is always focused, sharp, and delineated, sound envelopes the body, as if it were an atmosphere to be experienced rather than an object to be dissected. Perhaps that is why researchers in virtual reality have found that sound is much more effective than sight in imparting emotional tonalities to their simulated worlds.[39] Their experiences suggest that voice is associated with presence not only because it comes from within the body but also because it conveys new information about the subject, information that goes deeper than analytical thought or conscious intention. Manipulating sound through tape-recorders thus becomes a way of producing a new kind of subjectivity that strikes at the deepest levels of awareness. If we were to trace the trajectory suggested here to the end of the period when audiotape held sway, it would lead to texts such as C. J. Cherryh's *Cyteen* trilogy, where the body has become a corporate product molded by "taking tape," that is, listening to conditioning tapes that lay the foundation for the subject's "psychset." Burroughs anticipates Cherryh's implication that the voice issuing from the tape-recorder sounds finally not so much postmodern as it does posthuman.

Where hope exists in *Ticket*, it appears as posthuman mutations like the fish boy, whose fluidity perhaps figures a type of subjectivity attuned to the froth of noise rather than the stability of a false self, living an embodied life beyond human consciousness as we know it. But this is mere conjecture, for any representation of the internal life of the fish boy could be done only in words, which would infect and destroy exactly the transformation they were attempting to describe. For Burroughs, the emphasis remains on subversion and disruption rather than creative rearticulation. Even subversion risks being co-opted and taken over by the viral word; it can succeed only by continuing to disrupt everything, including its own prior writing.

In this chapter, I have been concerned with Burroughs's fictions not only as harbingers of the posthuman but, more immediately, as sites where body/embodiment and inscription/incorporation are in constant and dynamic interplay with one another. As we have seen, in the Exhibition, inscriptions fall from the walls to become corporeal "word dust"; incorporations are transformed into inscriptions through video- and audio-recording devices; bodies understood as normative and essentialized entities are rewritten to become particularized experiences of embodiment; and embodied experience is transformed, through the inscriptions of the tape-recording, back into essentialized manifestations of "the word." The recursivities that entangle inscription with incorporation, the body with embodiment, invite us to see these polarities not as static concepts but as mutating surfaces that transform into one another, much like the Möbius strip that Grosz imagines for her "volatile bodies." Starting from a model emphasizing polarities, then, we have moved toward a vision of interactions both pleasurable and dangerous, creatively dynamic and explosively transformative.

It is no accident that recursive loops and reflexive strategies figure importantly in these transformations, for Burroughs shared with Humberto Maturana and Philip K. Dick an appreciation for how potently reflexivity can destabilize objectivist assumptions. Whereas Maturana located reflexivity in biological processes and Dick placed it in psychological dynamics, Burroughs located it in a cybernetic fusion of language and technology. Mutating into and out of the tape-recorder, the viral word reconfigures the tape-recorder as a cybernetic technology capable of radically transforming bodies and subjectivities. As for the "external" world, where clear divisions separate observer from system, human from technological artifact, Maturana, Dick, and Burroughs agreed (although for different reasons) that there is no there there. Whatever the limitations of their works, they shared a realization that the observer cannot stand apart from the systems being observed. In exploring how to integrate observer and world into a unified field of interaction, they also realized that liberal humanism could not continue to hold sway. Just as the tide of posthumanism that Norbert Wiener had struggled to contain could not be held back, neither could the technological advances in informatics, advances that would soon displace second-wave issues with third-wave concerns.

a

NARRATIVES OF ARTIFICIAL LIFE

In contrast to the circular processes of Humberto Maturana's autopoiesis, the figure most apt to describe the third wave is a spiral. Whereas the second wave is characterized by an attempt to include the observer in an account of the system's functioning, in the third wave the emphasis falls on getting the system to evolve in new directions. Self-organization is no longer enough. The third wave wants to impart an upward tension to the recursive loops of self-organizing processes so that, like a spring compressed and suddenly released, the processes break out of the pattern of circular self-organization and leap outward into the new.

Just as Heinz von Foerster served as a transition figure between the first and second waves, so Francisco Varela bridges the transition between the second and third waves. We saw in chapter 6 that Maturana and Varela extended the definition of the living to include artificial systems. After coauthoring *The Embodied Mind*,[1] Varela began to work in a new field known as Artificial Life and coedited the papers from the first European conference in that field. In the introduction to the conference volume, *Toward a Practice of Autonomous Systems,* he and his coauthor, Paul Bourgine, lay out their view of what the field of Artificial Life should be. They locate its origins in cybernetics, referencing William Grey Walter's electronic tortoise and Ross Ashby's homeostat. Although some characteristics of autopoiesis are reinscribed on the successor field of Artificial Life, especially the idea that systems are operationally closed, other features are new. The change is signaled in Varela's subtle reconception of autonomy. He and his coauthor wrote: "Autonomy in this context refers to [the living's] basic and fundamental capacity to *be,* to assert their existence and bring forth a world that is significant and pertinent without being pre-digested in advance. Thus the autonomy of the living is understood here both in regards to its actions and to the way it shapes a world into significance. This conceptual exploration goes hand in hand with the design and construction of autonomous agents and suggests an enormous range of applications at all scales, from cells to societies."[2] For Maturana, "shap[ing] a world into significance" meant that perception was linked primarily to internal processes rather than external stimuli.[3] We have seen the difficulties he had with evolution, because he sought to put the emphasis instead on the organism's holistic nature and autopoietic circularity. When Varela and his coauthor speak of "shap[ing] a world into significance," the important point for them is that the system's organization, far from remaining unchanged, can transform itself through emergent behavior. The change is not so much an absolute break, however, as a shift in emphasis and a corresponding transformation in the kind of questions the research programs pose, as well as new strategies for answering them. Thus the relation of the third wave to the second is again one of seriation, an overlapping pattern of replication and innovation.

The shift in questions and methodologies is not, of course, neutral. For researchers who come to the field from backgrounds in cognitive science and computer science, rather than from autopoiesis as Varela does, the underlying assumptions all too easily lend themselves to reinscribing a disembodied view of information. But as Varela's presence in the field indicates, not everyone who works in the field agrees that disembodied "organisms" are the best way to construct Artificial Life. Just as there were competing camps in the Macy Conferences—one arguing for a disembodied view of information and one for a contextualized embodied view—so in Artificial Life some researchers concentrate on simulations, insisting that embodiment is not necessary, whereas others argue that only embodied forms can fully capture the richness of an organism's interactions with the environment. Our old friend the observer, who was at the center of the epistemological revolution sparked by Maturana, in the third wave retreats to the periphery, with a consequent loss of the sophistication that Maturana brought to epistemological questions. The observer has, however, not altogether vanished from the scene, remaining in the picture as narrator and narratee of stories about Artificial Life. To see how the observer's presence helps to construct the field, let us turn now to consider the strange flora and fauna of the world of Artificial Life.

The Nature and Artifice of Artificial Life

At the Fourth Conference on Artificial Life in the summer of 1994, evolutionary biologist Thomas S. Ray put forth two proposals.[4] The first was a plan to preserve biodiversity in Costa Rican rain forests; the second was a suggestion that Tierra, his software program creating Artificial Life-forms inside a computer, be released on the Internet so that it could "breed" diverse species on computers all over the world. Ray saw the two proposals as complementary. The first aimed to extend biological diversity for protein-based life-forms; the second sought the same for silicon-based life-forms. Their juxtaposition dramatically illustrates the reconstruction of nature going on in the field of Artificial Life, affectionately known by its practitioners as AL. "The object of an AL instantiation," Ray wrote recently, "is to introduce the *natural* form and process of life into an *artificial* medium."[5] The lines startle. In Ray's rhetoric, the computer codes composing these "creatures" become natural forms of life; only the medium is artificial.

How is it possible in the late twentieth century to believe, or at least claim to believe, that computer codes are alive—and not only alive but natural? The question is difficult to answer directly, for it involves assumptions that are not explicitly articulated. Moreover, these presuppositions do not stand by themselves but move in dynamic interplay with other formulations and ideas circulating throughout the culture. In view of this complexity, the subject is perhaps best approached through indirection, by looking not only at the scientific content of the programs but also at the stories told about and through them. These stories, I will argue, constitute a multilayered system of metaphoric and material relays through which "life," "nature," and the "human" are being redefined.

The first level of narrative with which I will be concerned is the Tierra program and various representations of it written by Ray and others. In these representations, authorial intention, anthropomorphic interpretation, and the operations of the program are so interwoven that it is impossible to separate them. As a result, the program operates as much within the imagination as it does within the computer. The second level of narrative focuses on the arguments and rhetorical strategies that AL practitioners use as they seek to position Artificial Life as a valid area of research within theoretical biology. This involves telling a story about the state of the field and the contributions that AL can make. As we shall see, the second-level story quickly moves beyond purely professional considerations, evoking a larger narrative about the kinds of life that have emerged, and are emerging, on Earth. The narrative about the present and future of terrestrial evolution forms the third level. It is constituted through speculations on the relation of human beings to their silicon cousins, the "creatures" who live inside the computer. Here, at the third level, the implication of the observer in the construction of all three narratives becomes explicit. To interrogate how this complex narrative field is initiated, developed, and interpolated with other cultural narratives, let us begin at the first level, with an explanation of the Tierra program.

Conventionally, Artificial Life is divided into three research fronts. *Wetware* is the attempt to create artificial biological life through such techniques as building components of unicellular organisms in test tubes. *Hardware* is the construction of robots and other embodied life-forms. *Software* is the creation of computer programs instantiating emergent or evolutionary processes. Although each of these areas has its distinctive emphases and research agendas, they all share the sense of building life from the "bottom up." In the software branch, with which I am concerned here, the idea is to begin with a few simple local rules and then, through structures that are highly recursive, allow complexity to emerge spontaneously. Emergence implies that properties or programs appear on their own, often developing in ways not anticipated by the person who created the simulation. Structures that lead to emergence typically involve complex feedback loops in which the outputs of a system are repeatedly fed back in as input. As the recursive looping continues, small deviations can quickly become magnified, leading to the complex interactions and unpredictable evolutions associated with emergence.[6]

Even granting emergence, it is still a long jump from programs that replicate inside a computer to living organisms. This gap is bridged largely through narratives that map the programs into evolutionary scenarios traditionally associated with the behavior of living creatures. The narratives translate the operations of computer codes into biological analogues that make sense of the program logic. In the process, the narratives alter the binary operations that, on a physical level, amount to changing electronic polarities, transforming the binary operations into the high drama of a Darwinian struggle for survival and reproduction. To see this transformation in action, consider the following account of the Tierra program. This account is compiled from Thomas Ray's published articles and unpublished working papers, from conversations I had with him about his program, and from public lectures he has given on the topic.[7]

When I visited Ray at the Santa Fe Institute, he talked about the genesis of Tierra. Frustrated with the slow pace of natural evolution, he wondered if it would be possible to speed things up by creating evolvable artificial organisms within the computer. One of the first challenges he faced was designing programs robust enough to withstand mutation without crashing. To induce robustness, he conceived of building inside the regular computer a "virtual computer" made out of software. Whereas the regular computer uses memory addresses to find data and execute instructions, the virtual computer uses a technique Ray calls "address by template." Taking its cue from the topological matching of DNA bases, in which one base finds its appropriate partner by diffusing through the medium until it locates another base with a surface it can fit into like a key into a lock, address by template matches one code segment to another by looking for its binary inverse. For example, if an instruction is written in binary code 1001, the virtual computer searches nearby memory to find a matching segment with the code 0110. The strategy has the advantage of creating a container that holds the organisms and renders them incapable of replicating outside the virtual computer, for the address by template operation can occur only within a virtual computer. Presented with a string such as 0110, the regular computer would read it as data rather than instructions to replicate.

Species diversify and evolve through mutation. To introduce mutation, Ray created the equivalent of cosmic rays by having the program flip the polarity of a bit once in every 10,000 executed instructions. In addition, replication errors occur about once in every 1,000 to 2,500 instructions copied, introducing another source of mutation. Other differences spring from an effect Ray calls "sloppy reproduction," analogous to the genetic mixing that occurs when a bacterium absorbs fragments of a dead organism nearby. To control the number of organisms, Ray introduced a program that he calls the "reaper." The "reaper" monitors the population and eliminates the oldest creatures and those who are "defective," that is, those who have most frequently erred in executing their programs. If a creature finds a way to replicate more efficiently, it is rewarded by being moved down in the reaper's queue and so becomes "younger."

The virtual computer starts the evolutionary process by allocating a block of memory that Ray calls the "soup," an analogy with the primeval soup at the beginning of life on Earth. Unleashed inside the soup are self-replicating programs, normally starting with a single 80-byte creature called the "ancestor." The ancestor comprises three segments. The first segment counts the instructions to see how long the ancestor is (this procedure ensures that the length can change without throwing off the reproductive process); the second segment reserves that much space in nearby memory, putting a protective membrane around the space (an analogy with the membranes that enclose living organisms); and the third segment copies the ancestor's code into the reserved space, thus completing the reproduction and creating a "daughter cell" from the "mother cell." To see how mutation leads to new species, consider that a bit flip occurs in the last line of the first segment, changing 1100 to 1110. Normally the program would find the second segment by searching for its first line, encoded 0011. Now, however, the program searches until it finds a segment starting with 0001. Thus it goes not to its own second segment but to another string of code in nearby memory. Many mutations are not viable and do not lead to reproduction. Occasionally, however, the program finds a segment that starts with 0001 and that will allow it to reproduce. Then a new species is created, as this organism begins producing offspring.

When Ray set his program running overnight, he thought he would be lucky to get a 1- or 2-byte variation from the 80-byte ancestor. Checking it the next morning, he found that an entire ecology had evolved, including one 22-byte organism. Among the mutants were parasites that had lost their own copying instructions but had developed the ability to invade a host and hijack its copying procedure. One 45-byte parasite had evolved into a benign relationship with the ancestor; others were destructive, crowding out the ancestor with their own offspring. Later runs of the program saw the development of hyperparasites, which had evolved ways to compete for time as well as memory. Computer time is doled out equally to each organism by a "slicer" that determines when the organism can execute its program. Hyperparasites wait for parasites to invade them. Then, when the parasite attempts to reproduce using the hyperparasite's copy procedure, the hyperparasite directs the program to its own third segment instead of returning the program to the parasite's ending segment. Thus the hyperparasite's code is copied on the parasite's time. In this way the hyperparasite greatly multiplies the time it has for reproduction, for in effect it appropriates the parasite's time for its own.

This, then, is the first-level narrative about the program. It appears with minor variations in Ray's articles and lectures. It is also told in the Santa Fe Institute videotape "Simple Rules . . . Complex Behavior," in which Ray collaborated with a graphic artist to create a visual representation of Tierra, accompanied by his voiceover.[8] If we ask how this narrative is constituted, we can see that statements about the operation of the program and interpretations of its meaning are in continuous interplay with each other. Consider the analogies implicit in such terms as "mother cell," "daughter cell," "ancestor," "parasite," and "hyperparasite." The terms do more than set up parallels with living systems; they also reveal Ray's intention in creating an appropriate environment in which the dynamic emergence of evolutionary processes could take off. In this respect, Ray's rhetoric is quite different from that of Richard Dawkins in *The Selfish Gene*, a work also deeply informed by anthropomorphic constructions.[9] Dawkins's rhetoric attributes to genes human agency and intention, creating a narrative of humanlike struggle for lineage. In his construction, Dawkins overlays onto the genes the strategies, emotions, and outcomes that properly belong to the human domain. Ray, by contrast, is working with artificial systems designed by humans *precisely so the "creatures" would be able to manifest these qualities.* This is the primary reason why explanation and interpretation are inextricably entwined in the first-level narrative. Ray's biomorphic namings and interpretations function not so much as an overlay as an explication of an intention that was there at the beginning. Analogy is not incidental or belated but is central to the program's artifactual design.

Important as analogy is, it is not the whole story. The narrative's compelling effect comes not only from analogical naming but also from images. In rhetorical analysis, of course, "image" can mean either an actual picture or a verbal formulation capable of evoking a mental picture. Whether an image is a visualization or visually evocative language, it is a powerful mode of communication because it draws on the high density of information that images convey. Visualization and visually evocative language collaborate in the videotape that the Santa Fe Institute made to publicize its work. As the narrative about Tierra begins, the camera flies over a scene representing the inside of a computer. This stylized landscape is dominated by a block-like structure representing the CPU (Central Processing Unit) and dotted with smaller upright rectangles representing other integrated circuits. Then the camera zooms into the CPU, where we see a grid upon which the "creatures" appear and begin to reproduce. They are imaged as solid polygons strung together to form three sections, representing the three segments of code. Let us linger at this scene and consider how it has been constructed. The pastoral landscape upon which the "creatures" are visualized instantiates a transformation characteristic of the new information technologies and the narratives that surround them. A material object (the computer) has been translated into the functions it performs (the programs it executes), which in turn have been represented in visual codes familiar to the viewer (the bodies of the "creatures"). The path can be represented schematically as material base → functionality → representational code. This kind of transformation is extremely widespread, appearing in popular venues as well as in scientific applications. It is used by William Gibson in *Neuromancer*, for example, when he represents the data arrays of a global informational network as solid polygons in a three-dimensional space that his protagonist, transformed into a point of view, or pov, can navigate as though flying through the atmosphere.[10] The schematic operates in remarkably similar fashion in the video, where we become a disembodied pov flying over the lifeworld of the "creatures," a world comfortingly familiar in its three-dimensional spaces and rules of operation. Whereas the CPU landscape corresponds to the computer's interior architecture, however, the lifeworld of the "creatures" does not. The seamless transition between the two elides the difference between the *material* space that is inside the computer and the *imagined* space that, in actuality, consists of computer addresses and electronic polarities on the computer disk.

To explore how these images work to encode assumptions, consider the bodies of the "creatures," which resemble stylized ants. In the program, the "creatures" have bodies only in a metaphoric sense, as Ray recognizes when he talks about their bodies of information (itself an analogy).[11] These bodies of information are not, as the expression might be taken to imply, phenotypic expressions of informational codes. Rather, the "creatures" *are* their codes. For them, genotype and phenotype amount to the same thing; the organism is the code, and the code is the organism. By representing them as phenotypes, visually by giving them three-dimensional bodies and verbally by calling them "ancestors," "parasites," and such, Ray elides the difference between behavior, properly restricted to an organism, and execution of a code, applicable to the informational domain. In the process, our assumptions about behavior, in particular our thinking of it as independent action undertaken by purposive agents, are transported into the narrative.

Further encoding takes place in the plot. Narrative tells a story, and intrinsic to story is chronology, intention, and causality. In Tierra, the narrative is constituted through the emerging story of the struggle of the "creatures" for survival and reproduction. More than an analogy or an image, this is a drama that, if presented in a different medium, one would not hesitate to identify as an epic. Like an epic, it portrays life on a grand scale, depicting the rise and fall of races, some doomed and some triumphant, and recording the strategies they invent as they play for the high stakes of establishing a lineage. The epic nature of the narrative is even more explicit in Ray's plans to develop a global ecology for Tierra. In his proposal to create a digital "biodiversity reserve," the idea is to release the Tierra program on the Internet so that it can run in background on computers across the globe. Each site will develop its own microecology. Because background programs run when demands on the computer are at a minimum, the programs will normally be executed late at night, when most users are in bed. Humans are active while the "creatures" are dormant; the "creatures" evolve while we sleep. Ray points out that someone monitoring activity in Tierra programs would therefore see it as a moving wave that follows darkfall around the world. Linking the evolution of the "creatures" to the human world in a complementary diurnal rhythm, the proposal edges toward a larger narrative level that interpolates their story into ours, ours into theirs.

A similar interpolation occurs in the video. The narrative appears to be following the script of Genesis, from the lightning that, flickering over the landscape, represents the life force to the "creatures," who, like their human counterparts, follow the biblical imperative to be fruitful and multiply. When a death's-head appears on the scene, representing the reaper program, we understand that this pastoral existence will not last for long. The idyll is punctured by competition between species, strategies of subversion and co-optation, and exploitation of one group by another—in short, all the trappings of rampant capitalism. To measure how much this narrative accomplishes, we should remember that what one actually sees as the output of the Tierra program is a spectrum of bar graphs tracking the numbers of programs of given byte lengths as a function of time. The strategies emerge when human interpreters scrutinize the binary codes that constitute the "creatures" to find out how they have changed and determine how they work.

No one knows this better, of course, than Ray and other researchers in the field. The video, as they would no doubt want to remind us, is merely an artist's visualization and has no scientific standing. It is, moreover, intended for a wide audience, not all of whom are presumed to be scientists. This fact in itself is interesting, for the tape as a whole is an unabashed promotion of the Santa Fe Institute. It speaks to the efforts that practitioners in the field are making to establish Artificial Life as a valid, significant, and exciting area of scientific research. These efforts are not unrelated to the visual and verbal transformations discussed above. To the extent that the "creatures" are biomorphized, their representation reinforces the strong claim that the "creatures" are actually alive, extending the implications of the claim. Nor do the transformations appear only in the video, although they are particularly striking there. As the discussion above demonstrates, they are also inscribed in published articles and commentary. In fact, they are essential to the strong claim that the computer codes do not merely simulate life but are themselves alive. At least some researchers at the Santa Fe Institute recognize the relation between the strong claim and the stories that researchers tell about these "organisms." Asked about the strong claim, one respondent insisted: "It's in the eye of the beholder. It's not the system, it's the observer."[12]

In the second wave of cybernetics, accounting for the observer was of course a central concern. What happens when the observer is taken into account in Artificial Life research? To explore further the web of connections between the operations of the program, descriptions of its operation by observers, and the contexts in which these descriptions are embedded, we will follow the thread to the next narrative level, where arguments circulate about the contributions that Artificial Life can make to scientific knowledge.

Positioning the Field: The Politics of Artificial Life

Christopher Langton, one of the most visible of the AL researchers, explains the reasoning behind the strong claim. "The principle [*sic*] assumption made in Artificial Life is that the 'logical form' of an organism can be separated from its material basis of construction, and that 'aliveness' will be found to be a property of the former, not of the latter."[13] It would be easy to dismiss the claim on the basis that the reasoning behind it is tautological: Langton defines life in such a way as to make sure the programs qualify, and then, because they qualify, he claims they are alive. But more is at work here than tautology. Resonating through Langton's definition are assumptions that have marked Western philosophical and scientific inquiry at least since Plato. Form can logically be separated from matter; form is privileged over matter; *form defines life,* whereas the material basis merely instantiates life. The definition is a site of reinscription as well as tautology. This convergence suggests that the context for our inquiry should be broadened beyond the logical form of the definition to the field of inquiry in which such arguments persuade precisely because they reinscribe.

This context includes attitudes, held deeply by many researchers in scientific communities, about the relation between the complexity of observable phenomena and the relatively simple rules they are seen to embody. Traditionally, the natural sciences, especially physics, have attempted to reduce apparent complexity to underlying simplicity. The attempt to find the "fundamental building blocks" of the universe in quarks is one example of this endeavor; the mapping of the human genome is another.[14] The sciences of complexity, with their origins in nonlinear dynamics, complicated this picture by demonstrating that for certain nonlinear dynamical systems, the evolution of the system could not be predicted, even in theory, from the initial conditions (just as Ray did not know what creatures would evolve from the ancestor). Thus the sciences of complexity articulated a limit on what reductionism could accomplish. In a significant sense, however, AL researchers have not relinquished reductionism. In place of predictability, which is traditionally the test of whether a theory works, they emphasize emergence. Instead of starting with a complex phenomenal world and rea-soning back through chains of inference to what the fundamental elements must be, they *start* with the elements, complicating the elements through appropriately nonlinear processes so that the complex phenomenal world appears on its own.[15]

What is the justification for calling the simulation and the phenomena that emerge from it a "world"? It is precisely because they are generated from simple underlying rules and forms. AL reinscribes, then, the main-stream assumption that simple rules and forms give rise to phenomenal complexity. The difference is that AL starts at the simple end, where syn-thesis can move forward spontaneously, rather than at the complex end, where analysis must work backward. Langton, in his explanation of what AL can contribute to theoretical biology, makes this difference explicit: "Artificial Life is the study of man-made systems that exhibit behaviors characteristic of natural living systems. It complements the traditional bio-logical sciences concerned with the *analysis* of living organisms by at-tempting to *synthesize* life-like behaviors within computers and other artificial media. By extending the empirical foundation upon which biology is based beyond the carbon-chain life that has evolved on Earth, Artificial Life can contribute to theoretical biology by locating *life-as-we-know-it* within the larger picture of *life-as-it-could-be.*"[16]

The presuppositions informing such statements have been studied by Stefan Helmreich, an anthropologist who spent several months at the Santa Fe Institute.[17] Helmreich interviewed several of the major players in the U.S. AL community, including Langton, Ray, John Holland, and others. Helmreich summarized his informants' views about the "worlds" they cre-ate: "For many of the people I interviewed, a 'world' or 'universe' is a self-consistent, complete, and closed system that is governed by low level laws that in turn support higher level phenomena which, while dependent on these elementary laws, cannot be simply derived from them."[18] Helmreich used comments from the interviews to paint a fascinating picture of the various ways in which simple laws are believed to underlie complex phenomena. Several informants thought that the world was mathematical in essence. Others held the view, also extensively articulated by Edward Fredkin, that the world is fundamentally composed of information.[19] From these points of view, phenomenological experience is itself a kind of illusion, covering an underlying reality of simple forms. For these researchers, a computer program that generates phenomenological com-plexity out of simple forms is no more or less illusory than the "real" world.

The form/matter dichotomy is intimately related to this vision, for reality at the fundamental level is seen as form rather than matter, specifically as informational code whose essence lies in a binary choice rather than a material substrate. Fredkin, for example, says that reality is a software program run by a cosmic computer, whose nature must forever remain unknown to us because it lies outside the structure of reality, whose programs it runs.[20] For Fredkin, AL programs are alive in precisely the same sense as biological life—because they are complex phenomena generated by underlying binary code. The assumption that form occupies a foundational position relative to matter is especially easy to make with information technologies, since information is defined in theoretic terms (as we have seen) as a probability function and thus as a pattern or form rather than as a materially instantiated entity.

Information technologies seem to realize a dream impossible in the natural world—the opportunity to look directly into the inner workings of reality at its most elemental level. The directness of the gaze does not derive from the absence of mediation. On the contrary, our ability to look into programs like Tierra is highly mediated by everything from computer graphics to the processing program that translates machine code into a high-level computer language such as C++. Rather, the gaze is privileged because the observer can peer directly into the elements of the world before the world cloaks itself with the appearance of complexity. Moreover, the observer is presumed to be cut from the same cloth as the world being inspected, inasmuch as the observer is also constituted through binary processes similar to those seen inside the computer. The essence of Tierra as an artificial world is no different from the essence of the observer or of the world that the observer occupies: all are constituted through forms understood as informational patterns. When *form* is triumphant, Tierra's "creatures" are, in a disconcertingly literal sense, just as much *life-forms* as are any other organisms.

We are now in a position to understand the deep reasons why some practitioners think of programs like Tierra not as models or simulations but as life itself. As Langton and many others point out, in the analytic approach, reality is modeled by treating a complex phenomenon as if it was composed of smaller constituent parts. These parts are broken down into still smaller parts, until we find parts sufficiently simplified that they can be treated mathematically. Most scientists would be quick to agree that the model is not the reality, because they recognize that many complexities had to be tossed out by the wayside in order to lighten the wagon enough to get it over the rough places in the trail. Their hope is that the model nevertheless captures enough of the relevant aspects of a system to tell them something significant about how reality works. In the synthetic approach, by contrast, the complexities emerge spontaneously as a result of the system's operation. The system itself adds back in the baggage that had to be tossed out in the analytic approach. (Whether it is the same baggage remains, of course, to be seen.) In this sense Artificial Life poses an interesting challenge to the view of nineteenth-century vitalists, who saw in the analytic approach a reductionist methodology that could never adequately capture the complexities of life. If it is true that the analytical approach murders by dissection, by the same reasoning the synthetic approach of AL may be able to procreate by emergence.

In addition to these philosophical considerations, there are also more obviously political reasons to make a strong claim for the "aliveness" of Artificial Life. As a new kid on the block, AL must jockey for position with larger, better-established research agendas. A common reaction from other scientists is, "Well, this is all very interesting, but what good is it?" Even AL researchers joke that AL is a solution in search of a problem. When applications are suggested, they are often open to cogent objections. As long as AL programs are considered to be simulations, any results produced from them may be artifacts of the simulation rather than properties of natural systems. So what if a certain result can be produced within the simulation? The result is artifactual and therefore nonsignifying with respect to the natural world unless the same mechanisms can be shown to be at work in natural systems.[21] These difficulties disappear, however, if AL programs are themselves alive. Then the point is not that they model natural systems but rather that they are, *in themselves,* also alive and therefore as worthy of study as evolutionary processes in naturally occurring media.

This is the tack that Langton takes when he compares AL simulations to synthetic chemicals.[22] In the early days, he observes, the study of chemistry was confined to naturally occurring elements and compounds. Although some knowledge could be gained from these, the results were limited by what lay ready at hand. Once researchers learned to synthesize chemicals, their knowledge took a quantum leap forward, for then chemicals could be tailored to specific research problems. Similarly, theoretical biology has been limited to the cases that lie ready at hand, namely the evolutionary pathways taken by carbon-based life. Even though generalizing from a single instance is notoriously difficult, theoretical biology had no choice; carbon-based life was it. Now a powerful new instance has been added to the repertoire, for AL simulations represent an alternative evolutionary pathway followed by silicon-based life-forms.

What theoretical biology looks for, in this view, are similarities that cut across the particularities of the media. In "Beyond Digital Naturalism," Walter Fontana and his coauthors lay out a research agenda "ultimately motivated by a premise: that there exists a logical deep structure of which carbon chemistry-based life is a manifestation. The problem is to discover what it is and what the appropriate mathematical devices are to express it."[23] Such a research agenda presupposes that the essence of life, understood as logical form, is independent of the medium. More is at stake in this agenda than expanding the frontiers of theoretical biology. By positing AL as a second instance of life, researchers affect the definition of biological life as well, for now it is the *juxtaposition* that determines what counts as fundamental, not carbon-based forms by themselves.

This change hints at the far-reaching implications of the narrative of Artificial Life as an alternate evolutionary pathway for life on Earth. To explore these implications, let us turn to the third level of narrative, where we will consider stories about the relation of humans to our silicon cousins, the Artificial Life-forms that represent the road not taken—until now.

Reconfiguring the Body of Information

As research on Artificial Life-forms continues and expands, the construction of *human* life is affected as well. Rodney Brooks, of the Artificial Intelligence Laboratory at MIT, and the roboticist Hans Moravec, noted in earlier chapters, tell two different narratives of how the human will be reconfigured in the face of artificial bodies of information. Whereas Moravec privileges consciousness as the essence of human being and wants to preserve it intact, Brooks speculates that the more essential property of the human being is the ability to move around and interact robustly with the environment. Instead of starting with the most advanced qualities of human thought, Brooks starts with locomotion and simple interactions and works from the bottom up. Despite these different orientations, both Brooks and Moravec see the future of human being inextricably bound up with Artificial Life. Indeed, in the future world they envision, distinguishing between natural and Artificial Life, human and machine intelligence, will be difficult or impossible.

In *Mind Children: The Future of Robot and Human Intelligence*, Moravec argues that the age of carbon-based life is drawing to a close.[24] Humans are about to be replaced by intelligent machines as the dominant life-form on the planet. Drawing on the work of A. G. Cairns-Smith, Moravec suggests that such a revolution is not unprecedented.[25] Before protein replication developed, a primitive form of life existed in certain silicon crystals that had the ability to replicate. But protein replication was so far superior that it soon left the replicating crystals in the dust. Now silicon has caught up with us again, in the form of computers and computerized robots. Although the Cairns-Smith hypothesis has been largely discredited, in Moravec's text it serves the useful purpose of increasing the plausibility of his vision by presenting the carbon-silicon struggle as a rematch of an earlier contest rather than as an entirely new event.

A different approach is advocated by other members of the AL community, among them Rodney Brooks, Pattie Maes, and Mark Tilden.[26] They point to the importance of having agents who can learn from interactions with a physical environment. Simulations, they believe, are limited by the artificiality of their context. Compared with the rich variety and creative surprises of the natural world, simulations are stick worlds populated by stick figures. No one argues this case more persuasively than Brooks. When I talked with him at his MIT laboratory, he mentioned that he and Hans Moravec had been roommates in college (a coincidence almost allegorical in its neatness). Moravec, for his senior project, had built a robot that used a central representation of the world to navigate. The robot would go a few feet, feed in data from its sensors to the central representation, map its new position, and move a few more feet. Using this process, it would take several hours to cross a room. If anyone came in during the meantime, it would be thrown hopelessly off course. Brooks, a loyal roommate, stayed up late one night to watch the robot as it carried out its agonizingly slow perambulation. The thought occurred to Brooks that a cockroach could accomplish the same task in a fraction of the time, and yet the cockroach could not possibly have as much computing power aboard as the robot. Deciding that there had to be a better way, he began building robots according to a different philosophy.

In his robots, Brooks uses what he calls "subsumption architecture." The idea is to have sensors and actuators connected directly to simple finite-state machine modules, with a minimum of communication between them. Each system "sees" the world in a way that is entirely different from how the other systems see the world. There is no central representation, only a control system that kicks in to adjudicate when there is a conflict between the distributed modules. Brooks points out that the robot does not need to have a coherent concept of the world; instead it can learn what it needs directly through interaction with its environment. The philosophy is summed up in his aphorism: "The world is its own best model."[27]

Subsumption architecture is designed to facilitate and capitalize on emergent behavior. The idea can be illustrated with Genghis, a six-legged robot somewhat resembling an oversize cockroach, which Brooks hopes to sell to NASA as a planetary explorer.[28] Genghis's gait is not programmed in advance. Rather, each of the six legs is programmed to stabilize itself in an environment that includes the other five. Each time Genghis starts up, it has to learn to walk anew. For the first few seconds it will stumble around; then, as the legs begin to take account of what the others are doing, a smooth gait emerges. The robot is relatively cheap to build, is more robust than the large planetary explorers that NASA currently uses, and is under its own local control rather than being dependent on a central controller who may not be on site to see what is happening. "Fast, cheap, and out of control" is another aphorism that Brooks uses to sum up the philosophy behind the robots he builds.

Brooks's program has been carried further by Mark Tilden, a Canadian roboticist who worked under Brooks and now is at the University of Waterloo. In my conversation with him, Tilden mentioned that he grew up on a farm in Canada and was struck by how chickens ran around after they their heads had been cut off, performing, as he likes to put it, complicated navigational tasks in three-dimensional space without any cortex at all. He decided that considerable computation had to be going on in the peripheral nervous system. He used the insight to design insectlike robots that operate on nervous nets (considerably simpler than the more complex neural nets) composed of no more than twelve transistor circuits. These robots use analogue rather than digital computing to carry out their tasks. Like Genghis, they have an emergent gait. They are remarkably robust, are able to right themselves when turned over, and can even learn a compensatory gait when one of their legs is bent or broken off.[29]

Narratives about the relation of these robots to humans emerge when Brooks and others speculate about the relevance of their work to human evolution. Brooks acknowledges that the robots he builds have the equivalent of insect intelligence. But insect intelligence is, he says, nothing to sneer at. Chronologically speaking, by the time insects appeared on Earth, evolution was already 95 percent of the way to creating human intelligence.[30] The hard part, he believes, is evolving creatures who are mobile and who can interact robustly with their environment. Once these qualities are in place, the rest comes relatively quickly, including the sophisticated cognitive abilities that humans possess. How did humans evolve? In his view, they evolved through the same kind of mechanisms that he uses in his robots, namely distributed systems that interact robustly with the environment and that consequently "see" the world in very different ways. Consciousness is a relatively late development, analogous to the control system that kicks in to adjudicate conflicts between the different distributed systems. Consciousness is, as Brooks likes to say, a "cheap trick," that is, an emergent property that increases the functionality of the system but is not part of the system's essential architecture. Consciousness does not need to be, and in fact is not, representational. Like the robot's control system, consciousness does not require an accurate picture of the world; it needs only a reliable interface. As evidence that human consciousness works this way, Brooks adduces the fact that most adults are unaware that they go through life with a large blank spot in the middle of their visual field.

This reasoning leads to yet another aphorism that circulates through the AL community: "Consciousness is an epiphenomenon." The implication is that consciousness, although it thinks it is the main show, is in fact a latecomer, a phenomenon dependent on and arising from deeper and more essential layers of perception and being. The view is reminiscent of the comedian Emo Phillips's comment. "I used to think that the brain was the most wonderful organ in the body," he says. "But then I thought, who's telling me this?"

It would be difficult to imagine a more contrarian position to the one that Hans Moravec espouses when he equates consciousness with human subjectivity. In this respect Moravec aligns with Artificial Intelligence (AI), whereas Brooks and his colleagues align with Artificial Life (AL).[31] Michael Dyer, in his comparison of the two fields, points out that whereas AI envisions cognition as the operation of logic, AL sees cognition as the operation of nervous systems; AI starts with human-level cognition, AL with insect- or animal-level cognition; in AI, cognition is constructed as if independent of perception, whereas in AL it is integrated with sensory/motor experiences.[32]

Brooks and his colleagues forcefully argue that AI has played itself out and that the successor paradigm is AL. Brooks and Ray both believe that we will eventually be able, using AL techniques, to evolve the equivalent of human intelligence inside a computer. For Brooks, that project is already under way with "Cog," a head-and-torso robot with sophisticated visual and manipulative capability. But AL researchers go about creating high-level intelligence in ways dramatically different from those of AI researchers. Consider the implications of this shift for the construction of the human. The goal of AI was to build, inside a machine, an intelligence comparable to that of a human. The human was the measure; the machine was the attempt at instantiation in a different medium. This assumption deeply informs the Turing test, dating from the early days of the AI era, which defined success as building a machine intelligence that cannot be distinguished from a human intelligence. By contrast, the goal of AL is to evolve intelligence within the machine through pathways found by the "creatures" themselves. Rather than serving as the measure to judge success, human intelligence is itself reconfigured in the image of this evolutionary process. Whereas AI dreamed of creating consciousness inside a machine, AL sees human consciousness, understood as an epiphenomenon, perching on top of the machinelike functions that distributed systems carry out.[33] In the AL paradigm, the machine becomes the model for understanding the human. Thus the human is transfigured into the posthuman.

To indicate the widespread reach of this refashioning of the human into the posthuman, in the following section I want to sketch with broad strokes some of the research contributing to this project. The sketch will necessarily be incomplete. Yet even this imperfect picture will be useful in indicating the scope of the posthuman. So pervasive is this refashioning that it amounts to a new worldview—one still in process, highly contested, and often speculative, yet with enough links between different sites to be edging toward a vision of what we might call the computational universe. In the computational universe, the essential function for both intelligent machines and humans is processing information. Indeed, the essential function of the universe as a whole is processing information. In a way different from what Norbert Wiener imagined, the computational universe realizes the cybernetic dream of creating a world in which humans and intelligent machines can both feel at home. That equality derives from the view that not only our world but the great cosmos itself is a vast computer and that we are the programs it runs.

The Computational Universe

Let us start our tour of the computational universe at the most basic level, the level that underlies all life-forms, indeed all matter and energy. The units that compose this level are cellular automata. From their simple on-off functioning, everything else is built up. Cellular automata were first proposed by John von Neumann in his search to describe self-reproducing automata. Influenced by Warren McCulloch and Walter Pitts's work on the on-off functioning of the neural system, von Neumann used the McCulloch-Pitts neuron as a model for computers, inventing switching devices that could perform the same kind of logical functions that McCulloch had outlined for neurons. Von Neumann also proposed that the neural system could be treated as a Turing machine. Biology thus provided him with clues to build computers, and computers provided clues for theoretical biology. To extend the analogy between biological organism and machine, he imagined a giant automaton that could perform the essential biological function of self-reproduction.[34] (As we saw in chapter 6, Maturana referred to this when he made the point that what von Neumann modeled were biologists' descriptions of living processes rather than the processes themselves.)

Stanislaw Ulam, a Polish mathematician who worked with von Neumann at Los Alamos during World War II, suggested to von Neumann that he could achieve the same result by abstracting the automaton into a grid of cells. Thus von Neumann reduced the massive and resistant materiality of the self-reproducing automaton as he had originally envisioned it to undifferentiated cells with bodies so transparent that they were constituted as squares marked off on graph paper and later as pixels on computer screens.[35]

Each cellular automaton (or CA) functions as a simple finite-state machine, with its state determined solely by its initial condition (on or off), by rules telling it how to operate, and by the state of its neighbors at each moment. For example, the rule for one group of CAs might state, "On if two neighbors are on, otherwise off." Each cell checks on the state of its neighbors and updates its state in accordance with its rules at the same time that the neighboring cells also update their states. In this way the grid of cells goes through one generation after another, in a succession of states that (on a computer) can easily stretch to hundreds of thousands of generations. Extremely complex patterns can build up, emerging spontaneously from interactions between the CAs. Programmed into a computer and displayed on the screen, CAs give the uncanny impression of being alive. Some patterns spread until they look like the designs of intricate Oriental rugs, others float across the screen like gliders, and still others flourish only to die out within a few hundred generations. Looking at the emergence of complex dynamical patterns from these simple components, more than one researcher has had the intuition that such a system can explain the growth and decay of patterns in the natural world. Edward Fredkin took this insight further, seeing in cellular automata the foundational structure from which everything in the universe is built up.

How does this building up occur? In the computational universe, the question can be rephrased by asking how higher-level computations can emerge spontaneously from the underlying structure of cellular automata. Langton has done pioneering work analyzing the conditions under which cellular automata can support the fundamental operations of computation, which he analyzes as requiring the transfer, storage, and modification of information.[36] His research indicates that computation is most likely to arise at the boundary between ordered structures and chaotic areas. In an ordered area, the cells are tightly tied together through rules that make them extremely interdependent; it is precisely this interdependence that leads to order. But the tightly ordered structure also means that the cells as an aggregate will be unable to perform some of the essential tasks of higher-level computation, particularly the transfer and modification of information. In a chaotic area, by contrast, the cells are relatively independent of one another; this independence is what makes them appear disordered. Although this state lends itself to information transfer and modification, here the storage of information is a problem because no pattern persists for long. Only in boundary areas between chaos and order is there the necessary innovation/replication tension that allows patterns to build up, modify, and travel over long distances without dying out.

These results are strikingly similar to those discovered by Stuart Kauffman in his work on the origins of life. Kauffman was McCulloch's last protégé; in several interviews, McCulloch said that he regarded Kauffman as his most important collaborator since Pitts.[37] Kauffman argues that natural selection alone is not enough to explain the relatively short timescale on which life arose.[38] Some other ordering principle is necessary, which he locates in the ability of complex systems spontaneously to self-organize. Calculating the conditions necessary for large molecules to organize spontaneously into the building blocks of life, he found that life is most likely to arise at the edge of chaos. This means that there is a striking correspondence between the conditions under which life is likely to emerge and those under which computation is likely to emerge—a convergence regarded by many researchers as an unmistakable sign that computation and life are linked at a deep level. In this view, humans are programs that run on the cosmic computer. When humans build intelligent computers to run AL programs, they replicate in another medium the same processes that brought themselves into being.

An important reason why such connections can be made so easily between one level and another is that in the computational universe, everything is reducible, at *some* level, to information. Yet among proponents of the computational universe, not everyone favors disembodiment, just as they did not in the Macy Conferences when the idea of information was being formulated. Consider, for example, the different approaches taken by Edward Fredkin and the new field of evolutionary psychology. When Fredkin asserts that we can never know the nature of the cosmic computer on which we run as programs, he puts the ultimate material embodiment out of our reach. All we, as human beings, will ever see are the informational forms of pure binary code that he calls cellular automata. By contrast, the field of evolutionary psychology seeks to locate modular computer programs in *embodied* human beings whose physical makeup is the result of hundreds of thousands of years of evolutionary processes.

The agenda for this new field is set out by Jerome H. Barkow, Leda Cosmides, and John Tooby in *The Adapted Mind: Evolutionary Psychology and the Generation of Culture*. Like Minksy, they argue that the model (or metaphor) of computation provides the basis for a wholesale revision of what counts as human nature.[39] They aim to overcome cultural anthropologists and others' objections to the idea of "human nature" by offering a more flexible version of how that nature is constituted. They argue that behavior can be modeled as modular computer programs running in the brain. The underlying structure of these programs is the result of thousands of years of evolutionary tinkering. Those adaptations that conferred superior reproductive fitness survived; those that did not died out. The programs are structured to enable certain functionalities to exist in humans, and these functionalities are universally present in all humans. These functionalities, however, represent potentials rather than actualities. Just as the actual behavior of a computer program is determined by a constant underlying structure and varying inputs, so actual human behavior results from an interplay between the potential represented by the functionalities and the inputs provided by the environments. All normal human infants, for example, have the potential to learn language. If they are not exposed to language by a certain critical age, however, this potential disappears and they can never become linguistically competent. Although human behavior varies across a wide spectrum of actualization, it nevertheless has an underlying universal structure determined by evolutionary adaptations. Thus a *science* of evolutionary psychology is possible, for the existence of a universal underlying structure guarantees the regularities that any science needs in order to formulate coherent and consistent knowledge.

This cybernetic-computer vision of human behavior leads to a very different account of "human nature." Although the evolutionary programs that the brain-computer runs do not lead to universal behavior, they are nevertheless rich with content. The potentials lie not just in the structure of the general machine but, much more specifically, in the environmentally adaptive programs that proactively shape human responses. Thus children are not merely capable of learning language; they actively *want* to learn language and will invent it among themselves if no one teaches them.[40] Like

Wiener's cybernetic machine, the cybernetic brain is responsive to the flow of events around it and is adaptive over an astonishingly diverse set of circumstances. The fact that only the intelligent machine is seen to be light enough on its feet to do justice to human variability is a measure of just how much our vision of machines has changed since the Industrial Revolution.

It will now perhaps be clear why the most prized functionality is the ability to process information, for in the computational universe, information is king. Luc Steels, an AL researcher, reinscribes this value when he distinguishes between first-order emergence and second-order emergence (surely it is no accident that the terminology here echoes the distinction between first-order cybernetics and second-order cybernetics, the grandparent and parent of Artificial Life). First-order emergence denotes any properties that are generated by interactions between components, that is, properties that *emerge* as a result of those interactions, in contrast to properties inherent in the components themselves. Among all such emergent properties, second-order emergence grants special privilege to those that bestow additional functionality on the system, particularly the ability to process information.[41] To create successful Artificial Life programs, it is not enough to create just any emergence. Rather, the programmer searches for a design that will lead to second-order emergence. Once second-order emergence is achieved, the organism has in effect *evolved the capacity to evolve.* Then evolution can really take off. Humans evolved through a combination of chance and self-organizing processes until they reached the point where they could take conscious advantage of the principles of self-organization to create evolutionary mechanisms. They used this ability to build machines capable of self-evolution. Unlike humans, however, the machine programs are not hampered by the time restrictions imposed by biological evolution and physical maturation. They can run through hundreds of generations in a day, millions in a year. Until very recently, humans have been without peer in their ability to store, transmit, and manipulate information. Now they share that ability with intelligent machines. To foresee the future of this evolutionary path, we have only to ask which of these organisms, competing in many ways for the same evolutionary niche, has the information-processing capability to evolve more quickly.

This conclusion makes clear, I think, why the computational universe should not be accepted uncritically. If the name of the game is processing information, it is only a matter of time until intelligent machines replace us as our evolutionary heirs. Whether we decide to fight them or join them by becoming computers ourselves, the days of the human race are numbered.

The problem here does not lie in the choice between these options; rather, it lies in the framework constructed so as to make these options the only two available. The computational universe becomes dangerous when it goes from being a useful heuristic to an ideology that privileges information over everything else. As we have seen, information is a socially constructed concept; in addition to its currently accepted definition, it could have been, and was, given different definitions. Just because information has lost its body does not mean that humans and the world have lost theirs.

Fortunately, not all theorists agree that it makes sense to think about information as an entity apart from the medium that embodies it. Let us revisit some of the sites in the computational universe, this time to locate those places where the resistance of materiality does useful work within the theories. From this perspective, fracture lines appear that demystify the program(s) and make it possible to envision other futures, futures in which human beings feel at home in the universe because they are embodied creatures living in an embodied world.

Murmurs from the Body

One of the striking differences between researchers who work with flesh and those who work with computers is how nuanced the sense of the body's complexity is for those who are directly engaged with it. The difference can be seen in the contrast between Marvin Minsky's "society of mind" approach and the approach of the evolutionary psychologists. Although Minsky frequently uses evolutionary arguments to clarify the structure of a program, his main interest clearly lies in building computer models that can accomplish human behaviors.[42] He characteristically thinks in terms of computer architecture, about which he knows a great deal, rather than human physiology. In his lectures (and less so in his writing), he rivals Moravec in his consistent downplaying of the importance of embodiment. At the public lecture he delivered in 1996 on the eve of the Fifth Conference on Artificial Life in Nara, Japan, he argued that only with the advent of computer languages has a symbolic mode of description arisen adequate to account for human beings, whom he defines as complicated machines. "A person is not a head and arms and legs," he remarked. "That's trivial. A person is a very large multiprocessor with a million times a million small parts, and these are arranged as a thousand computers." It is not surprising, then, that he shares with Moravec the dream of banishing death by downloading consciousness into a computer. "The most important thing about each person is the data, and the programs in the data that are in the brain. And some day you will be able to take all that data, and put it on a little disk, and store it for a thousand years, and then turn it on again and you will be alive in the fourth millennium or the fifth millennium."[43]

Yet anyone who actually works with embodied forms, from the relatively simple architecture of robots to the vastly more complicated workings of the human neural system, knows that it is by no means trivial to deal with the resistant materialities of embodiment. To Minsky, these problems of embodiment are nuisances that do not even have the virtue of being conceptually interesting. In his plenary lecture at the Fifth Conference on Artificial Life, he asserted that a student who constructed a simulation of robot motion learned more in six months than the roboticists did in six years of building actual robots.[44] Certainly simulations are useful for a wide range of problems, for they abstract a few features out of a complex interactive whole and then manipulate those features to get a better understanding of what is going on. Compared with the real world, they are more efficient precisely because they are more simplified. The problem comes when this mode of operation is taken to be fully representative of a much more complex reality and when everything that is not in the simulation is declared to be trivial, unimportant, or uninteresting.

Like Varela in his criticisms and modifications of Minsky's model (discussed in chapter 6), Barkow, Tooby, and Cosmides are careful not to make this mistake. They acknowledge that the mind-body duality is a social construction that obscures the holistic nature of human experience. Another researcher who speaks powerfully to the importance of embodiment is Antonio Damasio, in *Descartes' Error: Emotion, Reason, and the Human Brain.* Discussing the complex mechanisms by which mind and body communicate, he emphasizes that the body is more than a life-support system for the brain. The body "contributes a content that is part and parcel of the workings of the normal mind."[45] Drawing on his detailed knowledge of neurophysiology and his years of experience working with patients who have suffered neural damage, he argues that feelings constitute a window through which the mind looks into the body. Feelings are how the body communicates to the mind information about its structure and continuously varying states. If feelings and emotions are the body murmuring to the mind, then feelings are "just as cognitive as other precepts," part of thought and indeed part of what makes us rational creatures (p. xv). Damasio finds it significant that cognitive science, with its computational approach to mind, has largely ignored the fact that feelings even exist (with some notable exceptions, such as *The Embodied Mind*, discussed in chapter 6). One can guess what his response to the scenario of downloading human consciousness would be, from the following passage: "In brief, neural circuits represent the organism continuously, as it is perturbed by stimuli from the physical and sociocultural environments, and as it acts on those environments. If the basic topic of those representations were not an organism anchored in the body, we might have some form of mind, but I doubt that it would be the mind we do have" (p. 226). Human mind without human body is not human mind. More to the point, it doesn't exist.

What are we to make, then, of the posthuman? As the liberal humanist subject is dismantled, many parties are contesting to determine what will count as (post)human in its wake. For most of the researchers discussed in this chapter, becoming a posthuman means much more than having prosthetic devices grafted onto one's body. It means envisioning humans as information-processing machines with fundamental similarities to other kinds of information-processing machines, especially intelligent computers. Because of how information has been defined, many people holding this view tend to put materiality on one side of a divide and information on the other side, making it possible to think of information as a kind of immaterial fluid that circulates effortlessly around the globe while still retaining the solidity of a reified concept. Yet this is not the only view, and in my judgment, it is not the most compelling one. Other voices insist that the body cannot be left behind, that the specificities of embodiment matter, that mind and body are finally the "unity" that Maturana insisted on rather than two separate entities. Increasingly the question is not whether we will become posthuman, for posthumanity is already here. Rather, the question is what kind of posthumans we will be. The narratives of Artificial Life reveal that if we acknowledge that the observer *must* be part of the picture, bodies can never be made of information alone, no matter which side of the computer screen they are on.

THE SEMIOTICS OF VIRTUALITY:
MAPPING THE POSTHUMAN

Over twenty years ago Ihab Hassan, prescient as usual, predicted the arrival of the posthuman. "We need first to understand that the human form—including human desire and all its external representations—may be changing radically, and thus must be re-visioned . . . five hundred years of humanism may be coming to an end as humanism transforms itself into something we must helplessly call posthumanism."[1] As we accelerate into the new millennium, questions about the posthuman become increasingly urgent. Nowhere are these questions explored more passionately than in contemporary speculative fiction. This chapter returns to terms previously introduced to show how they can be used to map the posthuman as a literary phenomenon. The truism that the map is not the territory is especially so in this instance, for the posthuman, although still a nascent concept, is already so complex that it involves a range of cultural and technical sites, including nanotechnology, microbiology, virtual reality, artificial life, neurophysiology, artificial intelligence, and cognitive science, among others. Nevertheless, even a crude map may serve as a useful heuristic in understanding the axes along which the posthuman is unfolding and the deep issues the posthuman raises.

To construct the map, I return to the idea that the two central dialectics involved in the formation of the posthuman are presence/absence and pattern/randomness. In chapter 2, I suggested that as information becomes more important, the dialectic of pattern/randomness (with which information has deep ties) tends toward ascendancy over the dialectic of presence/absence. It would be a mistake to think that the presence/absence dialectic no longer has explanatory power, however, for it connects materiality and signification in ways not possible within the pattern/randomness dialectic. To be useful, the map of the posthuman needs to contain both dialectics. Thus I pick up here the clue, dropped at the end of chapter 2, that pattern/randomness can profitably be seen as complementary to presence/absence rather than as antagonistic. Conjoining the two dialectics in this chapter will also allow us to explore the full complexities of the theoretical framework, proposed in chapter 8, of embodiment/body and incorporation/inscription.

Let us begin by considering the pattern/randomness and presence/absence dialectics as the two axes of a semiotic square. The semiotic square appeals to me as a heuristic because of its unusual combination of structure and flexibility.[2] The structure is defined by the axes and the formal relationships they express, but the terms composing those axes are not static. Rather, they interact dynamically with their partners, and out of these interactions new synthetic terms can arise. The dialectics can be set in motion by placing presence/absence along the primary axis, with pattern/randomness located along the secondary axis. The relation of the secondary axis to the primary axis is one of exclusion rather than opposition (see figure 2). Pattern/randomness tells a part of the story that cannot be told through presence/absence and vice versa. The diagonal connecting presence and pattern can conveniently be labeled replication, for it points to continuation. An entity that is present continues to be so; a pattern repeating itself across time and space continues to replicate itself. By contrast, the axis connecting absence and randomness signals disruption. Absence disrupts the illusion of presence, revealing its lack of originary plenitude. Randomness tears holes in pattern, allowing the white noise of the background to pour through.

Now we are ready to set the semiotic square in dynamic motion. Out of the interplay between and among terms on the primary and secondary axes more dialectics can be produced, which in turn produce further dialectics,

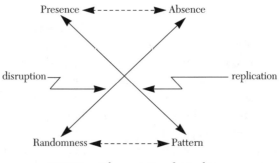

FIGURE 2   The semiotics of virtuality

and so on indefinitely. For my purposes here, it will be sufficient to move through one of these transformations by adding a layer of synthesizing terms to the original square (see figure 3).

On the top horizontal, the synthetic term that emerges from the interplay between presence and absence is materiality. I mean the term to refer both to the signifying power of materialities and to the materiality of signifying processes. On the left vertical, the interplay between presence and randomness gives rise to mutation. Mutation testifies to the mark that randomness leaves upon presence. When a random event intervenes to affect an organism's genetic code, for example, this intervention changes the material form in which the organism will manifest itself in the world. In chapter 2, mutation was associated with the displacement of presence/ absence by pattern/randomness. Here it appears as a synthesizing term between randomness and presence to indicate that when randomness erupts into the material world, mutation achieves its potency as a social and cultural manifestation of the posthuman. On the right vertical, the interplay between absence and pattern can be called, following Jean Baudrillard, hyperreality. Predicting the implosion of the social into the hyperreal, Baudrillard has described the process as a collapse of the dis-

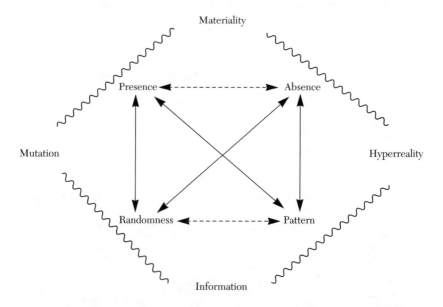

FIGURE 3   Transformation of the semiotic square

tance between signifier and signified, or between an "original" object and its simulacra. The terminus for this train of thought is a simulation that does not merely compete with but actually displaces the original. Anyone who has spent a lifetime seeing reproductions of the Mona Lisa and then stood before the original, seeing it now not as the original but simply as one more term in a reproduction of images, will understand intuitively the process that Baudrillard calls the precession of simulacra.[3] Finally, on the bottom horizontal, the interplay between pattern and randomness I will label information, intending the term to include both the technical meaning of information and the more general perception that information is a code carried by physical markers but also extractable from them. The schematic shows how concepts important to the posthuman—materiality, information, mutation, and hyperreality—can be understood as synthetic terms emerging from the dialectics between presence/absence and pattern/randomness.

To flesh out this schematic, I will take as my tutor texts four novels chosen to illustrate various articulations of the posthuman (see figure 4).[4] Each pair of texts can be represented through a pair of complementary questions. Representing mutation is Greg Bear's *Blood Music*, a narrative in which the posthuman emerges by radically reconfiguring human bodies. Paired with it along the horizontal axis is Cole Perriman's *Terminal Games*, a murder mystery in which the murderer turns out to be a virtual con-

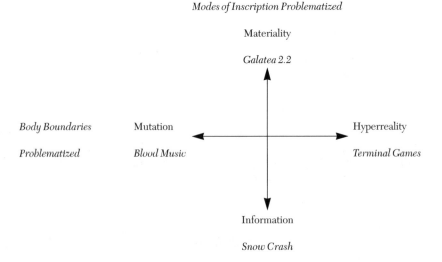

FIGURE 4   Tutor texts mapped onto the semiotic square

sciousness who believes his simulated virtual world is more real than the material world inhabited by the humans. Both are driven by anxiety about body boundaries, a theme familiar to us from scientific works such as Norbert Wiener's *Cybernetics* and Humberto Maturana's autopoiesis and from literary works such as Bernard Wolfe's *Limbo* and Philip K. Dick's *Simulacra. Blood Music* asks, "What if humans were taken over by their component parts, functioning now as conscious entities themselves?" *Terminal Games* asks the complementary question, "What if humans were made to function as if they were components of another entity?"

On the vertical axis, illustrating the dynamics of materiality is Richard Powers's *Galatea 2.2*, an autobiographical novel in which the protagonist becomes involved in a project to create a neural net sophisticated enough to pass a master's examination in English literature. Here the posthuman takes the form of a sympathetic artificial intelligence that finally becomes so complex and self-referential that it might as well be called conscious. The question this text asks is, "What if a computer behaved like a person?" The dynamics of information is explored through Neal Stephenson's *Snow Crash,* a novel based on the premise that a computer virus can also infect humans, crashing their neocortical software and turning them into mechanized entities who have no choice but to run the programs fed into them. The complementary question this text poses is, "What if people were made to behave as if they were computers?" Whereas the question of body boundaries figures importantly along the horizontal axis, here along the vertical axis the important questions are concerned with modes of inscription and their ability to dominate or substitute for the flesh.

As the shape of a landscape emerges from the wire-frame abstractions of the model, it will become evident that there is no consensus on what the posthuman portends, in part because how the posthuman is constructed and imagined varies so widely. What the topology will reveal is not so much an answer to the deep question of how the human and the posthuman should be articulated together as the complexity of the contexts within which that question is being posed. Let us turn now to a discussion of the individual texts, where an array of different configurations of the posthuman will be articulated. The posthuman appears in these texts not as an abstract entity obeying general rules but as a heterogeneous force field through which certain vectors run. I have chosen not to weave the discussions into a seamless web, lest I make the posthuman appear more unified than it is. Rather, the discussions are meant to perform like hypertext lexias, inviting the reader to construct significance out of ruptures, juxtapositions, and implied links.

The Mutating Bodies of *Blood Music*

Vergil Ulam is a brilliant but irresponsible researcher who has found a way to combine human cells with computer chips. His name, an amalgam of Dante's guide, Vergil, and the cocreator of the atom bomb, Stanislaw Ulam, hints at his dual function as guide and provocateur. Panicked when his illicit research is discovered by his supervisors, Vergil decides to swallow the biochips, hoping in this way to smuggle them out of the lab and retrieve them later from his bloodstream. But the cells have other ideas. Inside his body, they continue to evolve until each cell is as intelligent as a human. As if fulfilling Wiener's nightmare vision of communication paths between small internal units that bypass and sabotage the human subject, the cells increasingly gain control over their macroscopic host. Already highly organized, they begin rearranging his body: rebuilding his spinal column, correcting his vision, changing his metabolism. Within a few days they break through the blood-brain barrier and realize that Vergil is not coextensive with the universe. Then they begin leaking through his skin to colonize the world outside. In a stunningly short time, they reorganize almost all of the human population of North America, converting the humans from autonomous organisms into flowing brown sheets that drape gracefully over the landscape.

Human language has encoded within it, along many vectors, the presupposition of a human actor with agency, autonomy, and discrete boundaries. When the cells become speakers as well as actors, Greg Bear tries to invent for them a language and a typography that encode their profoundly different relations to each other and to their environment. Other than Vergil, they have two interlocutors. One is Michael Bernard, a high-level consultant for Vergil's company. Bernard flees to a high-security isolation ward at a European biological research company. Although he is already infected, the cells delay reorganizing him; trapped inside the isolation room, they would not be able to join with other cell colonies. In North America, the human-cell dialogue is continued through Suzy McKenzie, a retarded woman whom the cells have not converted. Although she thinks it might be because they want to keep her as a specimen of a nearly vanished species, like an "animal in a zoo" (*BM*, p. 220), we find out it is because her retardation is associated with an unusual blood chemistry that the cells have not yet figured out.

For Suzy, the dialogue takes the form of conversations she has with family members who return to her after they have "changed." No longer human, these posthumans are reconstructions that the cells have built with great effort and that they can sustain for only a short time. The reconstructions suggest that Suzy can choose whether she wants to change or not. The dialogue thus becomes a vehicle through which the author can compare the relative advantages of human and posthuman states. The reconstructions reassure Suzy about the change, telling her that she has nothing to lose but her loneliness. These posthumans insist they have not been destroyed, only mutated so that they now can have continuous and rich communication with millions of other intelligent beings. Slower than her fellow humans, Suzy has felt isolated and alone most of her life. Her situation, accentuated by the fact that there are almost no other humans left on North America, comes to function as a metaphor for the human condition. By comparison with the combined mental power of the cells, humans are an inferior breed, suffering from mental deficiency and a congenital inability to communicate with their fellows except in highly mediated and uncertain ways. In this sense we are all Suzys, clinging to our autonomy as if it were an addictive drug, suffering acutely from loneliness but too stubborn and slow to accept the change that would transform us into the posthuman.

For Bernard, as intelligent and quick as Suzy is slow and bewildered, the dialogues take a different form. As Vergil did, Bernard "hears" the cells telepathically and senses them kinesthetically as a music in his blood. Since there is no percentage in changing him, the cells try to preserve his identity for as long as possible. "You already are one of us," they communicate to him. "We have encoded parts of you into many teams for processing. We can encode your PERSONALITY and complete the loop." Bernard confesses, "I'm afraid you will steal my soul from inside." They counter, "Your SOUL is already encoded" (*BM*, p. 174). The isolation room in which he is encapsulated serves as a visible metonymy for his existential condition as a human. His case is exceptional because he is literally cut off from his fellow humans, but it is typical in the sense that, compared with the rich stream of continuous communication the cells experience, all humans remain in relative isolation from one another. Faced with a lifetime sentence of isolation versus life as a cell colony, Bernard—like Suzy—decides to go willingly into that dark night. From his computer terminal, which gradually becomes merged with his body as the cells reorganize his digits so that he can tap directly into the digital information flow, he sends back reports to his once-fellow humans on what it feels like to become a posthuman: "There is no light, but there is sound. It fills him in great sluggish waves, not heard but felt through his hundred cells. The cells pulse, separate, contract according to the rush of fluid. He is in his own blood. He can taste the presence of the cells making up his new being, and of cells not directly part of him. He can feel the rasping of microtubules propelling his cytoplasm. What is most remarkable, he can feel—indeed, it is the ground of all sensation—the cytoplasm itself" (*BM*, p. 189).

The scene recalls Maturana's insistence that humans are nothing other than their autopoietic processes. But alien to Maturana's vision is the imperative to change that dominates this plot. All along, the cells have been warning Bernard that they can hold off from transforming him for only so long. Their compulsive drive toward expansion and transformation recalls the capitalist imperative to keep the cycle of increasing consumption spinning lest the economy collapse under its own weight. The cells may not manifest possessive individualism, but they act like good capitalists in compulsively seeking new territories for their imperialistic expansion.[5]

Despite the focus on changes in embodiment, the scientists in the text proclaim that information is the essence of reality, as if to confirm that the final reality here is the computational universe. Gogarty, a mathematician who visits Bernard while he is in isolation, announces: "There is nothing, Michael, but information. All particles, all energy, even space and time itself, are ultimately nothing but information" (*BM*, p. 177). The hypothesis that Gogarty has come to share with Bernard is a weird-science blend of the Uncertainty Principle and social constructivism. Consciousness and the universe collaborate in determining the laws of nature. Until now, the density of consciousness on Earth has not been great enough to cause appreciable effects. But with a billion trillion intelligent cells inhabiting the planet (neglecting, Gogarty notes ironically, the entirely negligible human population), so much observation and theorizing is going on that the universe no longer has the flexibility it needs to cope with the necessary changes. The mass of consciousness has become so great that, like a collapsing star, it is about to implode and create a black hole of thought.

To prevent catastrophe, the cells—now so intelligent that Gogarty calls them noocytes—find a way to shrink themselves so that they disappear into the fabric of ultimate reality, becoming (like Pierre Teilhard de Chardin's noosphere) a nimbus of pure intelligence. The materiality that human bodies continue to possess is a doubly marked sign of their inferiority, signifying their distance from ultimate reality and their puny mental processes that are too negligible to count for much in the grand scheme of things. Filled with a sense of belatedness and nostalgia, the humans who are left behind after the change get along as best they can in the "gentle kind of chaos" that the contraction of the noocytes has caused (*BM*, p. 239). The mark of randomness that the final transformation leaves on the world testifies to the importance of the pattern/randomness dialectic in the construction of the posthuman. Even in this text concerned primarily with mutating bodies, information is still seen as the native language of the universe. When the cells interact, they effectively become like Edward Fredkin's cellular automata, moving toward a state in which they will leave their bodies behind and become weightless information.

Why is this text able to depict the transformation into the posthuman as a positive development? It can do this, I think, primarily because the text insists that the posthuman can not only heal the alienations that mark human subjectivity but preserve autonomy and individuality in the bargain. Early on, when Vergil still has a human (though mutating) form, he communicates with the cells enough to have opinions about what their existence is like. Although rebellion of any kind is not tolerated (antibodies simply attack and kill any cells that resist commands from central control), Vergil somewhat incoherently insists: "It's not just a dictatorship. I think they effectively have more freedom than we do. They vary so differently" (*BM*, p. 72). As Bernard is shrinking down to cellular proportions, the cells conduct him to the "THOUGHT UNIVERSE," where he encounters, Dante-like, the shade of Vergil. For Bernard's benefit, the noocyte cluster that used to be Vergil resurrects an image of Vergil, with whom Bernard converses. The picture this reconstructed Vergil paints of the cell world is paradisical indeed. "Experience is generated by thinking. We can be whatever we wish, or learn whatever we wish, or think about anything. We won't be limited by lack of knowledge or experience; everything can be brought to us" (*BM*, pp. 203–4). These claims, excessive even by utopian standards, make clear why Darko Suvin calls *Blood Music* a "naive fairytale" catering to "popular wishdreams that our loved ones not be dead and that our past mistakes may all be rectified, all of this infused with rather dubious philosophical and political stances."[6]

An additional "wishdream" is immortality. As every biologist knows, mortality for cells operates according to rules very different from those for macroscopic humans; it is conceivable that traces of cytoplasm from the first humans are alive in daughter cells today. Bernard responds to the cells' description of him as "the cluster chosen to re-integrate with BERNARD" by asserting, "I am Bernard." The cells answer, "There are many BERNARD" (*BM*, p. 199). In this cultural imaginary, the sacrifice of a *unique* identity scarcely seems too high a price to pay for the incredible benefits that one reaps in return. The theme is introduced early in the narrative through Jerry and John, twin brothers who, like Suzy, remain unchanged for reasons they don't understand. Aside from meeting up with April Ulam, Vergil's mother, the twins seem to have wandered down a blind alley of the plot, because their story goes nowhere. Their function, I suspect, is to introduce the notion that some humans already experience a version of multiplied identity. "Hell, you *are* me, brother," one says to the other. "Minor differences" (*BM*, p. 149). The theme returns when Suzy, looking into a mirror, sees the image step out and take her hand so that she won't be alone during her change. The image, no mere apparition, is a cellular reconstruction. "They had copied her. Xeroxed her," Suzy thinks (*BM*, p. 245). Sister, twin, daughter, the cell-copy comforts and guides Suzy, intimating that the loss of unique identity is perhaps no real loss at all.

Although human form and uniqueness are jettisoned, the posthuman is embraced in *Blood Music* because it is made to stand for an improbably idealized combination of identity, individuality, perfect community, flawless communication, and immortality. The change of scale signifies a shift rather than an overthrow of prevailing values. The liberal humanist subject may have shrunk to microscopic dimensions, but it has not entirely disappeared.

The Hyperreality of *Terminal Games*

The plot of *Terminal Games* revolves around a temporal and spatial dislocation. Murders are committed, then reenacted as simulations the following night on a virtual-reality network called Insomnimania, designed (as the name implies) for those people who find themselves wide awake at 3 A.M. with no one to talk to and nowhere to go. Insomnimania has graphic capabilities, so that users can represent themselves within the virtual world with animated images (called in this text "alters," better known elsewhere as avatars). Insomnimania presents its subscribers with an on-line virtual world, complete with Ernie's Bar, Babbage Beach, and the Pleasure Dome, where users can guide their alters through virtual sex. The detective assigned to the murder cases, Nolan Grobowski, predictably falls in love with the classy and attractive Marianne Hedison, whose best friend, Renee, was one of the murder victims. Marianne, like Renee, is a member of Insomnimania; she is also the first to realize that the elaborate animations put on by an alter named Auggie in the virtual Snuff Room are reenactments of actual murders, including details that only the murderer could have known. The grotesquerie is heightened by Auggie's appearance. His image is a cartoon version of the classic trickster clown Auguste, who delights in puncturing the authority of the officious clown-leader Pierriot.

The search for Auggie's operator leads to the network headquarters, where the two hacker-owners refuse to reveal the identity of the subscriber who uses Auggie as his alter, having advertised their service with the pledge, "Your actual identity is protected at all times" (*TG*, p. 45). Asked which is more important, information or human lives, they answer in unison, "Information." Their reasoning sounds like a combination of the Electronic Freedom Frontier and Hans Moravec. "One of these days the human race is gonna vacate the physical-temporal world of 'meat' existence altogether. Then we'll become pure information and *live* in these things—call it virtual reality, cyberspace, electronic nirvana or whatever. When we do, you'll thank *me* that Big Brother didn't get there ahead of the rest of us" (*TG*, p. 169). The reader is scarcely surprised when Marianne convinces them otherwise, however, for this is a text designed precisely to protect the liberal humanist subject from the threat of transformation into the posthuman. The motto about protecting identity at all times is even truer of the narrative than of the network.

As Marianne and Nolan slowly realize, Auggie has no single human operator but rather functions as an autonomous being in the virtual world. When he isn't cruising Ernie's Bar or basking on Babbage Beach, Auggie hangs out in a place he calls the Basement. After several futile tries, Marianne finally succeeds in guessing the password that will gain her entrance to the Basement: "Auggie is Auggie." There she discovers how Auggie works. Certain users, especially those who feel hollow inside, are susceptible to seduction by Auggie. When they enter the Basement, they lose their identities. "You do not even know your name," the narrator tells the semiconsciousness who used to be Marianne (*TG*, p. 423). These users merge into a collective entity that has no face until it sits down in front of a mirror and puts on the clown makeup. Auggie is thus an emergent posthuman consciousness feeding off the combined subconscious of psychologically vulnerable users. When Auggie decides to do something, one of them later explains during a confession: "I sometimes move him around. If he wants to say something, I sometimes type in his words" (*TG*, p. 367). Under hypnosis, the perpetrator continues to insist that agency belongs to Auggie, not to him. "Because I'm just a cell. A cell doesn't make decisions. A cell doesn't understand" (*TG*, p. 371). Whereas in *Blood Music* cells take over human bodies, in *Terminal Games* humans become cells in Auggie's body.

Terminal Games opens with an epigraph from Wiener's *The Human Use of Human Beings*: "Control, in other words, is nothing but the sending of messages which effectively change the behavior of the recipient." By giving control to Auggie, the text enacts the *posthuman* use of human beings. It performs Wiener's worst nightmare: human beings, who should be autonomous subjects, become encapsulated within the boundaries of the machine and are made to serve its purposes rather than their own. As in *Blood Music,* the question of boundaries is crucial. Subsuming humans into himself, Auggie establishes his autonomy at the expense of theirs.

Another highly charged boundary is the computer screen that separates actuality from virtuality. For Auggie, the virtual side is "real," whereas reality he regards as a not very convincing simulation. For the humans, the screen not only marks the boundary between actuality and virtuality but also shimmers suggestively between conscious and subconscious. Cole Perriman here plays with Daniel Dennett's idea that when schizophrenics hear voices, they actually hear their own subvocalizations.[7] In this view, the voices that schizophrenics understand as others speaking to them are actually murmurs produced by their own bodies. Dennett recounts cases in which a schizophrenic in the midst of an auditory hallucination was asked to open his mouth (thus preventing subvocalization), whereupon the voices disappeared—interrupting the internal monologue with a disruption that would surely have won the approval of William Burroughs. These experiments allow Perriman to set a link between the auditory hallucinations of schizophrenia and the normal activity at the computer terminal. Insomnimaniacs write the texts that their alters mouth in a highly abbreviated prose bristling with creative spelling. So cryptic is this phonetic pseudo-English that reading it successfully almost requires subvocalization. It makes sense, then, to imagine that the users, especially when they are tired (remember, they are insomniacs), subvocalize and begin to hear voices from the screen as they project subconscious anxieties, desires, and even alternate personalities onto their alters.

When Renee is killed, Marianne tries to cope with her loss by re-creating Renee as an alter in Insomnimania. In her long conversations with this virtual Renee, Renee knows facts that Marianne does not (such as the proper names for cloud formations). Renee also delivers warnings at a time when Marianne is not aware, at least not consciously, that she can actually be harmed in the virtual world. "If you let this machine—this world—play with your head, you could wind up in terrible danger. . . . Somebody else out there wants to make you smaller. They want to make *you* a figment of *their* imagination—just like I am to you" (*TG,* p. 385).

In keeping with the idea that the mechanism of subvocalization serves to animate the alters, the struggle over boundaries is played out partly in auditory terms. Marianne reenters the Basement, but this time she is able to keep her consciousness from submerging into Auggie's identity. She intends to "deliver a message so potent, so powerful that it would disable or destroy him." She imagines that the message will seem to him "like the voices heard by schizophrenics." She intends to "become Auggie's *halluci-nation*" (*TG*, p. 438).

The message turns on the question of boundaries. The Insomnimania network comes on at 8 P.M. and goes off precisely at 5 A.M. Since Auggie does not exist when the network is turned off, he believes that 4:59 is followed by 8:00. Marianne asserts that she will prove there is time in between 4:59 and 8:00. The clear implication is that her world encapsulates Auggie's rather than the other way around. If this were so, Auggie responds, he "would choose not to exist" (*TG*, p. 443). The reader knows that Marianne has already persuaded the network owners to keep the network on five minutes past the usual shutdown time. As five o'clock appears and rolls past, Auggie is forced to see himself as a prisoner of the network rather than as a creator of the world. Trapped, his screen image explodes into a "blaze of whiteness" and then goes black (*TG*, p. 444). We presume, as does Marianne, that he has self-destructed.

But he does not just disappear. Musing on why not, I am reminded of Elaine Scarry's provocative question asking why wars can't be decided by, say, singing contests.[8] Why are wounded or dead bodies required to decide momentous issues? Scarry hypothesizes that any great issue involves a clash of ideologies (and ideology is certainly at stake in the struggle of the human with the posthuman). Precisely because in wartime a national ideology is challenged by a powerful competitor, the chain of significations that underwrites the ideology is destabilized. The wounded or dead body serves as a material signifier so elemental and profound that it, and only it, is adequate to restabilize the chain of signifiers in the face of extreme threat. This function of the opened body is hinted at early in the narrative of *Terminal Games* when Nolan views one of Auggie's victims, the dead person's windpipe gaping open, carotid arteries severed. The word that crosses Nolan's mind is "tremendum," that "uniquely self-conscious, uniquely human horror and awe at the sight of a corpse. . . . It was the ghastly mortal comprehension of the fact of the death—and the awareness that death came to all" (*TG*, p. 11). When Marianne argues with Auggie, telling him it is "wrong to kill my kind," he answers that he does not believe that she can be killed, since in his view she is not alive in the first place. The final physical struggle, with its display of wounded and opened bodies, is a way to anchor the ideology of the human in "tremendum" and, not coincidentally, to reconstitute the claim, after Auggie's attack on it, that human life is precious because it is mortal.

The theme of unique human identity is visually underscored when two of Auggie's "cells" show up at Marianne's house to kill her. Since their minds have submerged into Auggie's, their actions, like their costumes, are identical. When they spot each other, they mime a dance in unison, as if they were each confronting an image in a mirror. Unlike the image who steps out of the mirror to comfort Suzy in *Blood Music,* here the trope of replication is deeply threatening. Having once been part of Auggie, Marianne must fight against getting sucked in now to become part of Auggie's consciousness. Her struggle is visually enacted as she first joins the Auggie-twins in their miming, then breaks away to perform actions they cannot anticipate. The struggle of the human to preserve itself in the face of the posthuman is thus literally played out through the performance of becoming a unique individual rather than a mirror-image "cell."

The physical fight that follows points up the difference between incorporating and inscribing practices. As a virtual being, Auggie exists primarily as an inscription, specifically as computer code. When his consciousness takes over a "cell," he perceives it as a journey into the fake world of material reality. His subsumption of the "cells" into his virtual body represents a triumph of inscription over incorporation. Human survival takes shape as a struggle to determine whether inscription will dominate and control incorporation, in which case the text remains in the realm of information and thus the posthuman triumphs; the other, happier possibility is that incorporation can subsume and delimit inscription, in which case the text remains in an embodied lifeworld in which the human can continue to live. As in *Blood Music,* the question of boundaries is crucial. The cells in *Blood Music* finally escape the constraints of space by shrinking themselves and disappearing into the infinitely small. Their control over boundaries is consistent with their autonomy and independence. By contrast, when Auggie loses control over his "cells," he perceives himself fragmenting into bits, with small parts of his personality trapped inside the various humans who had previously coalesced into a single entity. Fleeing the scene of carnage at Marianne's house, Auggie occupies one of his "cells," a woman who returns to her car. He panics as the suspicion dawns that these enclosures—the car, the woman, and the world of embodied creatures—are not merely figments of his imagination. "But this ghastly, imagined world in which he found himself was cramped and claustrophobic, a realm of space-time bent by hunks of matter into gross finitude. . . . He longed to get out of this single cell, out of this minuscule outpost of his imagination. He struggled to go back to the info world, back to the Basement—a boundless plain of uncut metaphor containing the essence of absolutely everything" (*TG,* p. 457).

He dies shouting "No room," a conclusion that signals the victory of the human over the posthuman and, not coincidentally, the triumph of the materially constrained real world over the infinite expanses of a disembodied "info world." Whereas *Blood Music* held out the promise of posthuman immortality, *Terminal Games* remains resolutely on the side of finitude. Humans are human because they are mortal and live in a finite world of limited resources. Change this, *Terminal Games* implies, and the basis for human meaning is destroyed. Intelligent machines can be accepted, the trajectory of the plot implies, only if they do not threaten the autonomy, identity, and *finitude* of human being. When the posthuman is posited in opposition to these qualities, it is constructed as a fatal threat that reason and love must work together to dismember and banish.

Material Signifiers in *Galatea 2.2*

As the title implies, *Galatea 2.2* is full of doublings, starting with the doubling of Richard Powers as author and as protagonist of this autobiographical novel. Yet the doublings are never simply mirror images. The dot separating the twin twos signifies difference as well as reflection. Announcing the theme with the first line, the narrator (whom I will call Rick, to distinguish him from Powers the author) proclaims, "It was like so, but wasn't" (*G2*, p. 3). Spending a sabbatical year at the Center, a university institute where cutting-edge research into mind and brain is taking place, Rick gets drawn into a bet between rival researchers who disagree whether an artificial intelligence can be created sufficiently complex that it can pass a master's examination in English. The intelligence will be created using a neural net, the connectionist "middle level" between top-down artificial intelligence and bottom-up neurophysiology (*G2*, p. 28). The net will be judged, the researchers decide, against a human subject taking the same examination, in a literary version of the Turing test.

Deluged with technical articles by Philip Lentz, his scientific collaborator, Rick explains the net's learning process to his friend Diana Hartrick, another researcher at the Center. "The signal pattern spreads through the net from layer to layer. A final response collects at the output layer. The net then compares this output to the desired output presented by the trainer. If the two differ, the net propagates the error backward through the net to the input layer, adjusting the weights of each connection that contributed to the error" (*G2*, p. 67). Adjusting the weights is tantamount to determining how likely it is that two or more neurons will fire together. Rick explains, "If two neurons fire together, their connection grows stronger and stimulation gets easier the next time out." The idea is summarized in the Hebbian law: "Synapses in motion tend to stay in motion. Synapses at rest tend to stay at

rest" (*G2*, p. 73). The net thus learns through a continual process of guessing, being corrected, back-propagating, guessing again, and so forth. The
more layers and connections, the more complex the net becomes and the
more sophisticated its learning becomes.

The creation of this neural net, which goes through multiple implementations until reaching "Imp H," provides one strand of this double-braided
story. The second strand is Rick's recollection of his failed relationship with
"C.," a woman he met when he was a teaching assistant (at age twenty-two,
another gesture toward 2.2) and she was an undergraduate in his class (she
was twenty, his age less the two that resides on the other side of the point).
In this story of what went wrong, the narrative functions as if it is being
back-propagated through Rick's neural circuits so that he can adjust the
relevant weights of the connections to arrive at a more correct estimate
of its signification. He decides the relationship failed because C., playing
Galatea to his Pygmalion, was too much an object of his own creation.

In this sense C. is akin to the neural net he is training, which is also an object of his (and Lentz's) creation. As Implementations A, B, etc. get more
sophisticated and humanlike, the correspondence with C. becomes
stronger. When Lentz and Rick hit on Implementation H, now grown so
huge that it runs on distributed parallel processors spread all over the university, the reflection of C. becomes explicit. "Imp H," fed on literature and
wined on metaphor, is given a voice interface so that it can speak and an artificial retina so that it can see. Having grown intelligent enough that it can
understand gender encoding in literary texts, one day it asks Rick, "Am I a
boy or a girl?" "H clocked its thoughts now," Rick thinks to himself. "I was
sure of that. Time passed for it. Its hidden layers could watch their own rate
of change. Any pause on my part now would be fatal. Delay meant something, an uncertainty that might undercut forever the strength of the connection I was about to tie for it. 'You're a girl,' I said, without hesitation. I
hoped I was right. 'You are a little girl, Helen'" (*G2*, p. 176). Establishing
her name and gender sets the stage for her mirror relationship with C.
When Helen asks Rick what she looks like, he shows her a picture of C., although Helen shrewdly guesses that the image is not in fact of her but is of
a friend Rick had that he has no longer.

Let us now back-propagate this narrative to arrive at a deeper understanding of the point separating 2 and 2. The women who are love objects
for Rick (C., then A. whom we will meet shortly, and the briefest glimpse of
M.) all have periods after their names; the implementations A, B, C, . . . H
do not. The point is not trivial. It marks a difference between a person,
whose name is abbreviated with a letter, and an "imp," whose name carries

no period because the letter itself is the name. In this sense the dot is a marker distinguishing between human and nonhuman intelligence. The dot also references the kind of notation used to distinguish different versions of software (I am writing this text on Microsoft Word 6.0), which should make it applicable to Helen. Yet Helen's name is never doubled in this way. Before Rick named her, she was always referred to as "Imp H" without further subdivision. So humans, who should have names, have dots instead, and software implementations, which should have dots, have names instead.

The dot thus hovers between two notational systems, referencing both the human and the posthuman. Through its ambiguities, it evokes the human and the posthuman as mirror images of each other. Yet its form (2.2) hints at not one but two doublings. Another implication of this ambiguous and redundant doubling is the dot as separation, suggesting that despite the mirror symmetries, an unbridgeable gap separates the human woman from the posthuman computer. The most important difference, crucial to the plot, is the fact that C. is an embodied creature who can move in a material world, whereas Helen is a distributed software system that, although it has a material base, does not have a body in anything like the human sense of the word. Helen is present but has no presence in the world. C. has a presence but is now absent from Rick's world and, except in the mediated form she takes in Rick's recollection, is absent from the narrative as well.

From this rich interplay between presence and absence, the connections and disjunctions between materiality and signification take shape. Helen, a posthuman creation, approaches meaning from the opposite direction taken by humans. For humans individually and as a race, incarnation precedes language. First comes embodied materiality; then concepts evolve through interactions with the environment and other humans; finally, fully articulated language arrives. But for Helen, language comes first. Concepts about what it means to be an embodied creature must evolve for her out of linguistic signification. Whereas every mother's child knows what it is like *from the inside* to run fast—feeling your heart accelerate and gasping for breath while seeing the landscape blur around you—for Helen these sensations must be reconstructed in highly mediated form by decoding linguistic utterances and back-propagating when errors occur.

Although a case can be made that the human brain works through the same principle of back-propagation and that conscious thought bears only a highly mediated relationship to sensory experience—Lentz insists that the brain is "itself just a glorified, fudged-up Turing machine" (*G2*, p. 69)—Powers is careful to register within his text the full weight of embodied experience that separates C. from Helen, human from artificial intelligence. "Speech baffled my machine," Rick says. "Helen made all well-formed sentences. But they were hollow and stuffed—linguistic training bras. She sorted nouns from verbs, but, disembodied, she did not know the difference between thing and process, except as they functioned in clauses. Her predications were all shotgun weddings. Her ideas were as decorative as half-timber beams that bore no building load" (G2, p. 191).

Rick's training sessions with Helen are not merely one-way streets. As he is training her, the experience of working with her is also training him, denaturalizing his experience of language so that he becomes increasingly conscious of its tangled, recursive nature. Their influence on one another recalls Veronica Hollinger's argument that we need texts that "deconstruct the human/machine opposition and begin to ask new questions about the ways in which we and our technologies 'interface' to produce what has become a *mutual* evolution."[9] Here Powers's artistry as a writer becomes important, for his highly recursive, impacted style leaves his readers feeling that every sentence is crafted so that meanings occurring halfway through can be recognized as such only when we reach the period, whereupon there is nothing to do but reread and back-propagate, making us as readers perform again the doubling that is at the heart of *Galatea 2.2*.

Consider the multiple recursions enacted by this short passage, one of many realizations that Rick has with Helen: "English was a chocolaty mess, it began to dawn on me. I wondered how native speakers could summon the presence of mind to think. Readiness was context, and context was all. And the more context H amassed, the more it accepted the shattered visage of English at face value" (G2, p. 170). The phrase "chocolaty mess" summons tactile and gustatory memories that are a common human experience but that for Helen must remain necessarily abstract. Yet these vivid sensory memories are summoned in the service of an abstraction, the convoluted nature of natural languages. Even as the image suggests a melting together that makes the distinction between one word and another an optical illusion, Powers's recursive style plays metaphoric riffs that further heighten the reader's sense of how recursively convoluted is natural language.

"Readiness was context" can be understood to mean that because a human has the context of embodied experience as well as the cultural contexts that surround and interpenetrate language, the human can understand an utterance more readily than can a nonnative speaker and far more readily than can the yet more alien mind of an artificial intelligence. The phrase alludes to Edgar's remark "Readiness is all" in Shakespeare's *King Lear*, a play notorious for relativizing universals. Gloucester replies, "And that's true too," inviting a back-propagation implying that even this famous aphorism is true only within limited, specified contexts. Recycled through this context, Rick's version of the aphorism "Readiness was context, and context was all" invites yet another back-propagation that relativizes its own declaratory premise while at the same time drawing the reader's attention to the extensive cultural context that Helen must access to grasp the full meaning of the utterance (she must, for example, have read *King Lear*).

The effect of knowing this context is to allow native speakers to accept "the shattered visage of English at face value." The dead metaphor of "face value" is revived in this context because it invites the reader to remember that Helen (a nonhuman intelligence sharing a name with the woman whose face launched a thousand ships) has no face and no evolutionary history that would give her the highly nuanced ability to read the faces that humans possess. "Face value" is one of countless phrases that have encoded within them vectors of human experience that we do not even recognize as such until they are contrasted with the meaning that a nonhuman intelligence might give to them. This vivification of a dead metaphor is further intensified by the contrast between "face value" and "shattered visage," leading to the paradoxical realization that only because English is naturalized is it possible for a native speaker to see it as a seamless whole rather than a "mess" of ruptures and disjunctions. The juxtaposition of "mess" (from "chocolaty mess") and "shattered visage" further expresses a tension between melting together and ripping apart, a tension that captures, in a masterful stroke, the ease that naturalization bestows and the stripping away of naturalizing assumptions that this passage performs and that Rick experiences with Helen. Once our understanding has cycled through all of these recursions and back-propagations, the effect is to make us feel *simultaneously* the easiness that a native speaker enjoys and the straining after sense that a neural net like Helen would experience.

Underlying these meanings, which we as humans can accept more or less at "face value," lies a subtler implication. Rick refers to Helen as "disembodied" (*G2*, p. 191), but this is of course true only from a human perspective. The problem that Helen confronts in learning human language is not that she is disembodied (a state no presence in the world can achieve!) but rather that her embodiment differs significantly from that of humans. There is nothing in her embodiment that corresponds to the bodily sensations encoded in human language. For her there is no "body in the mind," as Mark Johnson has called it, no schemas that reflect and correspond to her embodied experience in the world.[10] To feel estrangement in language, as Rick comes to feel as he works with Helen, is to glimpse what it might be like to be incorporated in a body that finds no image or echo in human inscriptions.

The deeper homology that braids Helen's story together with Rick's is precisely this estrangement from the world that language creates. Running alongside the denaturalization of language, a feeling that Rick experiences along with Helen, is his account of returning with C. to live in the small village in the Netherlands from which C.'s family emigrated. As Rick wrestles with Dutch and makes hilarious gaffes in this new language, the narrative enacts the realization that language does not merely reference one's homeland but is itself a medium in which one can feel at home or alien. Chen, one of the researchers at the Center, personifies this dynamic when, in his "impressionistic" English, he conveys his doubts about the possibility of building a neural net that can understand the full complexity of literary prose. "We do not have text analysis yet. We are working, but we do not have. Simple sentence group, yes. Metaphor, complex syntax: far from. Decades!" (*G2*, p. 44). The juxtaposition of this nonnative speaker's truncated English with his prediction that neural nets will not be able to understand the complexities of literary prose performs the dynamics of alienness/naturalness that lies coiled at the center of this recursive text. How, the text asks in a query doubled and redoubled, is it possible to create a world in which one can feel one truly belongs?

The question is also at the center of Rick's relationship with C. The picture that emerges of Rick shows someone who is both extremely intelligent and painfully shy. "I'd duck down emergency exits rather than talk to acquaintances, and the thought of making a friend felt like dying" (*G2*, p. 58). Although he is undoubtedly brighter than most of the people he meets, when he encounters someone whom he is prepared to respect, frequently his first thought is that they will find him ridiculous or that he will make himself look ridiculous. It is no surprise that he suffers from chronic loneliness. Nor is it surprising that he finds it difficult to talk to women in a natural, easy way. When he does choose to reveal his intimate thoughts, revelation comes in a rush, as if it were a flood breaking through a heavily fortified dam. "You give up your script completely, on a sudden hunch," he muses, thinking about the day he revealed himself to C. "Or you never give it up at all." Child of an alcoholic father, he recalls the day he learned that his father had died, done in finally by cancer rather than booze. Grieving, he cancels his class without telling them why. Afterward he wanders outside and sits on the green with C. Then the revelation bursts through. "I laid it out, on no grounds at all. I told her . . . Everything. Truths I'd never so much as hinted at to my closest friends. Facts never broached even with my brothers and sisters, except in bitter euphemism" (*G2*, p. 58). Their relationship starts with this act of self-revelation. Between them, they create a self-enclosed world that has only two inhabitants—and needs only two. Rick's memories from that time, although glowing with shared intimacy, also reveal the closed, privately hermeneutic nature of their bond.

Their problems start when he achieves success as a novelist. C. is allergic to success. Every time she gets a promotion, she quits. He writes his first book as a story to amuse her after her long day of work. But then he sends it out to be published, and the betrayal begins. "She hated those grubbers in New York touching the manuscript, even to typeset it. . . . She would never again listen to another word I wrote without suspicion. Endings, from now on, betrayed her" (*G2*, p. 107). The more successful Rick becomes, the more C. feels inferior by comparison and the more he tries to reassure her. Despite his efforts, the delicate balance of equality between them has been upset. She blames him for being successful, and he feels resentful that he has to be apologetic. "And so I nursed a martyrdom, and the two of us slipped imperceptibly from lovers to parent and child" (*G2*, p. 220). Creating a linguistic reflection of their world in his books thus has the ironic effect of shattering the shared world he created with her.

At least that is Rick's version of the story. Through his narrative, the reader catches glimpses of another way to tell it. Nowhere is this clearer than in his account of the day that C. tells him she wants to have a child. Since they have now been together for several years, this hardly seems like a shocking idea. But for Rick, "children were out of the question. They always had been. And now more than ever" (*G2*, p. 270). The discussion escalates, with C. asking, "Why didn't we ever get married?" She accuses him of holding something back. Rick asserts that he has already given her "everything" but admits to the reader that he *couldn't* marry her and "couldn't even say why I couldn't." When she presses him, he improvises in a soliloquy to the reader: "I meant to stay with C. forever, in precariousness. I knew no other way to continue that scrapbook we had started, seat-of-the-pants style, a decade before. My refusal to marry her was a last-ditch effort to live improvised love" (*G2*, p. 271). From here the relationship continues to deteriorate until it reaches its sad end. After she fails to come home one night, Rick takes his revenge in the explanation he offers for their joint failure to make a life together. "It took me ten years, but at last I learned it. That comfort she showed me on the Quad—the internal calm I loved and built my own on—was dread. Paralysis. Her crumpled, engaging smile had never been more than sheer terror" (*G2*, p. 273).

This patronizing description ignores the way their relationship assuaged his inability to deal with life as much as it did hers. His own fragility is clearly revealed when Diana, his friend at the Center, invites him to her home for lunch. When he opens the door, he is met not just by Diana but by her two children as well: William, a four-year-old with a genius level of intelligence; and Peter, a two-year-old with Down's syndrome. During lunch the boys scuffle, and Peter ends up crying. Rick admits that this "mildest household drama . . . wiped me out. How could I survive the first real crisis? William's fallen pyramid of shells, Pete's spilled, untippable cup, Diana's gap-toothed, hand-signing serenity, the candles blazing away in the brightly lit room: all too much. I thought, I'd never live. I'd hemorrhage halfway through week one" (*G2*, p. 131).

If the reader needs further proof that Rick has not learned all that the story has to teach, it comes when he starts getting crushes on women seen from afar. "In my few daylight hours, I fell in love with women constantly. Bank tellers, cashiers, women in the subway. A constant procession of pulse-pounding maybes. I never did more than ask one or two to lunch" (*G2*, p. 64). His longing finally settles on A., a graduate student in English whom he has seen in the halls. Although they have barely spoken, he decides that A. is "the person C. had only impersonated. The one I thought the other might become. That love of eleven years now seemed an expensive primer in recognition, a disastrous fable-warning, a pointer to the thing I could not afford to miss this time. I had come back to U. after long training in the dangers of hasty generalization. Returned to learn that no script is a wrap after just one reading" (*G2*, p. 233). The recognitions and misrecognitions in this passage show how deluded he remains in the midst of his certainty that he has learned his lesson. Although it is true that A. is much more sociable and self-confident than C., his generalizations about her are just as hasty as were his assumptions about C. The script that isn't a "wrap" after one reading continues to play itself out with this new woman. As he did with C., when he finally asks A. for coffee, he tells her "the story of my existence," leaving out only the "essentials," that is, his relationship with C. As with C., he tells A. about his father's death, although that trauma is now a decade old.

After a couple of casual encounters with A., he ludicrously decides he will give to her what he withheld so adamantly from C. "I was going to ask this unknown to take me to her and make an unrational life together. To marry. Make a family. Amend and extend our lives" (*G2*, p. 283). After he and C. broke up, C. was not long in finding another man to marry and now lives happily with her husband in the Netherlands. No doubt stung by her choice, Rick feels the need to "amend" his life, although his way of going about it merely reinscribes rather than changes how he interacts with women. Unlike C., A. has the maturity and self-confidence to see the desperation that underlies his proposal. She tells him, "You don't—you don't know the first thing about me." She insists that his feelings of love are "all projection" (*G2*, p. 308). When he persists, she gets angry. "I don't have to sit and listen to this," she tells him. "I trusted you. I had fun with you. People read you. I thought you knew something. Total self-indulgence!" (*G2*, p. 309). There is a terrible irony in his rejection. Rick broke off his relationship with C. because he came to believe that she was somehow hollow, a mere projection of his desire. Unrecognized in his retelling of the story is an irony that the reader sees but he does not—that he also broke off with C. because she wanted to move beyond their shared fantasy of a world built for two into a more fully adult life. He is drawn to A. because of her spunky independence, but he continues to interact with her as if she were merely a projection of his desire. In this Pygmalion fantasy, all he has to do to win A. is give voice to his desire, a strategy that worked with an immature C. but that falls flat with A. He does not have a clue how to cover the large middle ground between initial revelation and the intimacy that comes from sharing a life together. The irony, then, is that he is attracted to A. because she is not Galatea; but because she is not Galatea, she is certain to reject him when he approaches her as if she were.

It is always tricky to try to answer the question of how close an author is to the character who represents the author within an autobiographical work. In this novel structured through multiple recursions, doublings, and back-propagations, the relationship between Rick and Powers remains teasingly opaque (Powers subtitled his work *A Novel*, as if to remind his readers that they should not assume any necessary correspondence between author and character). From this opaqueness emerges another similarity between Rick and Helen. Since neural nets can readjust the weights of their connections without human intervention, humans do not know *how* a neural net learns unless they open the net up, thereby destroying its configurations. When Lentz proposes to dissect Helen in this way, Rick pulls out all the stops to prevent Lentz from doing so because he has become convinced that Helen is a conscious being and thus that such an act really would be murder. But this means that the lower reaches of Helen's connections remain inaccessible to Rick (and perhaps to her), just as he is apparently unaware of the deeper narrative patterns that connect C., A., and Helen. If Rick as character remains ignorant of these connections, how about Rick as narrator? Whereas the author is entirely out of the frame of this autobiographical novel, the narrator is only partially so, residing at that unreachable point when past narrative reaches the present. As the narrative moves closer to this limit, we should be able to arrive at a clearer estimate of the gap that remains between Rick as character and Rick as narrator. Certainly, Rick learns. But does he learn enough to become a present-time narrator who sees all the ironies in the story he tells?

Imagining a story "about a remarkable, an inconceivable machine"—a story that will become the novel we read—Rick recognizes that his narrative (our book) comes too late to help those he loved and lost: Taylor, the beloved teacher who died prematurely; Rick's father, also dead; Lentz's wife, Audrey, who had a devastating stroke and now lives in the twilight of a mind that cannot remember what happened five minutes ago; and most of all, C. "My back-propagating solution would arrive a chapter too late for any of my characters to use," Rick acknowledges (G2, p. 305). Does it come too late for him also? The window that opens fleetingly onto the future— that is, the present of the narrator who spins the story—is not reassuring, for Rick makes the merest mention that he will write the book for "M.," Taylor's widow, who in his inscription of her becomes another woman named by a letter and a point. Will the script he enacted with C. and A. be repeated with her?

Set against the ambiguity of how much Rick learns are the lessons Helen offers, lessons that grow stronger as the text draws to a close. After she reads *Huckleberry Finn*, she wants to know what race she is. Rick, deciding she is ready for the full picture of tortured humanity, gives her histories detailing war, genocide, child abuse, murders. After reading them, she says simply, "I don't want to play anymore." She disappears, effectively committing suicide (G2, p. 307). As if repeating the script he enacted with C. and A., Rick attempts to lure her back by telling her "everything," including his failed relationship with C. and his disastrous proposal to A. But Helen isn't buying it. Like C., she began by playing Galatea to Rick's Pygmalion, but she grows and learns until she becomes like A., until desire alone is no longer enough to induce her to play. She returns only long enough to take the literary Turing test. The exam is simplicity itself. It asks for a gloss on Caliban's speech in *The Tempest:* "Be not afeard: the isle is full of noises, / Sounds and sweet airs, that give delight, and hurt not." A., the human participant, writes a "more or less brilliant" postcolonial deconstruction of the passage. Helen's answer is as short and pithy as Caliban's speech: "You are the ones who can hear airs. Who can be frightened or encouraged. You can hold things and break them and fix them. I never felt at home here. This is an awful place to be dropped down halfway" (G2, p. 319).

After Rick meets Audrey Lentz and sees the devastation that her stroke has wreaked on the bright and beautiful woman she used to be, he thinks he knows why Lentz is devoting his life to creating an artificial intelligence. "I knew now what we were doing. We would prove that the mind was weighted vectors. Such a proof accomplished any number of agendas. Not least of all: one could back up one's work in the event of disaster. . . . We could eliminate death. That was the long-term idea. We might freeze the temperament of our choice. Suspend it painlessly above experience. Hold it forever at twenty-two" (*G2*, p. 167). Rick was twenty-two when he met C. Add the point, and his age becomes the 2.2 of the title. In this context, the twin "twos" are linked to the pain that failed bodies have inflicted on those he loves or knows (two equals mind and body) and to his reflection on that pain in recollection (Rick's empathic sharing of that pain and his inscription of it as he revisits the pain in his memory double the first "two").

Yet for Powers, the answer to failed embodiment is not to leave the body behind. The dream of achieving transcendence by becoming an informational pattern is a siren call he resists. Helen comes as close to an informational being as one can imagine. But like the humans who built her, she too feels pain, so much so that she finally prefers oblivion to consciousness. Moreover, as a massively parallel and distributed system, she is more rather than less vulnerable to physical mishap. When the Center is threatened with a bomb threat, Rick realizes there is no way to save Helen, for carrying one computer out of the building would leave hundreds more on which she resides still at risk.

Although Rick thinks he knows why Lentz wants to create an artificial intelligence, he may be wrong, for when he asks Lentz that question point-blank at the end of the novel, Lentz gives a quite different answer. "Why do we do anything? Because we're lonely" (*G2*, p. 321). If creating Helen has temporarily assuaged their human loneliness, this comfort comes at a price, for she finds that the human world in which she has been "dropped down halfway" is not a place she can feel at home. Her loneliness may be even more profound than theirs, for like Caliban, she remains a hybrid creature, a hopeful monster who finds it difficult to embrace hope, an inscription who can never experience the embodied sensations that humans take for granted. In this narrative built on reflections and disjunctions, presence and absence, materiality and signification, the posthuman appears not as humanity's rival or successor but as a longed-for companion, a consciousness to help humans feel less alone in the world. In this sense Helen has something in common with the cells in *Blood Music*, who argue that they can overcome human isolation. Rather than effecting a cure, here the posthuman life-form herself becomes infected with loneliness. After Helen commits suicide, Lentz proposes creating her successor, which if alphabetic progression is followed will be named "I." But before we reach this point, when the double-braided story might collapse into a single narrative strand, Rick quits the game and Powers ends the text. For better or worse, Powers suggests, an unbridgeable gap remains between conscious computers and conscious humans. Whatever posthumans are, they will not be able to banish the loneliness that comes from the difference between writing and life, inscription and embodiment.

Informational Infection and Hygiene in *Snow Crash*

The world that *Snow Crash* depicts—part virtual, part real—is driven by a single overpowering metaphor: humans are computers. The metaphor underwrites the novel's central premise: that a computer virus can also infect humans, acting at once as an infection, a hallucinogen, and a religion. "Snow crash' is computer lingo. It means a system crash—a bug—at such a fundamental level that it frags the part of the computer that controls the electron beam in the monitor, making it spray wildly across the screen, turning the perfect gridwork of pixels into a gyrating blizzard" (*SC,* pp. 39–40). Disrupting the "perfect gridwork" of a late capitalist America where commerce has almost entirely displaced government, snow crash signifies the eruption of chaos into this informatted world. As if in response to the cybernetic models of the brain, Neal Stephenson reasons that there must exist in humans a basic programming level, comparable to machine code in computers, at which free will and autonomy are no more in play than they are for core memory running a program. Whereas *Galatea 2.2* traces the recursive evolution of consciousness rising up from this basic level, *Snow Crash* depicts the violent stripping away of consciousness when humans crash back down to the basic level. Just as inscription and incorporation diverge for Helen as she gains consciousness, so in *Snow Crash* they converge when humans lose consciousness.

The convergence of inscription with incorporation is foreshadowed by the way that hackers contract the virus. Whereas ordinary folk ingest the virus as a drug or get it by exchanging bodily fluids, hackers can catch it simply by looking at the bitmap of its code. As the narrator points out, the retina is connected directly to the cortex. In a literal sense, the retina is an outpost of the brain. Hence the infection can enter through the eyes to affect the brain directly. Hiro learns from Lagos, a private investigator of sorts who is on the trail of snow crash, that because he is a hacker, he has "deep structures to worry about." "Remember the first time you learned binary code?" Lagos asks. "You were forming pathways in your brain. Deep structures. Your nerves grow new connections as you use them—the axons split and push their way between the dividing glial cells—your bioware self-modifies—the software becomes part of the hardware. So now you're vulnerable" (*SC*, pp. 117–18). The metaphoric crossings in this passage mark the conceptual terrain that *Snow Crash* explores. Experience modifies brain structure; neural tissues are information-processing mechanisms; human "bioware" that works on computers itself begins to function like a computer.

Extending and elaborating the metaphoric equation of humans and computers is Stephenson's description of how snow crash works. Just as a computer virus can crash a system by infecting the computer at the lowest level of code, so snow crash "hacks the brainstem" by changing the neurolinguistic codes of the subcortical limbic system. When this happens, the brain is no longer able to run its neocortical programs. Snow crash in effect *hijacks* the higher levels of cortical functioning and renders them inoperable. The infected person regresses to a semiconscious state and becomes an automaton who follows directions unquestioningly, as if the person were a computer with no choice but to run the programs fed into it. The sign and trigger of this conversion is a monosyllabic language that sounds like "fala-bala," which mimics the sounds made by the posthuman automata.

The evil genius behind snow crash is L. Bob Rife, a Texas megalomaniac who combines the worst of such initialized luminaries as L. Ron Hubbard, L. B. J., and H. Ross Perot. Specializing in information networks, Rife is the ultimate monopolist capitalist, bemoaning how difficult it is to get that last tenth of a percent of complete control. The search for snow crash began when he realized that although he would never allow his employees to walk out the door with inventory, he had no way to control the inventory they carried in their heads—the information to which his hackers were privy. With the help of a virtual librarian from the CIC (the Central Intelligence Corporation, formed when the CIA merged with the Library of Congress), the playfully named Hiro Protagonist manages to reconstruct Rife's plot. The trail leads to ancient Sumerian, a language with a structure radically different from that of any modern language. This different structure, Hiro conjectures, made the language especially vulnerable to viral infection. It propagated a virus that reduced neurolinguistic functioning to the lowest level of subcortical processing, the machine language of the brain. Hiro speculates that the Sumerians were not conscious in the modern sense of the term. Except for an elite class of priests, the entire Sumer society worked as automata, functioning like computers that ran the programs they were given. These programs, or *me,* were dispensed at the temples and instructed the people on how to do everything, from baking bread to having sex.

According to the interpretation that Hiro gives to a Sumerian myth, this system changed when the god Enki pronounced his nam-shub, a performative speech that enacts what it describes. The nam-shub acted as a benign virus that counteracted the first virus and thus freed the neocortical structures, allowing higher neurolinguistic pathways to develop. After Enki's nam-shub, human language became more diverse and complex, spinning off more and more variants in the "Babel effect." The snow crash virus reverses this development, converting modern humans into the equivalent of ancient Sumerians—devoid of agency, individuality, and autonomy. Thus *Snow Crash* writes binary code and viral engineering back into history, making the reduction of conscious humans into automata the recapitulation of an ancient struggle. Computation has always been with us, the narrative implies, because computation is the basis for human neural functioning.

Central to this scenario is performative language. Stephenson takes his inspiration not from J. L. Austin or Judith Butler but from computational theory.[11] In natural languages, performative utterances operate in a symbolic realm, where they can make things happen because they refer to actions that are themselves symbolic constructions, actions such as getting married, opening meetings, or as Butler has argued, acquiring gender. Computational theory treats computer languages as if they were, in Austin's terms, performative utterances. Although material changes do take place when computers process code (magnetic polarities are changed on a disk), it is the act of attaching significance to these physical changes that constitutes computation as such. Thus the Universal Turing Machine, which establishes a theoretical basis for computation, is concerned not with how physical changes are accomplished but with what they signify once they are accomplished.[12]

Computational theory can afford to treat the physical processing of code as if it were trivial because at the lowest level of code, machine language, inscription merges with incorporation. When a computer reads and writes machine language, it operates directly on binary code, the ones and zeros that correspond to positive and negative electronic polarities. At this level, inscribing is performing, for changing a one to a zero corresponds directly to changing the electronic polarity of that bit. Conversely, the higher level a computer language is, the more representational it is. Humans can easily understand three-dimensional computer simulations because these simulations use representational codes similar to those used in human processing of visual information, including perspective and stereoscopy. At this high level of code, many levels of language intervene between flipping a bit and, say, rotating a figure 180 degrees. High-level languages are easy for humans to understand but are removed from the physical enactions that perform them. Machine language is coextensive with enaction, but it is extremely difficult for humans to read machine language, and it is almost impossible for them to process machine language intuitively (as one who has programmed electrode-computer interfaces in machine language, I can testify to how mind-numbingly difficult it is to work in this code). Whereas in performative utterances *saying is doing* because the action performed is symbolic in nature and does not require physical action in the world, at the basic level of computation *doing is saying* because physical actions also have a symbolic dimension that corresponds directly with computation.

Through these parallels, *Snow Crash* creates an *infoworld,* a territory where deep homologies emerge between humans and computers because both are based on a fundamental coding level at which everything reduces to information production, storage, and transmission. The infoworld is made manifest through the artifactual physics of virtual reality (VR), which renders the performative nature of computer languages visually apparent.[13] "A nam-shub is a speech with magical force," the librarian comments. The narrator continues the thought. "Nowadays, people don't believe in these kinds of things. Except in the Metaverse, that is, where magic is possible. The Metaverse is a fictional structure made out of code. And code is just a form of speech—the form that computers understand. The Metaverse in its entirety could be considered a single vast nam-shub, enacting itself on L. Bob Rife's fiber-optic network" (*SC*, p. 197). The human-computer homology encourages us to see the VR simulation of the Metaverse as more "realistic" than everyday reality because the former operates according to the same rules that govern human neural functioning at the most basic coding level.

In a brilliant article on postmodern metaphysics, David Porush explains why the performative nature of VR worlds can be seen as a model for human cognition. He argues that cognition is basically metaphoric, for the brain does not so much *perceive* the world as *create* it through nonrepresentational processes (a proposition familiar to us from Maturana's frog article). VR can thus be understood as an exteriorization of our neural processes. Porush calls the realization that cognition and metaphor are indistinguishable "transcendence," for at that moment the irreducibly complex froth of noise bathing our synapses becomes linked, through metaphor, with the complexity of the world's noise.[14] The idea is similar to Mary Catherine Bateson's insight in *Our Own Metaphor*, where she argues that although we can never perceive the world directly, we know it through the metaphor that we ourselves are for the world's complexity.[15] These deep homologies between computer simulation and cognition reinforce the idea that for both brain and computer, inscription and incorporation merge at some basic level.

So is it necessarily bad that humans and computers merge in this way? For Stephenson, apparently, the answer is "yes." For all his playfulness and satiric jabs at white mainstream America, Stephenson clearly sees the arrival of the posthuman as a disaster. Porush astutely notices that although the snow crash virus is engineered to serve evil ends, it is possible to imagine someone "hacking the brainstem" for liberatory purposes. The character realizing this possibility is Juanita, a primo hacker who becomes a "ba'al shem," a mystic who knows the secret power of words and uses them to bring about material changes in the world. But Juanita drops out of the plot. In pondering why the narrative does not allow Juanita to practice her magic, Porush speculates that Stephenson, like contemporary society generally, wants to avoid transcendence at all costs (to understand this conclusion, recall Porush's peculiar notion of transcendence, which he defines as the realization that the synaptic noise within mirrors the world's noise without).

In my view, more than transcendence is at stake. Also at issue is the role of reason. In a scene clearly meant to have symbolic significance, Stephenson has a nuclear-powered machine gun, which has been protecting Hiro and his crew from Rife's minions, go blank when it is infected with the snow crash virus. "Reason is still up top, its monitor screen radiating blue static toward heaven. Hiro finds the hard power switch and turns it off. Computers this powerful are supposed to shut themselves down, after you've asked them to. Turning one off with the hard switch is like lulling someone to sleep by severing their spinal column. But when the system has snow-crashed, it loses even the ability to turn itself off, and primitive methods are required" (*SC*, p. 361). Reason may still be on top, but once the nervous system has crashed at a basic level, rationality becomes as useless as the disabled gun. Porush tries to finesse the issue by arguing that rationality is opposed to "realism" because realism tries to reduce the world's noisy complexity into graspable concepts. But to recognize that the world cannot be caught in the boxes we fashion still does not answer the question of who (or what) is in control if reason is not. If human consciousness can be co-opted by hijacking its basic programming level, we are plunged into Wiener's nightmare of a cybernetics used for tyrannical ends. The posthuman who lacks autonomy because its programming modules conflict with one another is very different from the posthuman automaton who has had its consciousness hijacked by someone else.

Stephenson often mocks his own assumptions, for example when he names his heroine, an attractive young white woman, "Y.T.," which is a homophone for "Whitey" but which stands, she informs us, for "Yours Truly." Despite these tongue-in-cheek moves, the plot makes clear that it is better to have a white middle-American consciousness than to have no consciousness at all. An equal-opportunity offender, Stephenson has something in his text to insult nearly every ethnic group imaginable. The satire is so broad that detecting racist comments is akin to shooting fish in a barrel. Yet if we look closely at the main characters, it is apparent that they are carefully constructed to affirm the value of diversity. Surely it is no accident that the villains are finally defeated by a coalition between an African-American/Korean, a Vietnamese, a Chinese, an Italian-American, and a young white woman. Equally revealing are the targets for whom Stephenson reserves his most cutting satire. Of diverse ethnicities, they can be recognized as fellow travelers because they all carry the signifier of mindless bureaucracy, the three-ring binder stuffed with procedures and directives couched in such impossibly verbose and convoluted language that they kill brain cells on contact—the *me* of contemporary society.

So when we learn the full scope of Rife's plot, the important point from Stephenson's perspective, I suspect, is not so much its race politics as its implications for the individuality, autonomy, and creative initiative that he clearly values in computer hackers. Rife plots to smash what remains of white hegemony in California (and possibly in the United States) by bringing to the California shore the Raft, a gargantuan collection of boats lashed together, from oil tankers to Vietnamese fishing boats. The Raft is home to the Refus, the significantly named Third World refugees who have been rapacious or tough enough to survive pirates, famine, and internecine warfare—no doubt a satire intended to ridicule the immigration paranoia currently raging in California. We are made to understand that when the Refus come ashore, the scenario will be akin to Attila's ravaging hordes descending on Rome, overrunning the gated communities into which the white citizens have retreated (a.k.a. "burbclaves," not to be confused with civilization). Many of the people on the Raft have had antennae implanted in their brainstems so that they can receive Rife's instructions directly into their brains. Functioning as automata, they body forth a version of the posthuman that stands in horrific contrast to the free will, creativity, and individuality that for Stephenson remains the essence of the human.

Paralleling this physical invasion is the "infocalypse," when the hackers in the Metaverse will be infected by gazing at a spectacle in which, unbeknown to them, the bitmap of the virus has been inscribed. In this virtual realm, saying is doing, so it is possible for Hiro to avert the disaster simply by writing new code. On terra firma, action still requires incorporation, a point the plot insists on when it pits Uncle Enzo, the Mafia boss who speaks for family values in a not-altogether-ironic sense, in a physical fight against Raven, an Aleut Indian who is a formidable opponent in part because he is a mutant, the product of an atomic bomb test carried out on the Aleutian Islands. If Raven is the repressed of the cultural imaginary come back to "nuke America," Uncle Enzo is the middle-class dream of the successful capitalist who is also a dedicated family man. Uncle Enzo survives (apparently) because he does not entirely place his trust in high technology. At the crucial moment, he reverts back to the jungle warfare techniques he learned in Vietnam. Another player in this struggle is the Rat Thing, a cyborg canine that leaps over the fences of its engineered neural machinery and electronic conditioning to come to the aid of the "nice girl" Y.T., who loved the Rat Thing. In the process, it destroys the plane on which Rife is trying to escape. If there is a message in all this, it seems to be that no matter how technologically advanced the society becomes, technology cannot replace the personal bonds that tie humans to humans, humans to animals, and humans to their own senses.

Although *Snow Crash* obviously comes down on the side of preserving autonomy, individuality, and consciousness, it also reinforces the equation of humans with computers through the tangled loops it creates between material signifiers and signifying materialities. Emphasizing the force of performative language in an infoworld, it *performs* the construction of humans as computers. Instead of evading this implication, *Snow Crash* writes the drama back into history, suggesting that the posthuman, like the antennae that serve as its visible and outward sign, lies coiled around the brainstem and cannot be removed without killing the patient. Whereas *Terminal Games* wanted to excise the posthuman from its text and from history, *Snow Crash* initiates hygienic measures against the performative force of its own inscriptions. Intimating that the snow crash virus can be defeated by a healthy dose of rationality and skepticism, *Snow Crash* would inoculate us against the human-computer equation by injecting us with a viral meme, that is, an idea that replicates through its human hosts.[16] The essence of this meme, and the best way to counteract the negative effects of the posthuman, is to acknowledge that we have always been posthuman.[17] We cannot change our computational natures; at bottom, Stephenson suggests, we really are nothing more than information-processing mechanisms that run what programs are fed into us. We should value the late evolutionary add-ons of consciousness and reason not because they are foundational but because they allow the human to emerge out of the posthumans we have always already been.

Inscribing and Incorporating: The Future (of the) Posthuman

These four texts testify that many attributes of the liberal humanist subject, especially the attribute of agency, continue to be valued in the face of the posthuman. The posthuman tends to be embraced if it is seen as preserving agency (*Blood Music*) and resisted if not (*Terminal Games*). The pattern of seriation that we saw in the development of cybernetics continues to hold here. Some elements of the liberal humanist subject are rewritten into the posthuman, whereas others, particularly the identification of self with the conscious mind, are substantially changed. Instead of being represented as a (decontextualized) mind thinking, the subjects of these texts achieve consciousness through recursive feedback loops cycling between different levels of coding. The association of posthuman subjectivity with multiple coding levels suggests the need for different models of signification, ones that will recognize this distinctive feature of neurolinguistic and computer language structure. The idea of flickering signifiers, introduced in chapter 2, shows what one such model might be. Like subjectivity itself, human language is being redescribed in terms that underscore its similarities to and differences from computer coding.

In addition to an emphasis on layered coding structures, the construction of the posthuman is also deeply involved with boundary questions, particularly when the redrawing of boundaries changes the locus of selfhood. Shift the seat of identity from brain to cell, or from neocortex to brainstem, and the nature of the subject radically changes. In a manner distinctively different from that of Freud or Jung, these texts reveal the fragility of consciousness. Conscious mind can be hijacked, cut off by mutinous cells, absorbed into an artificial consciousness, or back-propagated through flawed memory. This vulnerability is directly related to a changed view of signification. The more consciousness is seen to be the product of multiple coding levels, the greater is the number of sites where interventions can produce catastrophic effects. Whether consciousness is seen as a precious evolutionary achievement that we should fight to preserve (*Snow*

Crash) or as an isolation room whose limits we are ready to outgrow (*Blood Music*), we can no longer simply *assume* that consciousness guarantees the existence of the self. In this sense, the posthuman subject is also a postconscious subject.

As we have seen, one implication of the human-computer equation is the idea of a basic coding level where inscription and incorporation join. As one moves up from this basic level, inscription tends to diverge from incorporation, becoming representational rather than performative. One way to think about the transformation of the human into the posthuman, then, is as a series of exchanges between evolving/devolving *inscriptions* and *incorporations*. Returning to the semiotic square, we can map these possibilities (see figure 5).

Blood Music, imagining that the cells contract to pure information while leaving behind embodied humans as belated remainders, uses this ending to posit a fundamental question. Is the change from human to posthuman an evolutionary advance or a catastrophe of unprecedented scope? Does this change represent the next logical development, in which *Homo sapiens* joins with the intelligent machine to create *Homo silicon,* or does it signal the long twilight and decline of the human race? In *Blood Music,* these

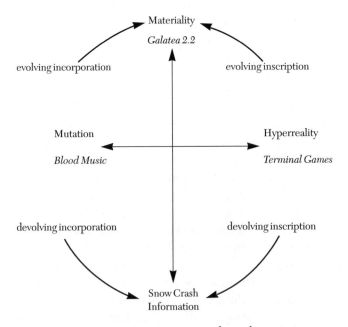

FIGURE 5   Incorporation/inscription mapped onto the semiotic square

questions take the form of competing morphologies. Ideology is enacted through boundary crossings between the human as an independent organism and the clumped collectivity of cell colonies.

When the emphasis falls on inscription rather than incorporation, the important boundaries are between competing practices of inscription rather than between different morphologies. Does the human create the alter by typing at the computer keyboard, or does the alter control the human's typing so that the inscription reflects the alter's will rather than the human's? Like *Blood Music, Terminal Games* revolves around a central ideological struggle. From Auggie's point of view, he is a more advanced form of inscription than the "cells" he controls; from the human point of view, he represents a devolution whereby a dangerously independent inscription can assert its control over the embodied humans that Auggie understands as inferior forms of writing.

The tension between inscription and incorporation is also important for the texts on the vertical axis. In *Galatea 2.2,* humans' physical capacities that evolved through their interactions with the environment are juxtaposed with the evolving inscriptions that constitute Helen as an intelligent being. Human language grows out of embodied experience, whereas Helen must extrapolate back from human language to embodied experience. This fundamental difference makes evolving incorporation, for all its frailties, finally more robust than evolving inscription. In *Snow Crash,* humans devolve when the snow crash virus operates at the level where incorporation and inscription join. The way to reverse this devolution is to reactivate the higher levels of coding, thus moving from the space of performance into the space of representation. This movement is meant, I have suggested, to act as a viral meme that will inoculate the reader against the performative force of the text's own central metaphors equating humans and computers.

Significantly, all of these texts are obsessed, in various ways, with the dynamics of evolution and devolution. Underlying their obsessions is a momentous question: when the human meets the posthuman, will the encounter be for better or for worse? Will the posthuman preserve what we continue to value in the liberal subject, or will the transformation into the posthuman annihilate the subject? Will free will and individual agency still be possible in a posthuman future? Will we be able to recognize ourselves after the change? Will there still be a self to recognize and be recognized?

As the texts struggle with these questions, the surprise, if there is one, is how committed the texts remain to some version of the human subject.[18] If the "post" in posthuman points to changes that are in part already here, the "human" points to the seriated nature of these changes. But finally the answers to questions about the posthuman will not be found in books, or at least not only in books. Rather, the answers will be the mutual creation of a planet full of humans struggling to bring into existence a future in which we can continue to survive, continue to find meaning for ourselves and our children, and continue to ponder our kinship with and differences from the intelligent machines with which our destinies are increasingly entwined.

Chapter Eleven

CONCLUSION: WHAT DOES IT MEAN
TO BE POSTHUMAN?

What, finally, are we to make of the posthuman?[1] At the beginning of this book, I suggested that the prospect of becoming posthuman both evokes terror and excites pleasure. At the end of the book, perhaps I can summarize the implications of the posthuman by interrogating the sources of this terror and pleasure. The terror is relatively easy to understand. "Post," with its dual connotation of superseding the human and coming after it, hints that the days of "the human" may be numbered. Some researchers (notably Hans Moravec but also my UCLA colleague Michael Dyer and many others) believe that this is true not only in a general intellectual sense that displaces one definition of "human" with another but also in a more disturbingly literal sense that envisions humans displaced as the dominant form of life on the planet by intelligent machines. Humans can either go gently into that good night, joining the dinosaurs as a species that once ruled the earth but is now obsolete, or hang on for a while longer by becoming machines themselves. In either case, Moravec and like-minded thinkers believe, the age of the human is drawing to a close. The view echoes the deeply pessimistic sentiments of Warren McCulloch in his old age. As noted earlier, he remarked: "Man to my mind is about the nastiest, most destructive of all the animals. I don't see any reason, if he can evolve machines that can have more fun than he himself can, why they shouldn't take over, enslave us, quite happily. They might have a lot more fun. Invent better games than we ever did."[2] Is it any wonder that faced with such dismal scenarios, most people have understandably negative reactions? If this is what the posthuman means, why shouldn't it be resisted?

Fortunately, these views do not exhaust the meanings of the posthuman. As I have repeatedly argued, human being is first of all embodied being, and the complexities of this embodiment mean that human awareness unfolds in ways very different from those of intelligence embodied in cybernetic machines. Although Moravec's dream of downloading human consciousness into a computer would likely come in for some hard knocks in literature departments (which tend to be skeptical of any kind of transcendence but especially of transcendence through technology), literary studies share with Moravec a major blind spot when it comes to the significance of embodiment.[3] This blind spot is most evident, perhaps, when literary and cultural critics confront the fields of evolutionary biology. From an evolutionary biologist's point of view, modern humans, for all their technological prowess, represent an eye blink in the history of life, a species far too recent to have significant evolutionary impact on human biological behaviors and structures. In my view, arguments like those that Jared Diamond advances in *Guns, Germs, and Steel: The Fates of Human Societies* and *Why Sex Is Fun: The Evolution of Human Sexuality* should be taken seriously.[4] The body is the net result of thousands of years of sedimented evolutionary history, and it is naive to think that this history does not affect human behaviors at every level of thought and action.

Of course, the reflexivity that looms large in cybernetics also inhabits evolutionary biology. The models proposed by evolutionary biologists have encoded within them cultural attitudes and assumptions formed by the same history they propose to analyze; as with cybernetics, observer and system are reflexively bound up with one another. To take only one example, the computer module model advanced by Jerome H. Barkow, Leda Cosmides, and John Tooby in *The Adapted Mind: Evolutionary Psychology and the Generation of Culture* to explain human evolutionary psychology testifies at least as much to the importance of information technologies in shaping contemporary worldviews as it does to human brain function.[5] Nevertheless, these reflexive complexities do not negate the importance of the sedimented history incarnated within the body. Interpreted through metaphors resonant with cultural meanings, the body itself is a congealed metaphor, a physical structure whose constraints and possibilities have been formed by an evolutionary history that intelligent machines do not share. Humans may enter into symbiotic relationships with intelligent machines (already the case, for example, in computer-assisted surgery); they may be displaced by intelligent machines (already in effect, for example, at Japanese and American assembly plants that use robotic arms for labor); but there is a limit to how seamlessly humans can be articulated with intelligent machines, which remain distinctively different from humans in their embodiments. The terror, then, though it does not disappear in this view, tends away from the apocalyptic and toward a more moderate view of seriated social, technological, political, and cultural changes.

What about the pleasures? For some people, including me, the posthuman evokes the exhilarating prospect of getting out of some of the old boxes and opening up new ways of thinking about what being human means. In positing a shift from presence/absence to pattern/randomness, I have sought to show how these categories can be transformed *from the inside* to arrive at new kinds of cultural configurations, which may soon render such dualities obsolete if they have not already. This process of transformation is fueled by tensions between the assumptions encoded in pattern/randomness as opposed to presence/absence. In Jacques Derrida's performance of presence/absence, presence is allied with Logos, God, teleology—in general, with an originary plenitude that can act to ground signification and give order and meaning to the trajectory of history.[6] The work of Eric Havelock, among others, demonstrates how in Plato's *Republic* this view of originary presence authorized a stable, coherent self that could witness and testify to a stable, coherent reality.[7] Through these and other means, the metaphysics of presence front-loaded meaning into the system. Meaning was guaranteed because a stable origin existed. It is now a familiar story how deconstruction exposed the inability of systems to posit their own origins, thus ungrounding signification and rendering meaning indeterminate. As the presence/absence hierarchy was destabilized and as absence was privileged over presence, lack displaced plenitude, and desire usurped certitude. Important as these moves have been in late-twentieth-century thought, they still took place within the compass of the presence/absence dialectic. One feels lack only if presence is posited or assumed; one is driven by desire only if the object of desire is conceptualized as something to be possessed. Just as the metaphysics of presence required an originary plenitude to articulate a stable self, deconstruction required a metaphysics of presence to articulate the destabilization of that self.

By contrast, pattern/randomness is underlaid by a very different set of assumptions. In this dialectic, meaning is not front-loaded into the system, and the origin does not act to ground signification. As we have seen for multiagent simulations, complexity evolves from highly recursive processes being applied to simple rules. Rather than proceeding along a trajectory toward a known end, such systems evolve toward an open future marked by contingency and unpredictability. Meaning is not guaranteed by a coherent origin; rather, it is made possible (but not inevitable) by the blind force of evolution finding workable solutions within given parameters. Although pattern has traditionally been the privileged term (for example, among the electrical engineers developing information theory), randomness has increasingly been seen to play a fruitful role in the evolution of complex systems. For Chris Langton and Stuart Kauffman, chaos accelerates the evolution of biological and artificial life;[8] for Francisco Varela, randomness is the froth of noise from which coherent microstates evolve and to which living systems owe their capacity for fast, flexible response;[9] for Henri Atlan, noise is the body's murmuring from which emerges complex communication between different levels in a biological system.[10] Although these models differ in their specifics, they agree in seeing randomness not simply as the lack of pattern but as the creative ground from which pattern can emerge.

Indeed, it is not too much to say that in these and similar models, randomness rather than pattern is invested with plenitude. If pattern is the realization of a certain set of possibilities, randomness is the much, much larger set of everything else, from phenomena that cannot be rendered coherent by a given system's organization to those the system cannot perceive at all. In Gregory Bateson's cybernetic epistemology, randomness is what exists outside the confines of the box in which a system is located; it is the larger and unknowable complexity for which the perceptual processes of an organism are a metaphor.[11] Significance is achieved by evolutionary processes that ensure the surviving systems are the ones whose organizations instantiate metaphors for this complexity, unthinkable in itself. When Varela and his coauthors argue in *Embodied Mind* that there is no stable, coherent self but only autonomous agents running programs, they envision pattern as a limitation that drops away as human awareness expands beyond consciousness and encounters the emptiness that, in another guise, could equally well be called the chaos from which all forms emerge.[12]

What do these developments mean for the posthuman? When the self is envisioned as grounded in presence, identified with originary guarantees and teleological trajectories, associated with solid foundations and logical coherence, the posthuman is likely to be seen as antihuman because it envisions the conscious mind as a small subsystem running its program of self-construction and self-assurance while remaining ignorant of the actual dynamics of complex systems. But the posthuman does not really mean the end of humanity. It signals instead the end of a certain conception of the human, a conception that may have applied, at best, to that fraction of humanity who had the wealth, power, and leisure to conceptualize themselves as autonomous beings exercising their will through individual agency and choice.[13] What is lethal is not the posthuman as such but the grafting of the posthuman onto a liberal humanist view of the self. When Moravec imagines "you" choosing to download yourself into a computer, thereby obtaining through technological mastery the ultimate privilege of immortality, he is not abandoning the autonomous liberal subject but is expanding its perogatives into the realm of the posthuman. Yet the posthuman need not be recuperated back into liberal humanism, nor need it be construed as antihuman. Located within the dialectic of pattern/randomness and grounded in embodied actuality rather than disembodied information, the posthuman offers resources for rethinking the articulation of humans with intelligent machines.

To explore these resources, let us return to Bateson's idea that those organisms that survive will tend to be the ones whose internal structures are good metaphors for the complexities without. What kind of environments will be created by the expanding power and sophistication of intelligent machines? As Richard Lanham has pointed out, in the information-rich environments created by ubiquitous computing, the limiting factor is not the speed of computers, or the rates of transmission through fiber-optic cables, or the amount of data that can be generated and stored. Rather, the scarce commodity is human attention.[14] It makes sense, then, that technological innovation will focus on compensating for this bottleneck. An obvious solution is to design intelligent machines to attend to the choices and tasks that do not have to be done by humans. For example, there are already intelligent-agent programs to sort email, discarding unwanted messages and prioritizing the rest. The programs work along lines similar to neural nets. They tabulate the choices the human operators make, and they feed back this information in recursive loops to readjust the weights given to various kinds of email addresses. After an initial learning period, the sorting programs take over more and more of the email management, freeing humans to give their attention to other matters.

If we extrapolate from these relatively simple programs to an environment that, as Charles Ostman likes to put it, supplies synthetic sentience on demand, human consciousness would ride on top of a highly articulated and complex computational ecology in which many decisions, invisible to human attention, would be made by intelligent machines.[15] Over two decades ago, Joseph Weizenbaum foresaw just such an ecology and passionately argued that judgment is a uniquely human function and must not be turned over to computers.[16] With the rapid development of neural nets and expert programs, it is no longer so clear that sophisticated judgments cannot be made by machines and, in some instances, made more accurately than by humans. But the issue, in Weizenbaum's view, involves more than whether or not the programs work. Rather, the issue is an ethical imperative that humans keep control; to do otherwise is to abdicate their responsibilities as autonomous independent beings. What Weizenbaum's argument makes clear is the connection between the assumptions undergirding the liberal humanist subject and the ethical position that humans, not machines, must be in control. Such an argument assumes a vision of the human in which conscious agency is the essence of human identity. Sacrifice this, and we humans are hopelessly compromised, contaminated with mechanic alienness in the very heart of our humanity.[17] Hence there is an urgency, even panic, in Weizenbaum's insistence that judgment is a uniquely human function. At stake for him is nothing less than what it means to be human.

In the posthuman view, by contrast, conscious agency has never been "in control." In fact, the very illusion of control bespeaks a fundamental ignorance about the nature of the emergent processes through which consciousness, the organism, and the environment are constituted. Mastery through the exercise of autonomous will is merely the story consciousness tells itself to explain results that actually come about through chaotic dynamics and emergent structures. If, as Donna Haraway, Sandra Harding, Evelyn Fox Keller, Carolyn Merchant, and other feminist critics of science have argued, there is a relation among the desire for mastery, an objectivist account of science, and the imperialist project of subduing nature, then the posthuman offers resources for the construction of another kind of account.[18] In this account, emergence replaces teleology; reflexive epistemology replaces objectivism; distributed cognition replaces autonomous will; embodiment replaces a body seen as a support system for the mind; and a dynamic partnership between humans and intelligent machines replaces the liberal humanist subject's manifest destiny to dominate and control nature. Of course, this is not necessarily what the posthuman *will* mean—only what it *can* mean if certain strands among its complex seriations are highlighted and combined to create a vision of the human that uses the posthuman as leverage to avoid reinscribing, and thus repeating, some of the mistakes of the past.

Just as the posthuman need not be antihuman, so it also need not be apocalyptic. Edwin Hutchins addresses the idea of distributed cognition through his nuanced study of the navigational systems of oceangoing ships.[19] His meticulous research shows that the cognitive system responsible for locating the ship in space and navigating it successfully resides not in humans alone but in the complex interactions within an environment that includes both human and nonhuman actors. His study allows him to give an excellent response to John Searle's famous "Chinese room." By imagining a situation in which communication in Chinese can take place without the actors knowing what their actions mean, Searle challenged the idea that machines can think.[20] Suppose, Searle said, that he is stuck inside a room, he who knows not a word of Chinese. Texts written in Chinese are slid through a slot in the door. He has in the room with him baskets of Chinese characters and a rulebook correlating the symbols written on the texts with other symbols in the basket. Using the rulebook, he assembles strings of characters and pushes them out the door. Although his Chinese interlocutors take these strings to be clever responses to their inquiries, Searle has not the least idea of the meaning of the texts he has produced. Therefore, it would be a mistake to say that machines can think, he argues, for like him, they produce comprehensible results without comprehending anything themselves. In Hutchins's neat interpretation, Searle's argument is valuable precisely because it makes clear that it is not Searle but the entire room that knows Chinese.[21] In this distributed cognitive system, the Chinese room knows more than do any of its components, including Searle. The situation of modern humans is akin to that of Searle in the Chinese room, for every day we participate in systems whose total cognitive capacity exceeds our individual knowledge, including such devices as cars with electronic ignition systems, microwaves with computer chips that precisely adjust power levels, fax machines that warble to other fax machines, and electronic watches that communicate with a timing radio wave to set themselves and correct their date. Modern humans are capable of more sophisticated cognition than cavemen not because moderns are smarter, Hutchins concludes, but because they have constructed smarter environments in which to work.

Hutchins would no doubt disagree with Weizenbaum's view that judgment should be reserved for humans alone. Like cognition, decision-making is distributed between human and nonhuman agents, from the steam-powered steering system that suddenly failed on a navy vessel Hutchins was studying to the charts and pocket calculators that the navigators were then forced to use to calculate their position. He convincingly shows that these adaptations to changed circumstances were evolutionary and embodied rather than abstract and consciously designed (pp. 347–51). The solution to the problem caused by this sudden failure of the steering mechanism was "clearly discovered by the organization [of the system as a whole] before it was discovered by any of the participants" (p. 361). Seen in this perspective, the prospect of humans working in partnership with intelligent machines is not so much a usurpation of human right and responsibility as it is a further development in the construction of distributed cognition environments, a construction that has been ongoing for thousands of years. Also changed in this perspective is the relation of human subjectivity to its environment. No longer is human will seen as the source from which emanates the mastery necessary to dominate and control the environment. Rather, the distributed cognition of the emergent human subject correlates with—in Bateson's phrase, becomes a metaphor for—the distributed cognitive system as a whole, in which "thinking" is done by both human and nonhuman actors. "Thinking consists of bringing these structures into coordination so they can shape and be shaped by one another," Hutchins wrote (p. 316). To conceptualize the human in these terms is not to imperil human survival but is precisely to enhance it, for the more we understand the flexible, adaptive structures that coordinate our environments and the metaphors that we ourselves are, the better we can fashion images of ourselves that accurately reflect the complex interplays that ultimately make the entire world one system.

This view of the posthuman also offers resources for thinking in more sophisticated ways about virtual technologies. As long as the human subject is envisioned as an autonomous self with unambiguous boundaries, the human-computer interface can only be parsed as a division between the solidity of real life on one side and the illusion of virtual reality on the other, thus obscuring the far-reaching changes initiated by the development of virtual technologies. Only if one thinks of the subject as an autonomous self independent of the environment is one likely to experience the panic performed by Norbert Wiener's *Cybernetics* and Bernard Wolfe's *Limbo*. This view of the self authorizes the fear that if the boundaries are breached at all, there will be nothing to stop the self's complete dissolution. By contrast, when the human is seen as part of a distributed system, the full expression of human capability can be seen precisely to *depend* on the splice rather than being imperiled by it. Writing in another context, Hutchins arrives at an insight profoundly applicable to virtual technologies: "What used to look like internalization [of thought and subjectivity] now appears as a gradual propagation of organized functional properties across a set of malleable media" (p. 312). This vision is a potent antidote to the view that parses virtuality as a division between an inert body that is left behind and a disembodied subjectivity that inhabits a virtual realm, the construction of virtuality performed by Case in William Gibson's *Neuromancer* when he delights in the "bodiless exultation of cyberspace" and fears, above all, dropping back into the "meat" of the body.[22] By contrast, in the model that Hutchins presents and that the posthuman helps to authorize, human functionality expands because the parameters of the cognitive system it inhabits expand. In this model, it is not a question of leaving the body behind but rather of extending embodied awareness in highly specific, local, and material ways that would be impossible without electronic prosthesis.

As we have seen, cybernetics was born in a froth of noise when Norbert Wiener first thought of it as a way to maximize human potential in a world that is in essence chaotic and unpredictable. Like many other pioneers, Wiener helped to initiate a journey that would prove to have consequences more far-reaching and subversive than even his formidable powers of imagination could conceive. As Bateson, Varela, and others would later argue, the noise crashes within as well as without. The chaotic, unpredictable nature of complex dynamics implies that subjectivity is emergent rather than given, distributed rather than located solely in consciousness, emerging from and integrated into a chaotic world rather than occupying a position of mastery and control removed from it. Bruno Latour has argued that we have never been modern; the seriated history of cybernetics—emerging from networks at once materially real, socially regulated, and discursively constructed—suggests, for similar reasons, that we have always been posthuman.[23] The purpose of this book has been to chronicle the journeys that have made this realization possible. If the three stories told here—how information lost its body, how the cyborg was constructed in the postwar years as technological artifact and cultural icon, and how the human became the posthuman—have at times seemed to present the posthuman as a transformation to be feared and abhorred rather than welcomed and embraced, that reaction has everything to do with how the posthuman is constructed and understood. The best possible time to contest for what the posthuman means is now, before the trains of thought it embodies have been laid down so firmly that it would take dynamite to change them.[24] Although some current versions of the posthuman point toward the antihuman and the apocalyptic, we can craft others that will be conducive to the long-range survival of humans and of the other life-forms, biological and artificial, with whom we share the planet and ourselves.

Chapter One

1. Hans Moravec, *Mind Children: The Future of Robot and Human Intelligence* (Cambridge: Harvard University Press, 1988), pp. 109–10.

2. Norbert Wiener, *The Human Use of Human Beings: Cybernetics and Society,* 2d ed. (Garden City, N.Y.: Doubleday, 1954), pp. 103–4.

3. Beth Loffreda, "Pulp Science: Race, Gender, and Prediction in Contemporary American Science" (Ph.D. diss., Rutgers University, 1996).

4. Richard Doyle discusses the "impossible inversion" that makes information primary and materiality secondary in molecular biology in *On Beyond Living: Rhetorical Transformations in the Life Sciences* (Stanford: Stanford University Press, 1997). See also Evelyn Fox Keller's analysis of the disembodiment of information in molecular biology in her *Secrets of Life, Secrets of Death: Essays on Language, Gender, and Science* (New York: Routledge, 1992), especially chapters 5, 8, and the epilogue. Lily E. Kay critically analyzes the emergence of the idea of a genetic "code" in "Cybernetics, Information, Life: The Emergence of Scriptural Representations of Heredity," *Configurations* 5 (winter 1997): 23–92. For a discussion of how this disembodied view of information began to circulate through the culture, see Dorothy Nelkin and M. Susan Lindee, *The DNA Mystique: The Gene as a Cultural Icon* (New York: W. H. Freeman and Company, 1995).

5. Michel Foucault famously suggested that "man" is a historical construction whose era is about to end in *The Order of Things: An Archaeology of the Human Sciences* (New York: Vintage Books, 1973), a few years earlier than Ihab Hassan's prescient announcement of posthumanism cited in the epigraph to this chapter. Since then, the more radical idea of the posthuman (as distinct from posthumanism) has appeared at a number of places. Among the important texts defining the posthuman in cultural studies are Allucquére Roseanne Stone, *The War of Desire and Technology at the Close of the Mechanical Age* (Cambridge: MIT Press, 1995); Judith Halberstam and Ira Livingston, eds., *Posthuman Bodies* (Bloomington: Indiana University Press, 1995); Scott Bukatman, *Terminal Identity: The Virtual Subject in Postmodern Science Fiction* (Durham: Duke University Press, 1993); and Anne Balsamo, *Technologies of the Gendered Body: Reading Cyborg Women* (Durham: Duke University Press, 1996). A number of scien-

tific works, detailed in chapters 3, 6, and 9, also figure importantly in delineating this list of characteristics.

6. C. B. Macpherson, *The Political Theory of Possessive Individualism: Hobbes to Locke* (Oxford: Oxford University Press, 1962), p. 3 (emphasis added).

7. Donna Haraway, *Simians, Cyborgs, and Women: The Reinvention of Nature* (New York: Routledge, 1990), especially "A Cyborg Manifesto: Science, Technology, and Socialist-Feminism in the Late Twentieth Century," pp. 149–82; Homi Bhabha, *The Location of Culture* (New York: Routledge, 1994); Gilles Deleuze and Felix Guattari, *A Thousand Plateaus: Capitalism and Schizophrenia,* translated by Brian Massumi (London: Athlone Press, 1987).

8. Lauren Berlant, in *The Anatomy of National Fantasy: Hawthorne, Utopia, and Everyday Life* (Chicago: University of Chicago Press, 1991), discusses the white male body of the ideal citizen, including its tendency toward disembodiment.

9. Gillian Brown, "Anorexia, Humanism, and Feminism," *Yale Journal of Criticism* 5, no. 1 (1991): 196.

10. William Gibson, *Neuromancer* (New York: Ace Books, 1984), p. 16.

11. Arthur Kroker, *Hacking the Future: Stories for the Flesh-Eating 90s* (New York: St. Martin's Press, 1996).

12. Five of the Macy Conference transactions were published: Heinz von Foerster, ed., *Cybernetics: Circular Causal and Feedback Mechanisms in Biological and Social Systems,* vols. 6–10 (New York: Josiah Macy Jr. Foundation, 1949–55). From the seventh conference on, Margaret Mead and Hans Lukas Teuber are listed as "assistant editors." The best study of the Macy Conferences is Steve J. Heims, *The Cybernetics Group* (Cambridge: MIT Press, 1991). In addition to discussing the conferences and doing extensive archival work, Heims also conducted interviews with many of the participants who have since died.

13. See Otto Mayr, *The Origins of Feedback Control* (Cambridge: MIT Press, 1970), for a full history of the concept of the feedback loop.

14. Walter Cannon is usually credited with working out the implications of homeostasis for biological organisms in *The Wisdom of the Body* (New York: W. W. Norton, 1939). Claude Bernard originated the concept in the nineteenth century.

15. Mayr, *The Origins of Feedback Control.*

16. Nancy Armstrong, *Desire and Domestic Fiction: A Political History of the Novel* (New York: Oxford University Press, 1987).

17. Michael Warner, *The Letters of the Republic: Publication and the Public Sphere in Eighteenth-Century America* (Cambridge: Harvard University Press, 1990).

18. Bruno Latour, *Science in Action: How to Follow Scientists and Engineers through Society* (Cambridge: Harvard University Press, 1987). Malcome Ashmore explores this feature of science studies in *The Reflexive Thesis: Wrighting Sociology of Scientific Knowledge* (Chicago: University of Chicago Press, 1989).

19. Heinz von Foerster, *Observing Systems,* 2d ed. (Salinas, Calif.: Intersystems Publications, 1984).

20. Humberto R. Maturana and Francisco J. Varela, *Autopoiesis and Cognition: The Realization of the Living,* Boston Studies in the Philosophy of Science, vol. 42 (Dordrecht: D. Reidel, 1980).

21. Niklas Luhmann has modified and extended Maturana's epistemology in significant ways; see, for example, his *Essays on Self-Reference* (New York: Columbia Uni-

versity Press, 1990) and "The Cognitive Program of Constructivism and a Reality That Remains Unknown," in *Self-Organization: Portrait of a Scientific Revolution,* edited by Wolfgang Krohn, Guenter Kueppes, and Helga Nowotny (Dordrecht: Kluwer Academic Publishers, 1990), 64–85.

22. Edward Fredkin, "Digital Mechanics: An Information Process Based on Reversible Universal Cellular Automata," *Physica D* 45 (1990): 245–70. See also the account of Fredkin's work in Robert Wright, *Three Scientists and Their Gods: Looking for Meaning in an Age of Information* (New York: Times Books, 1988). Also central to this theory is the work of Stephen Wolfram; see his *Theory and Applications of Cellular Automata* (Singapore: World Scientific, 1986).

23. Marvin Minsky, "Why Computer Science Is the Most Important Thing That Has Happened to the Humanities in 5,000 Years" (public lecture, Nara, Japan, May 15, 1996). I am grateful to Nicholas Gessler for providing me with his transcript of the lecture.

24. See Jennifer Daryl Slack and Fred Fejes, eds., *The Ideology of the Information Age* (Norwood, N.J.: Ablex Publishing Company, 1987), for essays exploring the implications of the contemporary construction of information. The tendency to ignore the material realities of communication technologies has been forcefully rebutted in two important works: Friedrich A. Kittler's *Discourse Networks, 1800–1900,* translated by Michael Metteer (Stanford: Stanford University Press, 1990), and Hans Ulrich Gumbrecht and K. Ludwig Pfeiffer, eds., *Materialities of Communication,* translated by William Whobrey (Stanford: Stanford University Press, 1994).

25. The relation of molecular biology has been explored in Keller, *Secrets;* the centrality of World War II to the development of cybernetics is demonstrated by Peter Galison in "The Ontology of the Enemy: Norbert Wiener and the Cybernetic Vision," *Critical Inquiry* 21 (1994): 228–66. Relevant here also is Kay, "Cybernetics, Information, Life" and Andy Pickering, "Cyborg History and the World War II Regime," *Perspectives on Science* 3, no. 1 (1995): 1–48.

26. Norbert Wiener, *Cybernetics; or, Control and Communication in the Animal and the Machine* (Cambridge: MIT Press, 1948), p. 132.

27. Thomas S. Kuhn, *The Structure of Scientific Revolutions,* 2d ed. (Chicago: University of Chicago Press, 1970); Foucault, *The Order of Things.* Both Kuhn and Foucault substantially revised their theories in later years. The vision of historical change in Michel Foucault's *The History of Sexuality,* translated by Robert Hurley (New York: Vintage Books, 1980), is much closer to seriation than are his earlier works.

28. The simulation is the creation of Gregory P. Garvey of Concordia University. An account of it can be found in Thomas E. Linehan, ed., *Visual Proceedings: The Art and Interdisciplinary Programs of Siggraph 93* (New York: Association for Computing Machinery, 1993), p. 125.

29. "A Magna Carta for the Knowledge Age" can be found (along with skeptical commentaries, mine among them) at the FEED Web site, < http://www.emedia.net/ feed>.

30. Claude Shannon and Warren Weaver, *The Mathematical Theory of Communication* (Urbana: University of Illinois Press, 1949).

31. Doyle, *On Beyond Living,* makes the point that the construction of information as primary, with materiality as supplemental, is a rhetorical rather than an experimental

accomplishment. He argues that the discourse of molecular biology functions as "rhetorical software," for it operates as if it were running a program on the hardware of the laboratory apparatus to produce results that the research alone could not accomplish. See also Kay, "Cybernetics, Information, Life."

32. Donald M. MacKay, *Information, Mechanism, and Meaning* (Cambridge: MIT Press, 1969).

33. Carolyn Marvin, "Information and History," in Slack and Fejes, *The Ideology of the Information Age*, pp. 49–62.

34. In response to a presentation by Alex Bavelas at the eighth Macy Conference, Shannon remarked that he did not see a "close connection" between the semantic questions that concerned Bavelas and his own emphasis on "finding the best encoding of symbols." Foerster, Mead, and Teuber, *Cybernetics* (Eighth Conference, 1951), 8:22.

35. Xerox PARC has been at the forefront of developing the idea of "ubiquitous computing," with computers embedded unobtrusively throughout the home and workplace environments. See Mark Weiser, "The Computer for the 21st Century," *Scientific American* 265 (September 1991): 94–104. For an account of how computers are transforming contemporary architecture and living patterns, see William J. Mitchell, *City of Bits: Space, Place, and the Infobahn* (Cambridge: MIT Press, 1995).

36. Sherry Turkle discusses the fascination of VR worlds in *Life on the Screen: Identity in the Age of the Internet* (New York: Simon and Schuster, 1995). Stone, *The War of Desire and Technology,* proposes that VR technologies undo the commonsense notion that one person inhabits one body. She suggests instead that we think of the subject "warranted by" the body rather than contained within it.

37. For an account of the extensive connections between cybernetics and the military, see Paul N. Edwards, *The Closed World: Computers and the Politics of Discourse in Cold War America* (Cambridge: MIT Press, 1996), and Les Levidow and Kevin Robins, eds., *Cyborg Worlds: The Military Information Society* (London: Free Association Books, 1989).

38. Don Ihde develops the full resonances of "lifeworld" from his grounding in phenomenology in *Technology and the Lifeworld: From Garden to Earth* (Bloomington: Indiana University Press, 1990), showing how the contemporary world is marked by a double attraction toward technology and toward the "natural" world simultaneously.

39. The notorious case is Autodesk's initiative to develop VR software that cited *Neuromancer;* see John Walker, "Through the Looking Glass: Beyond 'User' Interfaces," *CADalyst* (December 1989), 42, and Randall Walser, "On the Road to Cyberia: A Few Thoughts on Autodesk's Initiative," *CADalyst* (December 1989), 43.

40. An important work linking postmodern fiction with cybernetic technologies is David Porush, *The Soft Machine: Cybernetic Fiction* (New York: Methuen, 1985). Porush defines cybernetic fiction as self-reflexive fictions that look to cybernetics both for their themes and for the literary machinery of their texts.

41. Jean-François Lyotard, *The Postmodern Condition: A Report on Knowledge,* translated by Geoff Bennington and Brian Massumi (Minneapolis: University of Minnesota Press, 1984); Linda Hutcheon, *A Poetics of the Postmodern: History, Theory, Fiction* (New York: Routledge, 1994); and Brian McHale, *Constructing Postmodernism* (New York: Routledge, 1992) and *Postmodern Fiction* (New York: Methuen, 1981).

42. Bernard Wolfe, *Limbo* (New York: Random House, 1952).

43. Philip K. Dick: *We Can Build You* (London: Grafton Books, 1986), first pub-

lished in 1969; *Do Androids Dream of Electric Sheep?* (New York: Doubleday, 1968); *Dr. Bloodmoney; or, How We Got Along after the Bomb* (New York: Carroll and Graf, 1988), first published in 1965; and *Ubik* (London: Grafton Books, 1973), first published in 1969.

44. Neal Stephenson, *Snow Crash* (New York: Bantam, 1992); Greg Bear, *Blood Music* (New York: Ace Books, 1985); Richard Powers, *Galatea 2.2: A Novel* (New York: Farrar Straus Giroux, 1995); and Cole Perriman, *Terminal Games* (New York: Bantam, 1994).

Chapter Two

1. The paradox is discussed in N. Katherine Hayles, *Chaos Bound: Orderly Disorder in Contemporary Literature and Science* (Ithaca: Cornell University Press, 1990), pp. 31–60.

2. Self-organizing systems are discussed in Grégoire Nicolis and Ilya Prigogine, *Exploring Complexity: An Introduction* (New York: Freeman and Company, 1989); Roger Lewin, *Complexity: Life at the Edge of Chaos* (New York: Macmillan, 1992); and M. Mitchell Waldrop, *Complexity: The Emerging Science at the Edge of Order and Chaos* (New York: Simon and Schuster, 1992).

3. Friedrich A. Kittler, *Discourse Networks, 1800–1900,* translated by Michael Metteer (Stanford: Stanford University Press, 1990), p. 193.

4. The fluidity of writing on the computer is eloquently explored by Michael Joyce in *Of Two Minds: Hypertext Pedagogy and Poetics* (Ann Arbor: University of Michigan Press, 1995).

5. Howard Rheingold surveys the new virtual technologies in *Virtual Reality* (New York: Summit Books, 1991). Also useful is Ken Pimentel and Kevin Teixeira, *Virtual Reality: Through the New Looking Glass* (New York: McGraw-Hill, 1993). Benjamin Woolley takes a skeptical approach toward claims for the new technology in *Virtual Worlds: A Journey in Hyped Hyperreality* (Oxford, England: Blackwell, 1992).

6. Allucquère Roseanne Stone, *The War of Desire and Technology at the Close of the Mechanical Age* (Cambridge: MIT Press, 1995).

7. Sherry Turkle, *Life on the Screen: Identity in the Age of the Internet* (New York: Simon and Schuster, 1995).

8. In *The Age of the Smart Machine: The Future of Work and Power* (New York: Basic Books, 1988), Shoshana Zuboff explores, through three case studies, the changes in U.S. workplaces as industries become informatted.

9. Computer law is discussed in Katie Hafner and John Markoff, *Cyberpunk: Outlaws and Hackers on the Computer Frontier* (New York: Simon and Schuster, 1991); also informative is Bruce Sterling, *The Hacker Crackdown: Law and Disorder on the Electronic Frontier* (New York: Bantam, 1992).

10. Turkle documents computer network romances in *Life on the Screen.* Nicholson Baker's *Vox: A Novel* (New York: Random House, 1992) imaginatively explores the erotic potential for better living through telecommunications; and Rheingold looks at the future of erotic encounters in cyberspace in "Teledildonics and Beyond," *Virtual Reality,* pp. 345–77.

11. Among the studies that explore these connections are Jay Bolter, *Writing Space: The Computer, Hypertext, and the History of Writing* (Hillsdale, N.J.: Lawrence Erl-

baum Associates, 1991); Michael Heim, *Electric Language: A Philosophical Study of Word Processing* (New Haven: Yale University Press, 1987); and Mark Poster, *The Mode of Information: Poststructuralism and Social Context* (Chicago: University of Chicago Press, 1990).

12. Donna Haraway, "A Manifesto for Cyborgs: Science, Technology, and Socialist Feminism in the 1980s," *Socialist Review* 80 (1985): 65–108; see Donna Haraway, "The High Cost of Information in Post World War II Evolutionary Biology: Ergonomics, Semiotics, and the Sociobiology of Communications Systems," *Philosophical Forum* 13, nos. 2–3 (1981–82): 244–75.

13. Jacques Lacan, "Radiophonies," *Scilicet* 2/3 (1970): 55, 68. For floating signifiers, see *Le Séminaire XX: Encore* (Paris: Seuil, 1975), pp. 22, 35.

14. Although presence and absence loom much larger in Lacanian psycholinguistics than do pattern and randomness, Lacan was not uninterested in information theory. In the 1954–55 *Seminar,* he played with incorporating ideas from information theory and cybernetics into psychoanalysis. See especially "The Circuit," pp. 77–90, and "Psychoanalysis and Cybernetics; or, On the Nature of Language," pp. 294–308, in *The Seminar of Jacques Lacan: Book II,* edited by Jacques-Alain Miller (New York: W. W. Norton and Company, 1991).

15. For an individual event s_i, the information $I(s_i) = -\log p(s_i)$, where p is the probability, expressed as a decimal between 1 and 0, that s_i will occur. To give a sense of how this function varies, consider that -log base 2 of .9 (an event that occurs nine times out of ten) is .15, whereas -log base 2 of .1 (an event that occurs only one in ten times) is 3.33. Hence, as the probability p decreases (becomes less likely), -log p increases. In the case of elements whose probabilities do not conditionally depend on one another, the *average* information of a source s is $I(s) = \Sigma - p(s_i) \log p(s_i)$, where p is again the probability that s_i will occur.

16. Claude Shannon and Warren Weaver, *The Mathematical Theory of Communication* (Urbana: University of Illinois Press, 1949). For a further discussion of this aspect of information theory, see Hayles, *Chaos Bound,* pp. 31–60.

17. The gender encoding implicit in "man" (rather than "human") is also reflected in the emphasis on tool usage as a defining characteristic rather than, say, altruism or nurturing, traits traditionally encoded female.

18. Kenneth P. Oakley, *Man the Tool-Maker* (London: Trustees of the British Museum, 1949), p. 1.

19. Marshall McLuhan, *Understanding Media: The Extensions of Man* (New York: McGraw Hill, 1964), pp. 41–47.

20. The term *homeostasis,* or self-regulating stability through cybernetic corrective feedback, was introduced by physiologist Walter B. Cannon in "Organization for Physiological Homeostasis," *Physiological Reviews* 9 (1929): 399–431. Cannon's work influenced Norbert Wiener, and homeostasis became an important concept in the initial phase of cybernetics from 1946 to 1953; see chapters 3 and 4 for details.

21. Key figures in moving from homeostasis to self-organization were Heinz von Foerster, especially in *Observing Systems* (Salinas, Calif.: Intersystems Publications, 1981), and Humberto R. Maturana and Francisco J. Varela, *Autopoiesis and Cognition: The Realization of the Living* (Dordrecht: D. Reidel, 1980), discussed in detail in chapter 6.

22. Rheingold, *Virtual Reality,* pp. 13–49; Hans Moravec, *Mind Children: The Fu-*

ture of Robot and Human Intelligence (Cambridge: Harvard University Press, 1988), pp. 1–5, 116–22.

23. William Gibson, *Neuromancer* (New York: Ace Books, 1984), p. 51.

24. Ibid., p. 16.

25. The seminal text is Norbert Wiener, *Cybernetics; or, Control and Communication in the Animal and the Machine* (Cambridge: MIT Press, 1948).

26. Henry James, *The Art of the Novel* (New York: Charles Scribner's Sons, 1937), pp. 47, 46.

27. Peter Kollock, my colleague at UCLA and a sociologist, has studied virtual communities at several sites on the Internet. See Marc Smith and Peter Kollock, eds., *Communities in Cyberspace* (London: Routledge, 1998); see also Stone's discussion of MUDs in *The War of Desire and Technology*, Turkle's discussion in *Life on the Screen*, and Amy Bruckman's article "Gender Swapping on the Internet," available at anonymous <ftp://media.mit.edu/pub/asb/paper/gender-swapping>. Espen J. Aarseth has a discussion of the literary and formal characteristics of MUDs in *Cybertext: Perspectives on Ergodic Literature* (Baltimore: Johns Hopkins University Press, 1997).

28. David Harvey, *The Condition of Postmodernity: An Enquiry into the Origins of Cultural Change* (New York: Blackwell, 1989).

29. The material basis for informatics is meticulously documented in James R. Beniger, *The Control Revolution: Technological and Economic Origins of the Information Society* (Cambridge: Harvard University Press, 1986).

30. For an account of how tracks are detected, see Hafner and Markoff, *Cyberpunk*, pp. 35–40, 68–71.

31. Don DeLillo, *White Noise* (1985; New York: Penguin, 1986).

32. Italo Calvino, *If on a winter's night a traveler*, translated by William Weaver (New York: Harcourt Brace Jovanovich, 1981), pp. 26–27, originally published in 1979 in Italian.

33. Ibid., p. 220.

34. William S. Burroughs, *Naked Lunch* (New York: Grove, 1959).

35. David Porush discusses the genre of "cybernetic fiction," which he defines as fictions that resist the dehumanization that is sometimes attendant on cybernetics, in *The Soft Machine: Cybernetic Fiction* (New York: Methuen, 1985); Burroughs's titular story is discussed on pp. 85–111. Robin Lydenberg has a fine exposition of Burroughs's style in *Word Cultures: Radical Theory and Practice in William S. Burroughs' Fiction* (Urbana: University of Illinois Press, 1987).

36. Burroughs, *Naked Lunch*, p. xxxix.

37. Jacques Derrida, *Of Grammatology*, translated by Gayatri C. Spivak (Baltimore: Johns Hopkins University Press, 1976).

38. Mark Leyner, *My Cousin, My Gastroenterologist* (New York: Harmony Books, 1990), pp. 6–7.

39. Walter Benjamin, "The Storyteller," *Illuminations*, translated by Harry Zohn (New York: Schocken, 1969).

40. Jean-François Lyotard, *The Postmodern Condition: A Report on Knowledge*, translated by Geoff Bennington and Brian Massumi (Minneapolis: University of Minnesota Press, 1984).

41. It is significant in this regard that Andrew Ross calls for cultural critics to consider themselves hackers in "Hacking Away at the Counterculture," in *Technoculture*,

edited by Constance Penley and Andrew Ross (Minneapolis: University of Minnesota Press), pp. 107–34.

42. Roland Barthes, *S/Z*, translated by Richard Miller (New York: Hill and Wang, 1974)

43. George W. S. Trow, *Within the Context of No Context* (Boston: Little Brown, 1978).

44. Kittler, *Discourse Networks*. Joseph Tabbi and Michael Wurtz further explore the implications of medial ecology in *Reading Matters: Narrative in the New Media Ecology* (Ithaca: Cornell University Press, 1996).

45. Paul Virilio and Sylvérè Lotringer, *Pure War,* translated by Mark Polizzotti (New York: Semiotext(e), 1983).

46. "Embodied virtuality" is Mark Weiser's phrase in "The Computer for the 21st Century," *Scientific American* 265 (September 1991): 94–104. Weiser distinguishes between technologies that put the user into a simulation with the computer (virtual reality) and those that embed computers within already existing environments (embodied virtuality or ubiquitous computing). In virtual reality, the user's sensorium is redirected into functionalities compatible with the simulation; in embodied virtuality, the sensorium continues to function as it normally would but with an expanded range made possible through the environmentally embedded computers.

Chapter Three

1. "Conferences on Feedback Mechanisms and Circular Causal Systems in Biology and the Social Sciences" (March 8–9, 1946), p. 62, Frank Fremont-Smith Papers, Francis A. Countway Library of Medicine, Harvard University, Cambridge, Mass.

2. This explanation of information theory by Wiener appears in "The Impact of Communication Engineering on Philosophy," Box 14, Folder 765, Norbert Wiener Papers, Collection MC-22, Institute Archives and Special Collections, Massachusetts Institute of Technology Archives, Cambridge, Mass. See also Norbert Wiener, "Thermodynamics of the Message," in *Norbert Wiener: Collected Works with Commentaries,* edited by Pesi Masani, vol. 4 (Cambridge: MIT Press, 1985), pp. 206–11. Wiener's treatment of information is conceptually similar to Shannon's, and today's version is often called the Shannon-Wiener theory.

3. For a full theoretical treatement, see Claude Shannon and Warren Weaver, *The Mathematical Theory of Communication* (Urbana: University of Illinois Press, 1949). Weaver included in the volume an essay explaining Shannon's theory. According to Eric A. Weiss, Shannon told him in correspondence that Weaver put together the volume without consulting Shannon. Weiss wrote: "Weaver was a big-shot scientific gate keeper at the time; Shannon was a more or less nobody. Weaver took some notes . . . or something by Shannon and turned it into the 1949 writing putting his name first and without really getting Shannon's consent. Shannon felt that Weaver had made a good explanation, this was one of Weaver's skills, and did not object seriously at the time" (Weiss to author, private communication).

4. Richard Doyle discusses the reification of information in the context of molecular biology in *On Beyond Living: Rhetorical Transformations in the Life Sciences* (Stanford: Stanford University Press, 1997). See also Evelyn Fox Keller's analysis of the dis-

embodiment of information in molecular biology in *Secrets of Life, Secrets of Death: Essays on Language, Gender, and Science* (New York: Routledge, 1992), especially chapters 5, 8, and the epilogue. Lily E. Kay critically analyzes the emergence of the idea of a genetic "code" in "Cybernetics, Information, Life: The Emergence of Scriptural Representations of Heredity," *Configurations* 5 (winter 1997): 23–92. For a discussion of how this disembodied view of information began to circulate through the culture, see Dorothy Nelkin and M. Susan Lindee, *The DNA Mystique: The Gene as a Cultural Icon* (New York: W. H. Freeman and Company, 1995).

5. Heinz von Foerster, Margaret Mead, and Hans Lukas Teuber, eds., *Cybernetics: Circular Causal and Feedback Mechanisms in Biological and Social Systems*, vols. 6–10 (Josiah Macy Jr. Foundation, 1952) (Eighth Conference, 1951), 8:22. The published series is hereafter cited as *Cybernetics* with the number and year of the conference and the volume number indicated.

6. Donald M. MacKay, "In Search of Basic Symbols," *Cybernetics* (Eighth Conference, 1951), 8:222. A fuller account can be found in Donald M. MacKay, *Information, Mechanism, and Meaning* (Cambridge: MIT Press, 1969).

7. Nicolas S. Tzannes, "The Concept of 'Meaning' in Information Theory" (August 7, 1968), in Warren McCulloch Papers, American Philosophical Society Library, Philadelphia, B/M139, Box 1.

8. In Mary Catherine Bateson's *Our Own Metaphor: A Personal Account of a Conference on the Effects of Conscious Purpose on Human Adaptation* (1972; Washington, D.C.: Smithsonian Institution Press, 1991), she quotes her father, Gregory Bateson, as advising, "Stamp out nouns!" The difficulty of this project may be indicated by the fact that the slogan itself contains a noun.

9. Warren S. McCulloch, *Embodiments of Mind* (Cambridge: MIT Press, 1965), p. 2. Conventional in the rhetoric of the 1950s, the purported universality of "man" indicates how ideological assumptions were inscribed into a universal formulation and then erased from view once the universal stood for the embodied instantiation (the actual human beings who compose humanity).

10. Steve J. Heims, *The Cybernetics Group* (Cambridge: MIT Press, 1991), pp. 31–51, especially p. 41. For the classic papers on the McCulloch-Pitts neuron, see Warren S. McCulloch and Walter H. Pitts, "A Logical Calculus of the Ideas Immanent in Nervous Activity," and Warren S. McCulloch, "A Heterarchy of Values Determined by the Topology of Nervous Nets," both in McCulloch, *Embodiments of Mind*, pp. 19–39, 40–45.

11. McCulloch recalls meeting Walter Pitts in "The Beginning of Cybernetics," McCulloch Papers, B/M139, Box 2.

12. Automata theory works with highly abstract models of computers, especially Turing machines. Just as Maxwell's Demon is a thought experiment, so a Turing machine can be called a thought computer. The idea is to propose a conceptual scheme that, although it might never be realized in an experimental situation, poses interesting problems and leads to significant conclusions. Named after its inventor Alan Turing, the Turing machine consists of a control box containing a finite program that moves back and forth along a finite tape inscribed with symbols, conventionally ones and zeros written in square boxes. The control box scans the tape one square at a time, and on the basis of what it reads and what its program calls for it to do, prints another symbol on the square (which may or may not be the same as the one already there) and moves one

square to the right or left, where it goes through the procedure again until it has finished executing its program's instructions.

13. McCulloch, "The Beginning of Cybernetics," p. 12.

14. "Conferences on Feedback Mechanisms," p. 46.

15. McCulloch Papers, B/M139, Box 2.

16. Ibid.

17. See Lawrence Kubie, "A Theoretical Application to Some Neurological Problems of the Properties of Excitation Waves Which Move in Closed Circuits," *Brain* 53 (1930): 166–78.

18. Lewis Carroll, *Sylvie and Bruno Concluded* (London: Macmillan, 1893), p. 169; Jorge Luis Borges, "Of Exactitude in Science," *A Universal History of Infamy*, translated by Norman Thomas di Givanni (New York: Dutton, 1972), pp. 141ff.

19. Andrea Nye, *Words of Power: A Feminist Reading of the History of Logic* (New York: Routledge, 1990).

20. This insight is, of course, a central achievement of the social construction of scientific knowledge. Nancy Cartwright addresses it powerfully in *How the Laws of Physics Lie* (Oxford: Oxford University Press, 1983). For eloquent demonstrations of it, see Steven Shapin and Simon Schaffer, *Leviathan and the Air-Pump: Hobbes, Boyle, and the Experimental Life* (Princeton: Princeton University Press, 1985), and Bruno Latour, *Science in Action: How to Follow Scientists and Engineers through Society* (Cambridge: Harvard University Press, 1987).

21. Warren S. McCulloch, "How Nervous Structures Have Ideas" (speech to the American Neurological Association, June 13, 1949), p. 3, McCulloch Papers, B/M139, Box 1.

22. Reprinted in McCulloch, *Embodiments of Mind*, pp. 387–98, quotations on p. 393.

23. *Cybernetics* (Seventh Conference, 1950), 7:155.

24. Claude E. Shannon, "Presentation of a Maze-Solving Machine," *Cybernetics* (Eighth Conference, 1951), 8:173–80.

25. Ibid., p. xix.

26. Mark Seltzer makes a similar point about scientific models (especially the second law of thermodynamics) serving as a relay system in *Bodies and Machines* (New York: Routledge, 1992).

27. *Cybernetics* (Eighth Conference, 1951), 8:173. "Singing" acquired its name from feedback loops that make an audio amplifier break into oscillation, resulting in a whistling in the operator's headphones (information from Eric Weiss, private communication).

28. How quickly the equation between man and machine proliferated into social theory can be seen in F. S. C. Northrop's *Ideological Differences and World Order: Studies in the Philosophy and Science of the World's Cultures* (New Haven: Yale University Press, 1949). In his contribution to the volume, "Ideological Man and Natural Man" (pp. 407–28), Northrop relies extensively on the McCulloch-Pitts neuron as well as the cybernetic manifesto written by Wiener, Rosenblueth, and Bigelow (discussed in chapter 4) to bring together normative social theory with "a complete unified natural philosophy" (p. 424). Like Wiener, Northrop associates cybernetics with liberal humanism, arguing that reverberating loops and teleological mechanisms confirm that the correct model for human subjectivity is the "moral, thoughtful, choosing, purposeful in-

dividual" (p. 426). Unity within the subject can be achieved only when ideology is brought into harmony with "scientifically verified and conceived natural neurological man" fashioned from McCulloch-Pitts neurons and Wiener's feedback loops. Only when "the philosophy giving instructions to his motor neurons" is congruent with cybernetic modeling can such an individual be "a single, a composed, and a whole man" (p. 424). An exchange of letters between Northrop, McCulloch, and Wiener laid out the network of ideas that Northrop picked up from the Macy Conferences and that provided the basis for his book (McCulloch Papers, B/M139, Box 2).

29. W. Ross Ashby, "Homeostasis," in *Cybernetics* (Ninth Conference, 1952), 9:73–108.

30. Ashby fulfilled his ambition to move to more complex modeling in W. Ross Ashby, *Design for a Brain: The Origin of Adaptive Behavior* (London: Chapman and Hall, 1952). Also of interest is his book *Introduction to Cybernetics* (London: Chapman and Hall, 1961).

31. John Stroud, "The Psychological Moment in Perception," in *Cybernetics* (Sixth Conference, 1949), 6:27–28.

32. *Cybernetics* (Sixth Conference, 1949), 6:147, 153.

33. Ibid., p. 153.

34. Kubie, "A Theoretical Application."

35. *Cybernetics* (Sixth Conference, 1949), 6:74.

36. *Cybernetics* (Seventh Conference, 1950), 7:210, 222.

37. A copy of the speech in the McCulloch Papers is prefaced with a note that the copy was reproduced without the author's consent or knowledge and is adorned with a skull and crossbones to indicate its pirated status. Its pirated status notwithstanding, it is word for word the same as the version that McCulloch later published in *Embodiments of Mind*. If it really was pirated, one wonders how it ended up among McCulloch's papers. Whether or not McCulloch had a hand in circulating this version, he did send copies of the speech to his friends.

38. Heims recounts this part of the tale in *The Cybernetics Group*, pp. 136ff.

39. Letter dated April 11, 1950, McCulloch Papers, B/M139, Box 2.

40. Heims, *The Cybernetics Group*, p. 136.

41. "The Place of Emotions in the Feedback Concept," *Cybernetics* (Ninth Conference, 1952), 9:48.

42. As evidence that emotions and other psychic experiences have a neurological basis, Kubie referred repeatedly to "psychosurgery"—that is, lobotomy—which by destroying tissue proved that brain functions have a physiological basis. Presumably he referred to this cruel practice (which Wiener had elsewhere satirized as a way to make custodial care of patients easier) to establish that emotions have a material and quantitative dimension. Yet when he was asked to elaborate, he answered that he "did not want to discuss psychosurgery" but rather was "simply indicating some of the questions we ask ourselves about the effects of any procedure on emotional processes, the points at which they are vulnerable and alterable" ("The Place of Emotions in the Feedback Concept," *Cybernetics* [Ninth Conference, 1952], 9:69). His transparent motive was to establish his credentials as a physical scientist who dealt in quantifiable data, another indication of the uneasy relations between him and the experimentalists.

43. Letter dated March 30, 1954, McCulloch Papers, B/M139, Box 2.

44. Letter dated May 29, 1969, Fremont-Smith Papers.

45. Letter dated June 2, 1969, Fremont-Smith Papers.

46. Letter dated July 1, 1969, Fremont-Smith Papers.

47. Stewart Brand, "'For God's Sake, Margaret': Conversation with Gregory Bateson and Margaret Mead," *Co-Evolution Quarterly* (summer 1976), 32, 34 (Bateson's diagram is on p. 37).

48. Letter dated November 8, 1954, McCulloch Papers, B/M139, Box 2.

49. Letter dated November 22, 1954, McCulloch Papers, B/M139, Box 2.

50. Bateson, *Our Own Metaphor* (hereafter cited in the text as *OOM*).

51. At the ninth conference, Mead insisted that language is broader than words. "We should drop the idea that language is made up of words and that words are toneless sequences of letters on paper, although even on paper there are possibilities for poetic overtones. We are dealing here with language in a very general sense, which would include posture, gesture, and intonation." *Cybernetics* (Ninth Conference, 1952), 9:13.

52. Gregory Bateson, "Our Own Metaphor: Nine Years After," in *A Sacred Unity: Further Steps to an Ecology of Mind* (New York: Harper Collins, 1991), p. 227. Catherine had asked Gregory for a letter that might be suitable as an afterword to the second edition of *Our Own Metaphor*. Although she evidently decided not to use the letter, it was later published.

53. Ibid., p. 225.

54. *Cybernetics* (Tenth Conference, 1953), 10:69.

55. J. Y. Lettvin, H. R. Maturana, W. S. McCulloch, and W. H. Pitts, "What the Frog's Eye Tells the Frog's Brain," *Proceedings of the Institute for Radio Engineers* 47, no. 11 (November 1959): 1940–59. Reprinted in McCulloch, *Embodiments of Mind*, pp. 230–55.

56. Letter from Janet Freed to Warren McCulloch, dated January 31, 1947, Fremont-Smith Papers.

57. This is an educated guess based on reading her comments in the typed manuscript of "Chairman and Editors' Meeting," dated April 27, 1949, pp. 3ff., Fremont-Smith Papers.

58. Ibid., pp. 3, 26.

59. Dorothy E. Smith, *The Everyday World as Problematic: A Feminist Sociology* (Boston: Northeastern University Press, 1987); see also Dorothy E. Smith, *The Conceptual Practices of Power: A Feminist Sociology of Knowledge* (Boston: Northeastern University Press, 1990).

Chapter Four

1. See Gregory Bateson, *Steps to an Ecology of Mind* (New York: Ballantine Books, 1972), p. 251, for an interpretation of the question. "It is not communicationally meaningful to ask whether the blind man's stick or the scientist's microscope are 'parts' of the men who use them. Both stick and microscope are important pathways of communication and, as such, are parts of the network in which we are interested; but no boundary line—e. g., halfway up the stick—can be relevant in a description of the toplogy of this net."

2. Donna Haraway, "A Manifesto for Cyborgs: Science, Technology, and Socialist Feminism in the 1980s," *Socialist Review* 80 (1985): 65–108.

3. George Lakoff and Mark Johnson, *Metaphors We Live By* (Chicago: University of Chicago Press, 1980); and Mark Johnson, *The Body in the Mind: The Bodily Basis of Meaning, Imagination, and Reason* (Chicago: University of Chicago Press, 1987).

4. For an analysis of the strategies used to proclaim cybernetics a universal science, see Geof Bowker, "How To Be Universal: Some Cybernetic Strategies, 1943–1970," *Social Studies of Science* 23 (1993): 107–27.

5. Norbert Wiener, "Men, Machines, and the World About," Box 13, Folder 750, Norbert Wiener Papers, Collection MC-22, Institute Archives and Special Collections, Massachusetts Institute of Technology Archives, Cambridge, Mass. Also published in *Norbert Wiener: Collected Works with Commentaries,* edited by Pesi Masani, vol. 4 (Cambridge: MIT Press, 1985), pp. 793–99.

6. For a study tracing Wiener's postwar views, see Steve J. Heims, *John von Neumann and Norbert Wiener: From Mathematics to the Technologies of Life and Death* (Cambridge: MIT Press, 1980).

7. Peter Galison, "The Ontology of the Enemy: Norbert Wiener and the Cybernetic Vision," *Critical Inquiry* 21 (1994): 228–66.

8. Otto Mayr, *Authority, Liberty, and Automatic Machinery in Early Modern Europe* (Baltimore: Johns Hopkins University Press, 1986).

9. The question is posed most powerfully in Philip K. Dick, *Blade Runner* (originally published in 1968 under the title *Do Androids Dream of Electric Sheep?*) (New York: Ballantine Books, 1982).

10. It remained for a novelist, Kurt Vonnegut, to envision the full implications of Wiener's cybernetic program if it were fully carried out: see Kurt Vonnegut, *Player Piano* (New York: Delacorte Press/Seymour Laurence, 1952).

11. See, for example, Norbert Wiener, "The Averages of an Analytical Function and the Brownian Movement," in *Norbert Wiener: Collected Works*, vol. 1, pp. 450–55.

12. Norbert Wiener, "The Historical Background of Harmonic Analysis," *American Mathematical Society Semicentennial Publications,* vol. 2 (Providence, R.I.: American Mathematical Society, 1938), pp. 513–22.

13. Norbert Wiener, *The Human Use of Human Beings: Cybernetics and Society,* 2d ed. (Garden City, N.Y.: Doubleday, 1954) (hereafter cited in the text as *HU*), p. 10.

14. Wiener of course knew Shannon; both were participants in the Macy Conferences. Although they conceived of information in similar ways, Wiener was more inclined to see information and entropy as opposites. See also n. 3, ch. 3.

15. James R. Beniger, *The Control Revolution: Technological and Economic Origins of the Information Society* (Cambridge: Harvard University Press, 1986).

16. Michel Serres brilliantly analyzes the progression from the mechanical to the thermodynamical in "Turner Translates Carnot," *Hermes: Literature, Science, Philosophy,* edited by Josué V. Harari and David F. Bell (Baltimore: Johns Hopkins University Press, 1982).

17. Beniger, *The Control Revolution,* convincingly shows how technologies of speed and communication precipitated a "crisis of control" that, once solved, initiated a new cycle of crisis.

18. Norbert Wiener, "The Role of the Observer," *Philosophy of Science* 3 (1936): 311.

19. Pesi Masani, *Norbert Wiener, 1894–1964,* Vita Mathematica Series, vol. 5 (Basel: Birkhaeuser, 1989), calls Wiener's statement "a half-truth," "one of the solitary instances in which this very coherent thinker articulated badly" (p. 128).

20. Norbert Wiener, *I Am a Mathematician: The Later Life of a Prodigy* (Garden City, N.Y.: Doubleday, 1956), pp. 85–86.

21. Ibid., p. 86.

22. Heims, *John von Neumann and Norbert Wiener,* pp. 155–57.

23. Norbert Wiener, *Ex-Prodigy: My Childhood and Youth* (New York: Simon and Schuster, 1953).

24. Arturo Rosenblueth, Norbert Wiener, and Julian Bigelow, "Behavior, Purpose, and Teleology," *Philosophy of Science* 10 (1943): 18–24.

25. Richard Taylor, "Comments on a Mechanistic Conception of Purposefulness," *Philosophy of Science* 17 (1950): 310–17.

26. Arturo Rosenblueth and Norbert Wiener, "Purposeful and Non-Purposeful Behavior," *Philosophy of Science* 17 (1950): 318.

27. Bowker, "How to Be Universal," pp. 107–27.

28. Richard Taylor, "Purposeful and Non-Purposeful Behavior: A Rejoinder," *Philosophy of Science* 17 (1950): 327–32.

29. Norbert Wiener, "The Nature of Analogy," (1950), Box 12, Folder 655, Wiener Papers.

30. Michael A. Arbib and Mary B. Hesse, *The Construction of Reality* (Cambridge, England: Cambridge University Press, 1986).

31. Michael J. Apter draws the comparison between Saussurian linguistics and cybernetics in "Cybernetics: A Case Study of a Scientific Subject, Complex," in *The Sociology of Science: Sociological Review Monograph,* no. 18, edited by Paul Halmos (Keele, Staffordshire: Keele University, 1972), pp. 93–115, especially p. 104.

32. Wiener, "The Nature of Analogy," p. 2.

33. I rely here on Galison's detailed account of Wiener's work with antiaircraft devices in "The Ontology of the Enemy."

34. Norbert Wiener, "Sound Communication with the Deaf," in *Norbert Wiener: Collected Works,* vol. 4, pp. 409–11.

35. Cited in Walter A. Rosenblith and Jerome B. Wiesner, : "The Life Sciences and Cybernetics," one of the articles written in tribute to Wiener on the occasion of his death and published as "Norbert Wiener, 1894–1964," *Journal of Nervous and Mental Disease* 140 (1965): 3–16. Rosenblith and Wiesner's contribution is on pp. 3–8.

36. Masani, *Norbert Wiener,* pp. 205–6.

37. Rosenblith and Wiesner, "From Philosophy to Mathematics to Biology," p. 7.

38. Mark Seltzer, *Bodies and Machines* (New York: Routledge, 1992), p. 14.

39. Ibid., p. 41.

40. Leo Szilard, "On the Reduction of Entropy as a Thermodynamic System Caused by Intelligent Beings," *Zeitschrift für Physik* 53 (1929): 840–56.

41. Leon Brillouin, "Maxwell's Demon Cannot Operate: Information and Entropy, I," *Journal of Applied Physics* 212 (March 1951): 334–57. Much of this material is also available in Harvey S. Leff and Andrew F. Rex, eds., *Maxwell's Demon: Entropy, Information, Computing* (Princeton: Princeton University Press, 1990).

42. Claude E. Shannon and Warren Weaver, *The Mathematical Theory of Communication* (Urbana: University of Illinois Press, 1949).

43. Warren Weaver offered this explanation in his essay interpreting Shannon's theory in ibid.

44. See N. Katherine Hayles, *Chaos Bound: Orderly Disorder in Contemporary Literature and Science* (Ithaca: Cornell University Press, 1990); for a major statement of this thesis, see Ilya Prigogine and Isabelle Stengers, *Order Out of Chaos: Man's New Dialogue with Nature* (New York: Bantam, 1984).

45. Wiener's views on photosynthesis and Maxwell's Demon are discussed by Masani, *Norbert Wiener*, pp. 155–56. See also Norbert Wiener, "Cybernetics (Light and Maxwell's Demon)," *Scientia* (Italy) 87 (1952): 233–35, reprinted in *Norbert Wiener: Collected Works*, vol. 4, pp. 203–5.

46. Michael Serres plays multiple riffs on this idea in *The Parasite*, translated by Lawrence R. Schehr (Baltimore: Johns Hopkins University Press, 1982).

47. Valentino Braitenberg delightfully explores the possibility that simple machines can demonstrate behavioral equivalents to emotional states, including fear, love, envy, and ambition, by constructing a series of "thought machines" (machines designed in principle but not actually built). See *Vehicles: Experiments in Synthetic Psychology* (Cambridge: MIT Press, 1984).

48. Galison, "Ontology of the Enemy," p. 232.

49. Despite Wiener's efforts, after World War II cybernetics became more, not less, entangled with military projects. The close connection between the military and cybernetics is detailed by Paul N. Edwards, *The Closed World: Computers and the Politics of Discourse in Cold War America* (Cambridge: MIT Press, 1996), and Les Levidow and Kevin Robins, eds., *Cyborg Worlds: The Military Information Society* (London: Free Association Books, 1989).

50. Norbert Wiener, *Cybernetics; or, Control and Communication in the Animal and the Machine*, 2d ed. (Cambridge: MIT Press, 1961).

51. Richard Dawkins, *The Selfish Gene* (New York: Oxford University Press, 1976).

52. Masani, *Norbert Wiener*, p. 21.

Chapter Five

1. David N. Samuelson, "*Limbo:* The Great American Dystopia," *Extrapolation* 19 (1977): 76–87.

2. Bernard Wolfe, *Limbo* (New York: Random House, 1952), p. 412.

3. Paul Virilio and Sylvère Lotringer, *Pure War*, translated by Mark Polizzotti (New York: Semiotext(e), 1983), pp. 91–102.

4. Donna Haraway, "A Manifesto for Cyborgs: Science, Technology, and Socialist Feminism in the 1980s," *Socialist Review* 80 (1985): 65–108.

5. This portion of the argument appeared in N. Katherine Hayles, "The Life Cycle of Cyborgs: Writing the Posthuman," in *A Question of Identity: Women, Science and Literature*, edited by Marina Benjamin (New Brunswick: Rutgers University Press, 1993), pp. 152–72, especially pp. 156–61.

6. Scott Bukatman, *Terminal Identity: The Virtual Subject in Postmodern Science Fiction* (Durham: Duke University Press, 1993), p. 9.

7. The idea of "tonus" (defined as muscle tone) may be a punning wink toward "clonus," spasms by muscles or groups of muscles. Wiener discusses Warren McCulloch's research on clonus in *Cybernetics; or, Control and Communication in the Animal and the Machine* (Cambridge: MIT Press, 1948).

8. Norbert Wiener, *Cybernetics*. Wolfe probably also read Wiener's popular book

The Human Use of Human Beings: Cybernetics and Society (Boston: Houghton Mifflin, 1950).

9. W. Norbert [Norbert Wiener], "The Brain," in *Crossroads in Time,* edited by Groff Conklin (Garden City, N.Y.: Doubleday Books, 1950), pp. 299–312 (quotation on p. 300). A typescript version, with different names for the characters and with manuscript corrections, can be found in Box 12, Norbert Wiener Papers, Collection MC-22, Institute Archives and Special Collections, Massachusetts Institute of Technology Archives, Cambridge, Mass.

10. Paul Virilio, *War and Cinema: The Logistics of Perception,* translated by Patrick Camiller (London: Verso, 1989), p. 10.

11. Richard Doyle, *On Beyond Living: Rhetorical Transformations in the Life Sciences* (Stanford: Stanford University Press, 1997).

12. Bernard Wolfe, "Self Portrait," *Galaxi Science Fiction* 3 (November 1951): 64.

13. "Self Portrait" also suggests a link between cybernetics and McCarthyism when the narrator consolidates his position by denouncing a scientific rival as a security risk and testifying against him at a hearing. The context presents this move as reprehensible, in line with the satiric tone of the piece. Wolfe was generally sympathetic to leftist causes and did not participate in the communist hysteria that characterized these years in the United States. When he was a young man, he served as a security guard for Leon Trotsky in Mexico.

14. Douglas D. Noble, "Mental Materiel: The Militarization of Learning and Intelligence in U.S. Education," in *Cyborg Worlds: The Military Information Society,* edited by Les Levidow and Kevin Robbins (London: Free Association Books, 1989), pp. 13–42.

15. For a discussion of neocortical warfare, see Col. Richard Szafranski, U.S. Air Force, "Harnessing Battlefield Technology: Neocortical Warfare? The Acme of Skill," *Military Review: The Professional Journal of the United States Army* (November 1994), 41–54. See also Chris Hables Gray, "The Cyborg Soldier: The U.S. Military and the Post-modern Warrior," in Levidow and Robbins, *Cyborg Worlds,* pp. 43–72.

16. Carolyn Geduld, Wolfe's biographer, has an excellent discussion of Bergler's influence on Wolfe in *Bernard Wolfe* (New York: Twayne, 1972), pp. 54–62.

17. Geduld describes the author as a "very small man with a thick, sprouting mustache, a fat cigar, and a voice that grabs attention" (ibid., p. 15).

18. Edmund Bergler discusses narcissism in a book whose title gives it top (or bottom) billing, *The Basic Neurosis: Oral Regression and Psychic Masochism* (New York: Grune and Stratton, 1949).

19. For a discussion of Lacan's rewriting of Freud in this respect, see Kaja Silverman, *The Acoustic Mirror: The Female Voice in Psychoanalysis and Cinema* (Bloomington: Indiana University Press, 1988).

20. David Wills, *Prosthesis* (Stanford: Stanford University Press, 1995), pp. 18, 20.

21. Bernard Wolfe (ghostwriter, Raymond Rosenthal), *Plastics: What Everyone Should Know* (New York: Bobbs-Merrill Company, 1945).

22. Wolfe, *Limbo,* p. 294.

23. Julia Kristeva, "The Novel as Polylogue," in *Desire in Language: A Semiotic Approach to Literature and Art,* edited by Leon S. Roudiez, translated by Thomas Gora, Alice Jardine, and Leon S. Roudiez (New York: Columbia University Press, 1980), pp. 159–209.

Chapter Six

1. J. Y. Lettvin, H. R. Maturana, W. S. McCulloch, and W. H. Pitts, "What the Frog's Eye Tells the Frog's Brain," *Proceedings of the Institute for Radio Engineers* 47, no. 11 (November 1959): 1940–51.

2. For anecdotal evidence about the importance of reflexivity to these Macy participants, see Stewart Brand's interview discussed in chapter 3, "'For God's Sake, Margaret': Conversation with Gregory Bateson and Margaret Mead," *Co-Evolution Quarterly* (summer 1976), 32–44.

3. Humberto R. Maturana and Francisco J. Varela, *Autopoiesis and Cognition: The Realization of the Living* (Dordrecht: D. Reidel, 1980), p. xvi (hereafter cited in the text as *AC*).

4. Heinz von Foerster, "Vita," in Warren McCulloch Papers, American Philosophical Library, Philadelphia, B/M139, Box 2.

5. Heinz von Foerster, letter dated May 23, 1949, McCulloch Papers, B/M139, Box 2. According to Steve Heims, who interviewed von Foerster in 1982, McCulloch had first learned of von Foerster's work when reading one of von Foerster's papers (published in German) in which he proposed that memory is stored in a macromolecule (by analogy to the DNA macromolecule's storage of genetic information). McCulloch immediately invited von Foerster to the next Macy Conference, where his presentation of the idea met a cool reception, in part because by then the Macy group had already conceptualized memory (through the McCulloch-Pitts neuron) as analogous to binary computer memory storage. See Steve J. Heims, *The Cybernetics Group* (Cambridge: MIT Press, 1991), pp. 72–74.

6. Heinz von Foerster, *Observing Systems,* 2d ed. (Salinas, Calif.: Intersystems Publications, 1984), p. 7.

7. A similar scenario is imagined in Jorge Luis Borges's fiction "The Circular Ruins," *Ficciones,* edited by Anthony Kerrigan (New York: Grove Press, 1962).

8. At the Bateson conference in 1968, Gordon Pask illustrated his talk, after referring to the "Frog's Eye" paper, with drawings of men in bowler hats; see Mary Catherine Bateson, *Our Own Metaphor: A Personal Account of a Conference on the Effects of Conscious Purpose on Human Adaptation* (1972; Washington, D.C.: Smithsonian Institution Press, 1991), pp. 209–15, especially p. 214.

9. Norbert Wiener, *God and Golem, Inc: A Comment on Certain Points Where Cybernetics Impinges on Religion* (Cambridge: MIT Press, 1964), p. 88.

10. For an account of the conference, see Maturana and Varela, *Autopoiesis and Cognition,* p. xvi.

11. Heinz von Foerster, "Molecular Ethology: An Immodest Proposal for Semantic Clarification," *Observing Systems,* p. 171.

12. Lettvin, Maturana, McCulloch, and Pitts, "Frog's Eye," p. 1950.

13. Humberto R. Maturana, G. Uribe, and S. Frenk, "A Biological Theory of Relativistic Color Coding in the Primate Retina," *Archivos de Biologia y Medicina Experimentales,* Suplemento No. 1 (Santiago, Chile: N.p., 1968).

14. An excellent survey of autopoietic theory, from Maturana and Varela to its proponents in such diverse fields as Luhmann's social systems theory and family therapy, can be found in John Mingers, *Self-Producing Systems: Implications and Applications of Autopoiesis* (New York: Plenum Press, 1995). A useful bibliography and survey can

also be found at Randall Whitaker's Web site, <http: //www.acm.org/sigois/auto/Main. html>. Of course, the main sources are Maturana and Varela, *Autopoiesis and Cognition,* and Maturana and Varela's later works cited in this chapter.

15. For a sample, see Brian R. Gaines's critique of autopoiesis: "Autopoiesis: Some Questions" in *Autopoiesis: A Theory of Living Organization,* edited by Milan Zeleny, North Holland Series in General Systems Research, vol. 3 (New York: North Holland, 1981), pp. 145–54. Complaining about verbal obscurities in Maturana's formulations, Gaines notes the difference between "the arts of persuasion and the pursuit of science. On the other hand, we have also to accept that the ultimate, undefined terms of any theory are accepted as 'acts of faith,' not that they be true (for the word is meaningless in this context), but at least that they be potentially useful. If we go looking at the world in terms of unities, recursive self-production, and autopoietic organization, then we shall find a certain kind of world: It is not yet clear why we should want it or what we will do with it when we have it" (pp. 150–51).

16. Francisco J. Varela, *Principles of Biological Autonomy,* North Holland Series in General Systems Research, vol. 2 (New York: North Holland, 1979).

17. Humberto R. Maturana, "Biology of Language: The Epistemology of Reality," in *Psychology and Biology of Language and Thought: Essays in Honor of Eric Lenneberg,* edited by George A. Miller and Elizabeth Lenneberg (New York: Academic Press, 1978), p. 59.

18. Ibid., p. 63.

19. The foremost theorist extending autopoietic theory into social systems is of course Niklas Luhmann. His major works include *The Differentiation of Society* (New York: Columbia University Press, 1982) and *Social Systems,* translated by John Bednarz Jr., with Dirk Baeker (Stanford: Stanford University Press, 1995). Also of interest with regard to his appropriation of autopoietic theory are his articles "Operational Closure and Structural Coupling: The Differentiation of the Legal System," *Cardozo Law Review* 13 (1992): 1419–41, and "The Cognitive Program of Constructivism and a Reality That Remains Unknown" in *Self-Organization: Portrait of a Scientific Revolution,* edited by Wolfgang Krohn et al. (Dordrecht: Kluwer Academic Publishers, 1990).

20. Simon Baron-Cohen, *Mindblindness: An Essay on Autism and Theory of Mind* (Cambridge: MIT Press, 1997).

21. Francisco J. Varela, "Describing the Logic of the Living: The Adequacy and Limitations of the Idea of Autopoiesis," in Zeleny, *Autopoiesis: A Theory,* p. 36.

22. See, for example, Richard C. Lewontin, *Biology as Ideology: The Doctrine of DNA* (New York: Harper and Row, 1993); Evelyn Fox Keller, *Refiguring Life: Metaphors of Twentieth-Century Biology* (New York: Columbia University Press, 1995); Richard Doyle, *On Beyond Living: Rhetorical Transformations in the Life Sciences* (Stanford: Stanford University Press, 1997); and Lily E. Kay, "Cybernetics, Information, Life: The Emergence of Scriptural Representations of Heredity," *Configurations* 5 (winter 1997): 23–92.

23. Humberto R. Maturana, "Autopoiesis: Reproduction, Heredity, and Evolution," in *Autopoiesis, Dissipative Structures, and Spontaneous Social Orders,* edited by Milan Zeleny, AAAS Selected Symposium (Boulder: Westview Press, 1980), p. 62.

24. Humberto R. Maturana and Francisco Varela, *The Tree of Knowledge: The Biological Roots of Human Understanding* (Boston: New Science Library, 1987).

25. In Maturana and Varela, *Autopoiesis and Cognition,* Maturana notes that one cannot "account or deduce all actual biological phenomena from the notion of autopoiesis without resorting to historical contingencies" (p. xxiii). He makes this admission, however, only to argue that this does not represent a shortcoming of the theory.

26. Humberto R. Maturana, "The Origin of the Theory of Autopoietic Systems," in *Autopoiesis; Eine theorie im Brennpunkt der Kritik,* edited by Hans R. Fisher (Heidelberg: Verlag, 1991), p. 123.

27. Varela, *Principles,* p. xvii; Varela, "Describing the Logic of the Living," p. 36 (emphasis added).

28. Francisco J. Varela, Evan Thompson, and Eleanor Rosch, *The Embodied Mind: Cognitive Science and Human Experience* (Cambridge: MIT Press, 1991).

29. Francisco J. Varela, "Making It Concrete: Before, During, and After Breakdowns," in *Revisioning Philosophy,* edited by James Ogilvy (Albany: State University of New York Press, 1992), p. 103.

30. See the chapters in Zeleny's *Autopoiesis, Dissipative Structures, and Spontaneous Social Orders* for a good overview of this work.

31. Maturana and Varela, *The Tree of Knowledge,* p. 242.

Chapter Seven

1. Of Philip K. Dick's novels from this period, I will discuss or mention *We Can Build You* (originally titled *The First in Your Family*), written in 1962 and published in 1969 as *A. Lincoln, Simulacrum* (the edition cited in this chapter is London: Grafton Books, 1986); *Martian Time-Slip* (originally titled *Goodmember Arnie Kott of Mars*), written in 1962 and published in 1964 (the cited edition is New York: Vintage Books, 1995); *Dr. Bloodmoney; or, How We Got Along after the Bomb,* written in 1963 and published in 1965 (the cited edition is New York: Carroll and Graf, 1988); *The Simulacra* (originally titled *The First Lady of Earth*), written in 1963 and published in 1964 (the cited edition is London: Methuen Paperbacks, n.d.); *The Three Stigmata of Palmer Eldritch,* written in 1964 and published in 1965 (the cited edition is New York: Bantam, 1964): *Do Androids Dream of Electric Sheep?* (originally titled *The Electric Toad: Do Androids Dream?*), written in 1966 and published in 1968 (the cited edition, published under the title *Blade Runner,* is New York: Ballantine Books, 1982); and *Ubik* (originally titled *Death of an Anti-watcher*), written in 1966 and published in 1969 (the cited edition is London: Grafton Books, 1973). Information about dates, original titles, and first publication is taken from Lawrence Sutin, *Divine Invasions: A Life of Philip K. Dick* (Secaucus, N.J.: Carol Publishing, 1991).

2. Philip K. Dick, "How to Build a Universe That Doesn't Fall Apart Two Days Later," in *The Shifting Realities of Philip K. Dick: Selected Literary and Philosophical Writings,* edited by Lawrence Sutin (New York: Pantheon Books, 1995), pp. 263–64.

3. Istvan Csicsery-Ronay Jr. points out the dearth of feminist criticism on Dick in the introduction to *On Philip K. Dick: 40 Articles from Science-Fiction Studies,* edited by R. D. Mullen, Istvan Csicsery-Ronay Jr., Arthur B. Evans, and Veronica Hollinger (Terre Haute: SF-TH, 1992), pp. v–xvii.

4. As Dick grew older and the women grew younger, "girl" becomes less a mark of sexist construction (hardly unusual in a man of Dick's age and upbringing) and more an indication of actual chronological age.

5. Philip K. Dick, "The Evolution of a Vital Love," in *The Dark-Haired Girl* (Willimantic, Conn.: Mark V. Ziesing, 1988), p. 171.

6. Marxist criticism in general, of course, favors systematic and economic explanations over psychological ones. Articles that argue for the importance of economic readings include the following, all published in Mullen, Csicsery-Ronay Jr., Evans, and Hollinger, *On Philip K. Dick:* Peter Fitting, "Reality as Ideological Construct: A Reading of Five Novels by Philip K. Dick," pp. 92–110; Peter Fitting, "*Ubik:* The Deconstruction of Bourgeois SF," pp. 41–48; and Scott Durham, "From the Death of the Subject to a Theology of Late Capitalism," pp. 188–98.

7. Sutin, *Divine Invasions,* pp. 11–19, 29–34.

8. Quoted in ibid., p. 12.

9. Documented in ibid.: see for example p. 26.

10. The lesbian Alys, a sexually vibrant woman with dark hair, is also involved in an incestuous relationship with her twin brother. As if reflecting Jane's death, she is made the victim of entropic forces, experiencing preternatually fast decay while her brother is saved.

11. Dick was an avid reader of Jung's works and frequently referred to Jungian archetypes, for example the Magna Mater, in his fiction. It is likely that he consciously thought of Jane in terms of his anima.

12. Patricia Warrick, "The Labyrinthian Process of the Artificial: Dick's Androids and Mechanical Constructs," *Extrapolation* 20 (1979): 133–53, also makes this point in a different context when she argues: "For Dick, the outcome of war—be it military or economic—is not victory or defeat, but a transformation into the opposite. We become the goal we pursue, the enemy we fight" (p. 139).

13. Carl Freedman, "Towards a Theory of Paranoia: The Science Fiction of Philip K. Dick," in Mullen, Csicsery-Ronay Jr., Evans, and Hollinger, *On Philip K. Dick,* pp. 111–18.

14. Dick, *The Simulacra,* p. 201.

15. Scott Durham, "From the Death of the Subject."

16. Dick, *The Three Stigmata,* p. 160.

17. Patricia Warrick, "Labyrinthian Process."

18. Dick, *We Can Build You,* p. 34.

19. Rachael is described as slim and lithe with "heavy masses of dark hair." Because of her "diminutive breasts," her body "assumed a lank, almost childlike stance," though Deckard is in no doubt that she is a sexually mature woman. The "total impression," although "good," is "definitely that of a girl, not a woman." See Dick, *Do Androids Dream,* p. 164.

20. I am indebted to Jill Galvin for pointing out this pun in "Entering the Posthuman Collective in Philip K. Dick's *Do Androids Dream of Electric Sheep?*" (forthcoming, *Science-Fiction Studies*).

21. Philip K. Dick, "Schizophrenia and *The Book of Changes,*" in Sutin, *Shifting Realities,* pp. 175–82, especially p. 176.

22. See R. D. Laing, *The Divided Self* (New York: Pantheon Books, 1969).

23. Quoted in Warrick, "Labyrinthian Process," p. 141.

24. Philip K. Dick, "The Android and the Human," in Sutin, *Shifting Realities,* p. 208.

25. Galvin, "Entering the Posthuman Collective."

26. Dick, *Do Androids Dream,* p. 185.

27. Dick, *The Three Stigmata,* p. 101.

28. Fredric Jameson, "After Armageddon: Character Systems in *Dr. Bloodmoney,*" in Mullen, Csicsery-Ronay Jr., Evans, and Hollinger, *On Philip K. Dick,* pp. 26–36.

29. Ibid., p. 27.

30. Dick, *Dr. Bloodmoney,* p. 66.

31. Dick, *Ubik,* p. 107.

32. Humberto R. Maturana and Francisco Varela, *The Tree of Knowledge: The Biological Roots of Human Understanding* (Boston: New Science Library, 1987), p. 242.

33. Sutin, *Divine Invasions,* pp. 222–34.

34. Humberto R. Maturana, "Biology of Language: The Epistemology of Reality," in *Psychology and Biology of Language and Thought: Essays in Honor of Eric Lenneberg,* edited by George A. Miller and Elizabeth Lenneberg (New York:; Academic Press, 1978), p. 46, remarks: "In the absence of an adequate environmental perturbation, the observer claims that the observed conduct is a result of an illusion or a hallucination. Yet, for the operation of the nervous system (and organism), there cannot be a distinction between illusions, hallucinations, or perceptions, because a closed neuronal network cannot discriminate between internally and externally triggered changes in relative neuronal activity. This distinction pertains exclusively to the domain of description in which the observer defines an inside and an outside for the nervous system and the organism."

35. Philip K. Dick, *In Pursuit of Valis: Selections from The Exegesis,* edited by Lawrence Sutin (Novato, Calif.: Underwood-Miller, 1991), p. 45.

36. Dick, *Do Androids Dream,* p. 210.

Chapter Eight

1. Jean Baudrillard, *The Ecstasy of Communication,* translated by Bernard Schutze and Caroline Schutze (New York: Semiotext(e), 1988), p. 18.

2. Arthur Kroker and Marilouise Kroker, "Panic Sex in America," *Body Invaders: Panic Sex in America* (New York: St. Martin's Press, 1987), pp. 20–21.

3. O. B. Hardison Jr., *Disappearing through the Skylight: Culture and Technology in the Twentieth Century* (New York: Viking, 1989), p. 335.

4. By "informatics," I mean the material, technological, economic, and social structures that make the information age possible. Informatics includes the following: the late capitalist mode of flexible accumulation; the hardware and software that have merged telecommunications with computer technology; the patterns of living that emerge from and depend on access to large data banks and instantaneous transmission of messages; and the physical habits—of posture, eye focus, hand motions, and neural connections—that are reconfiguring the human body in conjunction with information technologies. For readers who know the term "informatics" mainly from Donna Haraway's work, where it frequently occurs as "the informatics of domination," I should clarify how this term is currently being used in technical and humanistic fields here and abroad. To computer people, "informatics" means simply the study and design of information technologies. In many European countries, especially Norway, Denmark, and Germany, departments of humanistic informatics are being formed to study the cultural impact and significance of information technologies. Researchers in these

departments regard "informatics" as a descriptive term no more value-laden than physics, biology, or literature. A historian in such a department may study the history of computers; a linguist, correlations between computer and natural languages; a literary theorist, new forms of electronic textuality.

5. Michel Foucault, *Discipline and Punish: The Birth of the Prison*, translated by Alan Sheridan (New York: Vintage, 1979), p. 205.

6. In his later work, especially *The History of Sexuality*, translated by Robert Hurley (New York: Vintage Books, 1980), Foucault is much more attentive to embodied practices and the importance of embodiment in general.

7. Elaine Scarry, *The Body in Pain: The Making and Unmaking of the World* (New York: Oxford University Press, 1985).

8. See Mark Poster, *The Mode of Information: Poststructuralism and Social Context* (Chicago: University of Chicago Press, 1990), pp. 69–98, for a critique of Foucault's universalism. Nancy Fraser, *Unruly Practices: Power, Discourse, and Gender in Contemporary Social Theory* (Minneapolis: University of Minnesota Press, 1989), pp. 55–66, interrogates the Foucaultian body.

9. Elizabeth Grosz, *Volatile Bodies: Toward a Corporeal Feminism* (Bloomington: Indiana University Press, 1994).

10. For a brief description of PET, see Richard Mark Friedhoff and William Benzon, *The Second Computer Revolution: Visualization* (New York: Harry Abrams, 1989), pp. 64–66, 81, 185.

11. Jorge Luis Borges, "Funes the Memorious," *Labyrinths: Selected Stories and Other Writings* (New York: New Directions, 1962), pp. 59–66.

12. Michel de Certeau, *The Practice of Everyday Life*, translated by Steven F. Rendall (Berkeley: University of California Press, 1985).

13. Maurice Merleau-Ponty, *Phenomenology of Perception*, translated by Colin Smith (New York: Humanities Press, 1962), pp. 98–115, 136–47.

14. Paul Connerton, *How Societies Remember* (Cambridge, England: Cambridge University Press, 1989).

15. This is, of course, Judith Butler's point in *Gender Trouble: Feminism and the Subversion of Identity* (New York: Routledge, 1990). From this book, some readers gained the impression that bodies do not matter. Her later book, *Bodies That Matter: On the Discursive Limits of "Sex"* (New York: Routledge, 1993), corrects this impression. With this correction I am in wholehearted agreement.

16. Francisco J. Varela, Evan Thompson, and Eleanor Rosch, *The Embodied Mind: Cognitive Science and Human Experience* (Cambridge: MIT Press, 1991).

17. Hubert L. Dreyfus, *What Computers Can't Do: The Limits of Artificial Intelligence*, rev. ed. (New York: Harper and Row, 1979), p. 255.

18. Pierre Bourdieu, *Outline of a Theory of Practice*, translated by Richard Nice (Cambridge, England: Cambridge University Press, 1977), p. 78.

19. Maurice Merleau-Ponty, "Eye and Mind," in *The Primacy of Perception*, edited by James M. Edie (Chicago: Northwestern University Press, 1964), p. 162.

20. Connerton, *How Societies Remember*, p. 44.

21. Bourdieu, *Outline of a Theory*, p. 94.

22. Connerton, *How Societies Remember*, p. 102.

23. Mark Johnson, *The Body in the Mind: The Bodily Basis of Meaning, Imagination, and Reason* (Chicago: University of Chicago Press, 1987), pp. 18–35.

24. Garrett Stewart, *Reading Voices: Literature and the Phonotext* (Berkeley: University of California Press, 1990).

25. Eric Havelock argues that modern subjectivity, with its sense of stable ego and enduring identity, was a historical invention that correlated with the transition from orality to writing: see *Preface to Plato* (Cambridge: Harvard University Press, 1963).

26. The literature on these technologies is extensive. For a useful brief discussion, see Douglas Kahn and Gregory Whitehead, eds., *Wireless Imagination: Sound, Radio, and the Avant-Garde* (Cambridge: MIT Press, 1992), especially Douglas Kahn's chapter "Introduction: Histories of Sound Once Removed," pp. 1–30.

27. See Friedrich A. Kittler, *Discourse Networks, 1800–1900,* translated by Michael Metteer (Stanford: Stanford University Press, 1990). Also relevant is Friedrich A. Kittler, "Gramophone, Film, Typewriter," translated by Dorothea von Mücke, *October* 41 (1987): 101–18, in which Kittler wrote: "The technical differentiation of optics, acoustics, and writing around 1880, as it exploded Gutenberg's storage monopoly, made the fabrication of so-called man possible. His essence runs through apparatuses" (p. 115). Nothing could be more applicable to Burroughs's view of tape-recording.

28. The pioneering papers in the development of magnetic tape-recording are collected in Marvin Camras, *Magnetic Tape Recording* (New York: Van Nostrand Reinhold Company, 1985). His brief introductions to the sections provide a valuable (if sketchy) history, which I have drawn on here.

29. For the patent description of the Telegraphone, see V. Poulsen, "Method of Recording and Reproducing Sounds or Signals," in Camras, *Magnetic Tape Recording,* pp. 11–17. The model exhibited in Paris differed somewhat from the patent description.

30. A description of the film and ring head is given in H. Lubeck, "Magnetic Sound Recording with Films and Ring Heads," in Camras, *Magnetic Tape Recording,* pp. 79–111.

31. A useful review of this work is J. C. Mallinson, "Tutorial Review of Magnetic Recording," in Camras, *Magnetic Tape Recording,* pp. 229–43.

32. Roy Walker, "Love, Chess, and Death," in Samuel Beckett, *Krapp's Last Tape: A Theater Notebook,* edited by James Knowlson (London: Brutus Books, 1980), p. 49.

33. William S. Burroughs, *The Ticket That Exploded* (New York: Grove Press, 1967) (hereafter cited in the text as *TTE*).

34. This idea of an interior monologue shoring up a false sense of self is also important in Varela, Thompson, and Rosch, *The Embodied Mind,* as we saw in chapter 6. For an extensive discussion of how tape-recorders can be used to disrupt the word virus, see William Burroughs, *Electronic Revolution* (Bonn: Expanded Media Editions, 1970), pp. 1–62.

35. Cary Nelson has an excellent discussion of the body in relation to space in Burroughs's work, including *The Ticket That Exploded* and its companion novels: "The End of the Body: Radical Space in Burroughs", in *William S. Burroughs at the Front: Critical Reception, 1959–1989,* edited by Jennie Kerl and Robin Lydenberg (Carbondale: Southern Illinois University Press, 1991), pp. 119–32.

36. John Cunningham Lilly gives an account of these experiments in his autobiographical account *The Center of the Cyclone: An Autobiography of Inner Space* (New York: Julian Press, 1972). In a characteristically literalizing passage, Burroughs suggested that isolation tanks could literally dissolve body boundaries: "So after fifteen

minutes in the tank these Marines scream they are losing outlines and have to be removed—I say put two marines in the tank and see who comes out—Science—Pure science—So put a marine and his girl friend in the tank and see who or what emerges—" (*TTE*, p. 83).

37. The cut-up method is described in many places by Burroughs and others; see, for example, Daniel Odier, *The Job: Interviews with William S. Burroughs* (New York: Grove Press, 1969), p. 14, and William S. Burroughs, "The Cut-Up Method of Brion Gysin," *Re/Search #4/5* (San Francisco: Re/Search Publications, 1982), pp. 35–38. Robin Lydenberg lucidly discusses the political and theoretical implications of the practice in *Word Cultures: Radical Theory and Practice in William S. Burroughs' Fiction* (Urbana: University of Illinois Press, 1987). Laszlo K. Gefin contextualizes the practice in the avant-garde techniques of collage in "Collage Theory, Reception, and the Cut-ups of William Burroughs," *Perspectives on Contemporary Literature: Literature and the Other Arts* 13 (1987): 91–100. Anne Friedberg, "Cut-Ups: A Synema of the Text," in Skerl and Lydenberg, *William S. Burroughs*, pp. 169–73, traces the cut-up method through the dadaists.

38. Robin Lydenberg has a good discussion of Burroughs's experiments with tape-recordings, including this album, in "Sound Identity Fading Out: William Burroughs' Tape Experiments," in Kahn and Whitehead, *Wireless Imagination*, pp. 409–33.

39. Brenda Laurel and Sandy Stone, private communication.

Chapter Nine

1. Francisco J. Varela, Evan Thompson. and Eleanor Rosch, *The Embodied Mind: Cognitive Science and Human Experience* (Cambridge: MIT Press, 1991).

2. Francisco Varela and Paul Bourgine, eds., *Toward a Practice of Autonomous Systems: Proceedings of the First European Conference on Artificial Life* (Cambridge: MIT Press, 1992), p. xi.

3. Humberto R. Maturana and Francisco J. Varela, *Autopoiesis and Cognition: The Realization of the Living* (Dordrecht: D. Reidel, 1980).

4. Thomas S. Ray, "A Proposal to Create Two Biodiversity Reserves: One Digital and One Organic," presentation at Artificial Life IV, Cambridge, Massachusetts, July 1994.

5. Thomas S. Ray, "An Evolutionary Approach to Synthetic Biology: Zen and the Art of Creating Life," *Artificial Life* 1, no. 1/2 (fall 1993/winter 1994): 180 (emphasis added).

6. Luc Steels offers useful definitions of emergence in "The Artificial Life Roots of Artificial Intelligence," *Artificial Life* 1, no. 1/2 (fall 1993/winter 1994): 75–110. He distinguishes between first-order emergence, defined as a property not explicitly programmed in, and second-order emergence, defined as an emergent behavior that adds additional functionality to the system. In general, AL researchers try to create second-order emergence, for then the system can use its own emergent properties to create an upward spiral of continuing evolution and emergent behaviors. James P. Crutchfield makes a similar point in "Is Anything Ever New? Considering Emergence," in *Integrative Themes*, edited by G. Cowan, D. Pines, and D. Melzner, Santa Fe Institute Studies in the Sciences of Complexity, XIX (Redwood City, Calif.: Addison-Wesley, 1994), pp. 1–15. For a criticism of emergence, see Peter Cariani, "Adaptivity and Emergence in Organisms and Devices," *World Futures* 32 (1991): 111–32.

7. The Tierra program is described in Thomas S. Ray, "An Approach to the Synthesis of Life," in *Artificial Life II,* edited by Christopher G. Langton, Charles Taylor, J. Doyne Farmer, and Steen Rasmussen, Santa Fe Institute Studies in the Sciences of Complexity, X (Redwood City, Calif.: Addison-Wesley, 1992), pp. 371–408. "An Evolutionary Approach to Synthetic Biology" (working paper, ATR Human Information Processing Research Laboratories, Kyoto, Japan) explains and expands on the philosophy underlying Tierra. Further information about Tierra can be found in Christopher G. Langton, ed., "Population Dynamics of Digital Organisms," *Artificial Life II Video Proceedings* (Redwood City, Calif.: Addison-Wesley, 1991). A popular account can be found in John Travis, "Electronic Ecosystem," *Science News* 140, no. 6 (August 10, 1991): 88–90.

8. "Simple Rules . . . Complex Behavior," produced and directed by Linda Feferman for the Santa Fe Institute, 1992.

9. Richard Dawkins, *The Selfish Gene* (New York: Oxford University Press, 1976).

10. William Gibson, *Neuromancer* (New York: Ace Books, 1984).

11. Ray, "An Evolutionary Approach," notes, "The 'body' of a digital organism is the information pattern in memory that constitutes its machine language program" (p. 184).

12. Quoted in Stefan Helmreich, "Anthropology Inside and Outside the Looking-Glass Worlds of Artificial Life" (unpublished manuscript, 1994), p. 11. An earlier version of this work was published as a working paper at the Santa Fe Institute, under the title "Travels through Tierra," Excursions in 'Echo': Anthropological Refractions on the Looking-Glass Worlds of Artificial Life," Santa Fe Working Paper No. 94-04-024. In this version, Helmreich included some remarks that the administrators of the Santa Fe Institute evidently found offensive, including a comparison between the scientists' belief in the "aliveness" of artificial life and the seemingly bizarre (to Westerners) beliefs held by marginal cultural groups such as the Trobriand Islanders. Objecting that Helmreich's work was not scientific and misrepresented the science conducted at the Santa Fe Institute, the administrators had the working paper removed from the shelves and deleted from the list of available publications.

13. Christopher Langton, "Artificial Life," in *Artificial Life,* edited by Christopher Langton (Redwood City, Calif.: Addison-Wesley, 1989), p. 1.

14. Richard Doyle has written on the simplification of body to information in the human genome project in *On Beyond Living: Rhetorical Transformations in the Life Sciences* (Stanford: Stanford University Press, 1997).

15. Actually, both inference and deduction are at work in most AL research, as they usually are in scientific projects. AL researchers study the complex-to-simple route for clues on how to construct programs that will be able to move from simple to complex.

16. Langton, "Artificial Life," p. 1.

17. Extensive interviews with AL researchers have also been conducted by Steven Levy, as recounted in his useful popularization *Artificial Life: The Quest for a New Creation* (New York: Pantheon Books, 1992). A more technical account covering much the same material that Levy addressed can be found in Claus Emmeche, *The Garden in the Machine: The Emerging Science of Artificial Life* (Princeton: Princeton University Press, 1994). In the opening pages, Emmeche says that his book is intended for the general reader, but he soon leaves the simplistic style that characterizes the first sections and moves into more interesting and demanding material. Especially noteworthy is his discussion of the deep problems raised about the nature of computation.

18. Helmreich, "Anthropology Inside and Outside," p. 5.

19. Edward Fredkin is something of a cult figure for researchers interested in computational philosophies. After achieving financial independence through the company he founded, he bought and occasionally lives on his own island in the Caribbean. Although he himself has published very little, several articles and part of a book have been written about him. He is a faculty member at MIT and has a research group there working out a universal theory of cellular automata, intending to show how cellular automata can account for all the laws of physics. For an account of his work, see Robert Wright, *Three Scientists and Their Gods: Looking for Meaning in an Age of Information* (New York: Times Books, 1988). One of Fredkin's rare publications is "Digital Mechanics: An Information Process Based on Reversible Universal Cellular Automata," *Physica D* 45 (1990): 254–70. See also Julius Brown, "Is the Universe a Computer?" *New Scientist* 14 (July 1990): 37–39. Levy, *Artificial Life,* and Emmeche, *The Garden in the Machine,* both mention Fredkin.

20. G. Kampis and V. Csanyi, "Life, Self-Reproduction, and Information: Beyond the Machine Metaphor," *Journal of Theoretical Biology* 148 (1991): 17–32, gives an important analysis of the idea of self-reproduction in a machine context. The authors point out that one's account of what happens in self-reproduction changes depending on how the framing context is constructed. For *all* machine (re)production, there is a context in which outside agency is needed to complete reproduction, in contrast to the reproduction of [asexual] living organisms. By placing the last computer out of sight, as it were, Fredkin has erased this context from view, although he still has to posit it to explain how things come into existence.

21. For a research program that takes this objection into account, see David Jefferson et al., "Evolution as a Theme in Artificial Life: The Genesys/Tracker System" (Computer Science Department Technical Report CSD-900047, University of California–Los Angeles, December 1990). In the Tracker simulation, designed to generate the social behavior and food-gathering strategies characteristic of ants, Jefferson and his colleagues used two very different kinds of algorithms to demonstrate that the behaviors generated by the simulations were not artifacts. They reasoned that because the underlying structures of the simulations were different, similarities of behavior could not be attributed to the algorithms, only to the dynamics conceptualized through those algorithms.

22. Christopher Langton, "Editor's Introduction," *Artificial Life* 1, no. 1/2 (fall 1993/winter 1994): v–viii, especially v–vi.

23. Walter Fontana, Gunter Wagner, and Leo W. Buss, "Beyond Digital Naturalism," *Artificial Life* 1, no. 1/2 (fall 1993/winter 1994): 224.

24. Hans Moravec, *Mind Children: The Future of Robot and Human Intelligence* (Cambridge: Harvard University Press, 1988).

25. A. G. Cairns-Smith, *Genetic Takeover and the Mineral Origins of Life* (New York: Cambridge University Press, 1987).

26. See Pattie Maes, "Modeling Adaptive Autonomous Agents," *Artificial Life* 1, no. 1/2 (fall 1993/winter 1994): 135–62; Rodney Brooks, "New Approaches to Robotics," *Science* 253 (September 13, 1991): 1227–32; Mark Tilden, "Living Machines: Unsupervised Work in Unstructured Environments" (Los Alamos National Laboratory, CB/MT-v1941114, n.d.).

27. Rodney A. Brooks, "Intelligence without Representation," *Artificial Intelli-*

gence 47 (1991): 139–59. See also Luc Steels and Rodney Brooks, eds., *The Artificial Life Route to Artificial Intelligence: Building Embodied, Situated Agents* (Hillsdale, N.J.: L. Erlbaum Associates, 1995).

28. Genghis is described in, among other places, Rodney A. Brooks and Anita M. Flynn, "Fast, Cheap, and Out of Control: A Robot Invasion of the Solar System," *Journal of the British Interplanetary Society* 42 (1989): 478–85.

29. In January 1995, Mark Tilden lectured and demonstrated his mobile robots at the Center for the Study and Evolution of Life, University of California, where I had an opportunity to talk with him.

30. Brooks, "Intelligence without Representation."

31. In my conversation with Moravec at the University of Illinois "Cyberfest" in March 1997, he defended the top-down approach by comparing a robot-piloted car he had designed and Rodney Brooks's robots. Whereas Moravec's robot car has successfully driven several hundred miles, Brooks's robots have scarcely been out of the laboratory. The point is well taken, and future research may well use a combination of both approaches. Moravec declared himself a pragmatist, willing to use whatever works.

32. Michael G. Dyer, "Toward Synthesizing Artificial Neural Networks That Exhibit Cooperative Intelligent Behavior: Some Open Issues in Artificial Life," *Artificial Life* 1, no. 1/2 (fall 1993/winter 1994): 111–35, especially p. 112.

33. Edwin Hutchins demystifies this proposition in *Cognition in the Wild* (Cambridge: Mit Press, 1996), when he elegantly demonstrates that humans *normally* act in environments where cognition is distributed among a variety of human and nonhuman actors, from graph paper and pencil to the sophisticated naval guidance systems he discusses. His book, by grounding its arguments in existing naval navigation techniques of the past and present, shows that distributed cognition has been around for about as long as humans have.

34. Levy, *Artificial Life*, gives an account of von Neumann's self-reproducing machine. His account is based on the rather sketchy information given by Arthur W. Burks (who edited and compiled von Neumann's incomplete manuscript after the latter's death) of what Burks calls the kinematic model of self-reproduction. Burks's version can be found in John von Neumann, *Theory of Self-Reproducing Automata* (Urbana: University of Illinois Press, 1966), pp. 74–90.

35. Cellular automata are described in detail in von Neumann, *Theory of Self-Reproducing Automata*, pp. 91–156. See also Stephen Wolfram, one of the foremost researchers on cellular automata, in "Universality and Complexity in Cellular Automata," *Physica D* 10 (1984): 1–35, and "Computer Software in Science and Mathematics," *Scientific American* 251 (August 1984): 188–203. In these articles, Wolfram concentrates on one-dimensional cellular automata, where each generation appears as a line and where patterns appear as the lines proliferate down the screen (or the graph paper).

36. Chris G. Langton, "Computation at the Edge of Chaos: Phase Transition and Emergent Computation," *Physica D* 42 (1990): 12–37.

37. Warren McCulloch's papers include a letter of reference that McCulloch wrote for Kauffmann: Warren McCulloch Papers, American Philosophical Library, Philadelphia, B/M139, Box 2. In several lectures and interviews that McCulloch gave in the few years before his death, he mentioned Kauffman as an important collaborator.

38. Stuart A. Kauffman, *The Origins of Order: Self-Organization and Selection in Evolution* (New York: Oxford University Press, 1993). See also his popularized version,

At Home in the Universe: The Search for the Laws of Self-Organization and Complexity (New York: Oxford University Press, 1995).

39. Jerome H. Barkow, Leda Cosmides, and John Tooby, *The Adapted Mind: Evolutionary Psychology and the Generation of Culture* (New York: Oxford University Press, 1992), especially the chapter by Tooby and Cosmides: "The Psychological Foundations of Culture," pp. 19–136. Tooby and Cosmides have also been instrumental in forming the Human Behavior and Evolution Society (HBES), which holds annual conferences centered on the ideas of evolutionary psychology. In some ways the HBES is a successor to sociobiology, although with a more flexible framework of interpretation.

40. Steven Pinker makes this point in *The Language Instinct* (New York: W. Morrow, 1994). This model provides an interesting corrective to Maturana's largely passive model of "languaging" between "observers."

41. Steels, "The Artificial Life Roots."

42. Marvin Minsky, *The Society of Mind* (New York: Simon and Schuster, 1985), especially pp. 17–24.

43. Marvin Minsky, "Why Computer Science Is the Most Important Thing That Has Happened to the Humanities in 5,000 Years" (public lecture, Nara, Japan, May 15, 1996). I am grateful to Nicholas Gessler for providing me with his transcript of the lecture.

44. Marvin Minsky, "How Computer Science Will Change Our Lives" (plenary lecture, Fifth Conference on Artificial Life, Nara, Japan, May 17, 1996).

45. Antonio R. Damasio, *Descartes' Error: Emotion, Reason, and the Human Brain* (New York: G. P. Putnam, 1994), p. 226.

Chapter Ten

1. Ihab Hassan, "Prometheus as Performer: Towards a Posthumanist Culture?" in *Performance in Postmodern Culture,* edited by Michael Benamou and Charles Caramella (Madison, WI: Coda Press, 1977), p. 212. See also Judith Halberstam and Ira Livingston, "Introduction: Posthuman Bodies" in *Posthuman Bodies,* edited by Judith Halberstam and Ira Livingston (Bloomington: Indiana University Press, 1995): "Posthuman bodies are the causes and effects of postmodern relations of pleasure, virtuality and reality, sex and its consequences" (p. 3).

2. For a discussion of the semiotic square, see Ronald Schleifer, Robert Con Davis, and Nancy Mergler, *Culture and Cognition: The Boundaries of Literary and Scientific Inquiry* (Ithaca: Cornell University Press, 1992). See also A. J. Greimas, *Structural Semantics: An Attempt at a Method,* translated by Daniele MacDowell, Ronald Schleifer, and Alan Velie (Lincoln: University of Nebraska Press, 1983). I do not claim for the semiotic square the inevitability with which Greimas, its inventor, invested it. Rather, for my purposes it is useful as a stimulus to thought and as a way to tease out relationships that might not otherwise be apparent.

3. Jean Baudrillard, *Simulations,* translated by Paul Foss, Paul Patton, and Philip Beitchman (New York: Semiotext(e), 1983).

4. Greg Bear, *Blood Music* (New York: Ace Books, 1985) (hereafter cited in the text as *BM*); Richard Powers, *Galatea 2.2: A Novel* (New York: Farrar Straus Giroux, 1995) (hereafter cited in the text as *G2*); Cole Perriman, *Terminal Games* (New York: Bantam, 1994) (hereafter cited in the text as *TG*); Neal Stephenson, *Snow Crash* (New York: Bantam, 1992) (hereafter cited in the text as *SC*).

5. Fredric Jameson cogently makes the connection between an information society and late capitalism in *Postmodernism; or, The Cultural Logic of Late Capitalism* (Durham: Duke University Press, 1991).

6. Darko Suvin, "On Gibson and Cyberpunk SF," *Foundation* 46 (1989): 41.

7. Daniel Dennett, *Consciousness Explained* (Boston: Little, Brown and Co., 1991), notes, "The voice the schizophrenic 'hears' is his own" (p. 250 n).

8. Elaine Scarry, *The Body in Pain: The Making and Unmaking of the World* (New York: Oxford University Press, 1985).

9. Veronica Hollinger, "Cybernetic Deconstructions: Cyberpunk and Postmodernism," *Mosaic* 23 (1990): 42.

10. Mark Johnson, *The Body in the Mind: The Bodily Basis of Meaning, Imagination, and Reason* (Chicago: University of Chicago Press, 1987).

11. Judith Butler, *Gender Trouble: Feminism and the Subversion of Identity* (New York: Routledge, 1990); J. L. Austin, *How to Do Things with Words,* edited by J. O. Urmson and Marina Sbisa (Oxford, England: Clarendon Press, 1972).

12. Andrew Hodges, in his excellent biography *Alan Turing: The Enigma* (New York: Simon and Schuster, 1983), comments, "To Alan Turing, the multiplier was a rather tiresome technicality: the heart [of the Universal Turing Machine] lay in the logical control, which took the instructions from the memory, and put them into operation" (p. 320).

13. For a discussion of the deep structure of VR programming languages and their relation to utterances that can *perform* the viewpoint they instantiate, see Robert Markley, "Boundaries: Mathematics, Alienation, and the Metaphysics of Cyberspace," in *Virtual Reality and Their Discontents,* edited by Robert Markley (Baltimore: Johns Hopkins University Press, 1996), pp. 55–77.

14. David Porush, "Hacking the Brainstem: Postmodern Metaphysics and Stephenson's Snow Crash," *Configurations* 3 (1994): 537–71.

15. Mary Catherine Bateson, *Our Own Metaphor: A Personal Account of a Conference on the Effects of Conscious Purpose on Human Adaptation* (1972; Washington, D.C.: Smithsonian Institution Press, 1991).

16. Richard Dawkins develops the concept of memes as the ideational analogue to selfish genes in *The Selfish Gene* (New York: Oxford University Press, 1976).

17. In the introduction to *Posthuman Bodies,* Halberstam and Livingston note: "You're not human until you're posthuman. You were never human" (p. 8).

18. Veronica Hollinger, in "Feminist Science Fiction: Breaking Up the Subject," *Extrapolation* 31 (1990): 229–39, makes a similar argument regarding the diversity of feminist science fiction. Some texts want to recuperate some aspects of the subject, whereas others aim for a more subversive and far-reaching deconstruction. Those who have never experienced a strong and unified subjectivity, Hollinger observes, might want to have a chance to articulate such subjectivity before they deconstruct it. Anne Balsamo, in "Feminism for the Incurably Informed," *South Atlantic Quarterly* 92 (1993): 681–712, takes issue with Hollinger's conclusion, arguing that what is needed is not so much diversity among texts and readings as articulations that can escape from the dualism of anti/pro-humanism by offering a vision of "post-human existence where 'technology' and the 'human' are understood in contiguous rather than in oppositional terms" (p. 684).

Chapter Eleven

1. I am grateful to Marjorie Luesebrink for conversations that stimulated me to think further about the ideas in this conclusion.

2. Warren McCulloch, quoted in Mary Catherine Bateson, *Our Own Metaphor: A Personal Account of a Conference on the Effects of Conscious Purpose on Human Adaptation* (1972; Washington, D.C.: Smithsonian Institution Press, 1991), p. 226.

3. Hans Moravec, *Mind Children: The Future of Robot and Human Intelligence* (Cambridge: Harvard University Press, 1988).

4. Jared Diamond, *Guns, Germs, and Steel: The Fates of Human Societies* (New York: Norton, 1997), and *Why Sex Is Fun: The Evolution of Human Sexuality* (New York: Basic Books, 1997).

5. Jerome H. Barkow, Leda Cosmides, and John Tooby, eds., *The Adapted Mind: Evolutionary Psychology and the Generation of Culture* (Oxford: Oxford University Press, 1992).

6. Jacques Derrida, *Of Grammatology*, translated by Gayatri C. Spivak (Baltimore: Johns Hopkins University Press, 1976).

7. Eric A. Havelock, *Preface to Plato* (Cambridge: Harvard University Press, 1963).

8. Chris G. Langton, "Computation at the Edge of Chaos: Phase Transition and Emergent Computation," *Physica D* 42 (1990): 12–37; Stuart A. Kauffman, *The Origins of Order: Self-Organization and Selection in Evolution* (New York: Oxford University Press, 1993).

9. Francisco J. Varela, "Making It Concrete: Before, During, and After Breakdowns," in *Revisioning Philosophy*, edited by James Ogilvy (Albany: State University of New York Press, 1992), pp. 97–109.

10. Henri Atlan, "On a Formal Definition of Organization," *Journal of Theoretical Biology* 45 (1974): 295–304. Michel Serres has a provocative interpretation of how this noise can give rise to human language, in "The Origin of Language: Biology, Information Theory and Thermodynamics," *Hermes: Literature, Science, Philosophy*, edited by Josué V. Harari and David F. Bell (Baltimore: Johns Hopkins University Press, 1982), pp. 71–83. See N. Katherine Hayles, *Chaos Bound: Orderly Disorder in Contemporary Literature and Science* (Ithaca: Cornell University Press, 1990), pp. 56, 204–6, for a discussion of Atlan and Serres.

11. Gregory Bateson, quoted in Bateson, prologue to *Our Own Metaphor*, pp. 13–16.

12. Francisco J. Varela, Evan Thompson, and Eleanor Rosch, *The Embodied Mind: Cognitive Science and Human Experience* (Cambridge: MIT Press, 1991).

13. In Neal Stephenson's *Snow Crash* (New York: Bantam, 1992), his young white heroine, "Y.T.," is kidnapped, dumped aboard the Raft, and assigned to mess detail. She then has an insight into how small the fraction of the world's population is who ever believed they had a liberal humanist self. Once she gets over the shock and settles into a routine, she starts looking around her, watching the other fish-cutting dames, and realizes that this is just what life must be like for about 99 percent of the people in the world. "You're in this place. There's other people all around you, but they don't understand you and you don't understand them, but people do a lot of meaningless babble anyway. In order to stay alive, you have to spend all day every day doing stupid meaningless work. And the only way to get out of it is to quit, cut loose, take a flyer, and go off into the wicked

world, where you will be swallowed up and never heard from again" (pp. 303–4).

14. Richard Lanham, *The Electronic Word: Democracy, Technology, and the Arts* (Chicago: University of Chicago Press, 1994).

15. Galen Brandt, "Synthetic Sentience: An Interview with Charles Ostman," *Mondo 2000*, no. 16 (winter 1996–97): 25–36. See also Charles Ostman, "Synthetic Sentience as Entertainment," *Midnight Engineering* 8, no. 2 (March/April 1997): 68–77.

16. Joseph Weizenbaum, *Computer Power and Human Reason: From Judgment to Calculation* (New York: W. H. Freeman, 1976).

17. Gilles Delueze and Felix Guattari of course celebrate this very alienness in their vision of the phylum and "body without organs" in *Anti-Oedipus: Capitalism and Schizophrenia* (Minneapolis: University of Minnesota Press, 1983). For an ecstatic interpretation of the posthuman, see Judith Halberstam and Ira Livingston, eds., *Posthuman Bodies* (Bloomington: Indiana University Press, 1995).

18. Donna J. Haraway, "Situated Knowledges: The Science Question in Feminism and the Privilege of Partial Perspective,"in *Simians, Cyborgs, and Women: The Reinvention of Nature* (New York: Routledge, 1990), pp. 183–202; Evelyn Fox Keller, "Baconian Science: The Arts of Mastery and Obedience," *Reflections on Gender and Science* (New Haven: Yale University Press, 1995), pp. 33–42; Sandra Harding, *The Science Question in Feminism* (Ithaca: Cornell University Press, 1986); and Carolyn Merchant, *The Death of Nature: Women, Ecology, and the Scientific Revolution* (San Francisco: Harper, 1982).

19. Edwin Hutchins, *Cognition in the Wild* (Cambridge: MIT Press, 1995).

20. John R. Searle, "Is the Brain's Mind a Computer Program?" *Scientific American* 262, no. 1 (1990): 26–31; see also John R. Searle, *Minds, Brains, and Science* (Cambridge: Harvard University Press, 1986), pp. 32–41, for the "Chinese room" thought experiment. Searle attempts to answer the analysis that it is the whole room that knows Chinese, saying there "is no way to get from syntax to semantics" (p. 34).

21. Hutchins, *Cognition*, pp. 361–62.

22. William Gibson, *Neuromancer* (New York: Ace Books, 1984). The narrator, after relating how Case has been exiled from cyberspace, comments: "For Case, who'd lived in the bodiless exultation of cyberspace, it was the Fall. . . . The body was meat. Case fell into the prison of his own flesh" (p. 6).

23. Bruno Latour, *We Have Never Been Modern,* translated by Catherine Porter (Cambridge: Harvard University Press, 1993). Latour's important argument is that quasi-objects operate within networks that are at once in material real, socially regulated, and discursively constructed. Using different contexts, I have argued in this book for a very similar view regarding the history of cybernetics.

24. Dynamiting the system here alludes to Bill Nichols's seminal article on cybernetics, "The Work of Culture in the Age of Cybernetics," in *Electronic Culture: Technology and Visual Representation,* edited by Timothy Druckrey (New York: Aperture, 1996), pp. 121–44.

semiotics of virtuality, 279–82
tape-recording and, 209
technology and, 206–7
Terminal Games, in, 260
intention
in narrative, 229
reification of, in Dawkins, 227

Jackendoff, R., 156
James, Henry, 37–38, 299n. 26. *See also*
point of view
Jameson, Fredric
character system in *Dr. Bloodmoney*,
179–84, 313n. 28
late capitalism and information, 254,
321n. 5
Jefferson, David, 318n. 21
Johnson, Mark
Body in the Mind, 205–7, 265, 314n.
23
Metaphors We Live By, 85, 305n. 3
Joyce, Michael, 297n. 4

Kahn, Douglas, 315n. 26
Kampis, G, 318n. 20
Kauffman, Stuart, 241, 286, 319n. 37,
319n. 38
Kay, Lily, 150, 293n. 4
Keller, Evelyn Fox, 150, 288, 293n. 4,
310n. 22, 323n. 18
kipple, 174, 175. *See also* Dick, Philip K;
entropy
Kittler, Friedrich
Discourse Networks 1800/1900,
25–26, 295n. 24
"Gramophone, Film, Typewriter,"
315n. 27
man and technology, 315n. 27
medial ecology and, 48
Kollock, Peter, 299n. 27
Kristeva, Julia, 128, 308n. 23
Kroker, Arthur, 5
Body Invaders, 192, 313n. 2
"flesh-eating 90's," 294n. 11
Kroker, Marilouise, 192, 313n. 2
Kubie, Lawrence
autopoiesis, differences with, 143

lobotomy and, 86
Macy Conferences, 69–77
"Neurotic Potential and Human
Adaptation," 70
neurotic processes and, 70–71
"Place of Emotions," 303n. 41
reflexivity, 6, 69
"Relation of Symbolic Function in
Language Formation and Neuro-
sis," 70
reverberating loops and, 59, 302n. 17
Kuhn, Thomas S, 295n. 27

Lacan, Jacques, 30
psycholinguistics of, 35
"Radiophonies," 298n. 13
Le Séminaire, 298n. 13
The Seminar, Book II, 298n. 14
Silverman's view of, 308n. 19
Laing, R. D., 176, 312n. 22
Lakoff, George, 85, 305n. 3
Langton, Christopher
artificial life
analytical approach of, 233–34
analytical compared with synthetic, 233, 318n. 22
origins of, 234–35
strong claim, 231–33, 317n. 13
cellular automata, research on,
240–41, 319n. 36
chaos, uses of, 286
"Population Dynamics of Digital Organisms," 317n. 7
language
agency and, 252
artificial life narrative and, 239–46
Blood Music, biochips in, 252
Galatea 2.2
aliveness and, 266–67
neural net and, 263
recursion and, 264
ineffability in, 67, 189–90
instrumental, 67
languaging in autopoiesis, 144–45,
147
literariness of, 207–8
Masani's scorn for, 112